Y0-BXQ-021

Biomathematics

Volume 16

Biomathematics

Volume 2
E. Batschelet
Introduction to Mathematics for Life Scientists

Volume 3
M. Iosifescu, P. Tăutu
Stochastic Processes and Applications in Biology and Medicine
Part 1: Theory

Volume 4
M. Iosifescu, P. Tăutu
Stochastic Processes and Applications in Biology and Medicine
Part 2: Models

Volume 5
A. Jacquard
The Genetic Structure of Populations
Translator: D. Charlesworth, B. Charlesworth

Volume 6
D. Smith, K. Keyfitz
Mathematical Demography
Selected Papers

Volume 7
E. R. Lewis
Network Models in Population Biology

Volume 8
A. T. Winfree
The Geometry of Biological Time

Volume 9
W. J. Ewens
Mathematical Population Genetics

Volume 10
A. Okubo
Diffusion and Ecological Problems: Mathematical Models

Volume 11
B. G. Mirkin, S. N. Rodin
Graphs and Genes
Translated from the Russian by H. L. Beus

Volume 12
R. Gittins
Canonical Analysis
A Review with Applications in Ecology

Volume 13
J. Impagliazzo
Deterministic Aspects of Mathematical Demography
An Investigation of the Stable Theory of Population including an Analysis of the Population Statistics of Denmark

Volume 14
C. J. Mode
Stochastic Processes in Demography and Their Computer Implementation

Volume 15
D. L. DeAngelis, W. M. Post, C. C. Travis
Positive Feedback in Natural Systems

Complexity, Language, and Life: Mathematical Approaches

Edited by
John L. Casti and Anders Karlqvist

Springer-Verlag
Berlin Heidelberg New York Tokyo

John L. Casti

International Institute for Applied Systems Analysis
2361 Laxenburg, Austria

Anders Karlqvist

Swedish Academy of Sciences
10110 Stockholm, Sweden

Mathematics Subject Classification (1980): 9202, 9302

ISBN 3-540-16180-5 Springer-Verlag Berlin Heidelberg New York Tokyo
ISBN 0-387-16180-5 Springer-Verlag New York Heidelberg Berlin Tokyo

Printed in Germany

Printing: Beltz, Hemsbach/Bergstraße
Binding: Schäffer, Grünstadt

2141/3145-543210

Preface

In May 1984 the Swedish Council for Scientific Research convened a small group of investigators at the scientific research station at Abisko, Sweden, for the purpose of examining various conceptual and mathematical views of the evolution of complex systems. The stated theme of the meeting was deliberately kept vague, with only the purpose of discussing alternative mathematically based approaches to the modeling of evolving processes being given as a guideline to the participants. In order to limit the scope to some degree, it was decided to emphasize living rather than nonliving processes and to invite participants from a range of disciplinary specialities spanning the spectrum from pure and applied mathematics to geography and analytic philosophy.

The results of the meeting were quite extraordinary; while there was no intent to focus the papers and discussion into predefined channels, an immediate self-organizing effect took place and the deliberations quickly oriented themselves into three main streams: conceptual and formal structures for characterizing system complexity; evolutionary processes in biology and ecology; the emergence of complexity through evolution in natural languages. The chapters presented in this volume are *not* the proceedings of the meeting. Following the meeting, the organizers felt that the ideas and spirit of the gathering should be preserved in some written form, so the participants were each requested to produce a chapter, explicating the views they presented at Abisko, written specifically for this volume. The results of this exercise form the volume you hold in your hand.

Special thanks for their help in various phases of organizations of the meeting and arrangement of the publication of this volume are due to M. Olson, P. Sahlstrom, and R. Duis.

December 1985

John Casti, Vienna
Anders Karlqvist, Stockholm

The International Institute for Applied Analysis

is a nongovernmental research institution, bringing together scientists from around the world to work on problems of common concern. Situated in Laxenburg, Austria, IIASA was founded in October 1972 by the academies of science and equivalent organizations of twelve countries. Its founders gave IIASA a unique position outside national, disciplinary, and institutional boundaries so that it might take the broadest possible view in pursuing its objectives:

To promote international cooperation in solving problems arising from social, economic, technological, and environmental change
To create a network of institutions in the national member organization countries and elsewhere for joint scientific research
To develop and formalize systems analysis and the sciences contributing to it, and promote the use of analytical techniques needed to evaluate and address complex problems
To inform policy advisors and decision makers about the potential application of the Institute's work to such problems

The Institute now has national member organizations in the following countries:

Austria – *The Austrian Academy of Sciences;* **Bulgaria** – *The National Committee for Applied Systems Analysis and Management;* **Canada** – *The Canadian Committee for IIASA;* **Czechoslovakia** – *The Committee for IIASA of the Czechoslovak Socialist Republic;* **Finland** – *The Finnish Committee for IIASA;* **France** – *The French Association for the Development of Systems Analysis;* **German Democratic Republic** – *The Academy of Sciences of the German Democratic Republic;* **Federal Republic of Germany** – *Association for the Advancement of IIASA;* **Hungary** – *The Hungarian Committee for Applied Systems Analysis;* **Italy** – *The National Research Council;* **Japan** – *The Japan Committee for IIASA;* **Netherlands** – *The Foundation IIASA – Netherlands;* **Poland** – *The Polish Academy of Sciences;* **Sweden** – *The Swedish Council for Planning and Coordination of Research;* **Union of Soviet Socialist Republics** – *The Academy of Sciences of the Union of Soviet Socialist Republics;* **United States of America** – *The American Academy of Arts and Sciences;*

List of Contributors

David Berlinski, Institut des Hautes Etudes Scientifiques, 35 Route de Chartres, 91440 Bures-sur-Yvette, France

John L. Casti, International Institute for Applied Systems Analysis, A-2361 Laxenburg, Austria

Peter Gould, Department of Geography, College of Earth and Mineral Sciences, The Pennsylvania State University, 302 Walker Building, University Park, Pennsylvania 16802, USA

Ulf Grenander, Mittag-Leffler Institute, Auravägen 17, S-10262 Djursholm, Sweden

Jeffrey Johnson, Centre for Configurational Studies, Design Discipline, Faculty of Technology, The Open University, Walton Hall, Milton Keynes MK7 6AA, England

Howard H. Pattee, Department of Systems Science, T. J. Watson School of Engineering, State University of New York at Binghampton, Binghampton NY 13901, USA

Robert Rosen, Department of Physiology and Biophysics, Dalhousie University, Nova Scotia, Canada, B3H 4H7

Karl Sigmund, Institute of Mathematics, University of Vienna, Strudelhofgasse 4, A-1090 Wien, Austria

Nils Chr. Stenseth, Zoological Institute, University of Oslo, PO Box 1050, Blindern, Oslo 3, Norway

René Thom, Institut des Hautes Etudes Scientifiques, 35 Route de Chartres, 91440 Bures-sur-Yvette, France

Table of Contents

Introduction

John L. Casti and *Anders Karlqvist*

Complexity and the Evolution of Living Systems

One of the most evident features distinguishing living from nonliving systems is the tendency for living processes to evolve ever more complex structural and behavioral modes during the course of time. This characteristic has been observed in lowly protozoa and in highly evolved linguistic and social communities, and has been variously labeled "negentropy", "variety", "information", or just plain *complexity*. The understanding and explanation of the trend toward the complexification of living things forms the heart of *any* research program in the biological, social, or behavioral sciences.

How does complexity emerge and what do we even mean when we speak of a system as being complex? In everyday language, complexity is associated with structural features such as large numbers of system components, high levels of connectivity between subsystems, feedback and feedforward data paths that are difficult to trace, and so on. Complexity is also associated with behavioral characteristics such as counterintuitive reactions, surprises, multiple modes of operation, fast and slow system time scales, and irreproducible surprises. While the structural features are, by and large, objective properties of the system *per se*, the behavioral components of complexity are decidedly subjective: what is counterintuitive, surprising, and so on is as much a property of the system doing the observing as it is a feature of the process under study. Thus, any mathematical formulations of the notion of complexity must respect the subjective, as well as the objective, aspects of the concept. In this volume, the chapters by Rosen and Casti address both aspects of the complexity question. Casti argues that complexity emerges from the *interaction* between the system and its observer, while Rosen introduces the idea of an activation–inhibition pattern to characterize the *informational interaction*, and then uses this concept to speak of complexity. Both of these chapters put forth strong arguments for the case that complexity of living systems is a contingent, rather than intrinsic, property of the system, and any theory of complexity management and control must start from this basis.

The explicit consideration of evolutionary processes in biology is taken up in the chapters by Sigmund and Stenseth. Stenseth reviews the essential components of the Darwinian theory: reproduction, variation, inheritance, and selection, and considers the issue of what kinds of mathematical models we need in order to capture the Darwinian view in operational form. His conclusion is that the numerous controversies surrounding the Darwinian paradigm in no way provide evidence for rejecting it, but rather point to the need to extend and improve our models in various directions. The chapter by Sigmund offers an intriguing mathematical structure suitable for capturing many of the most important features of evolving systems, the so-called "replicator" dynamics. The term is taken from Dawkins' work on general phenotypic evolution and here Sigmund shows how the general structure of the replicator equations covers a range of evolutionary phenomena stretching from population genetics and prebiotic evolution to population ecology and animal behavior. As a counterpoint to Darwinism, the chapter by Thom presents some speculative views on the way in which energetic constraints, together with a few genericity arguments, provide a key for understanding the relations of tools and organs within the context of embryology. Thom's argument is that constraints plus genericity force the embryo to develop along one of a very small number of paths characterized by archetypical geometric forms. As Thom emphasizes, such forms are imaginary entities offering no possibility for experimental prediction or verification; nonetheless, his claim is that they provide the basis for major theoretical advances bypassing the traditional Darwinian view.

The degree to which the evolutionary paradigm of biology mirrors the structure of a natural language is considered in the chapter by Berlinski. His point of view is to examine the claim that life, on some level, is a language-like system. He concludes that if life is a language-like system, then the neo-Darwinian theory is deficient in its repertoire of theoretical ideas. On the other hand, if life is not a language-like system, then the neo-Darwinian theory is singular in that it fails to explain or predict the properties of systems that are in some measure close to life. During the course of making these arguments, Berlinski touches on a number of ideas treated in the other chapters, such as complexity, randomness, information, pattern, and form.

The connection between the concept of measurement and the development of human language is explored in the chapter by Pattee. He argues that it is only by viewing measurement and language in an evolutionary context that we can appreciate how primitive and universal are the functional principles from which our highly specialized forms of measurement and language arose. Pattee's position is that the generalized functions of language and measurement form a semantically closed loop, which is a *necessary* condition for evolution. The chapter closes with some provocative arguments for why current theories of measurement and language do not satisfy the semantic closure requirement for evolution, and a suggestion is offered for a new approach to designing adaptive systems that have the possibility for greater evolutionary and learning potential than existing artificial intelligence models.

Mathematical and Human Affairs

The role of mathematics outside of mathematics itself has always been somewhat unclear and ill-defined. What has been clear is that every generation's problems, especially in the social and behavioral sciences, seem to far exceed the mathematical tools available, resulting in a continuing need to both develop new mathematical concepts and to examine ways in which the existing "abstract nonsense" of the pure mathematician can be recast in a form usable to the practitioner.

In the opening chapter of this volume, Peter Gould examines some of the problems of mathematicizing the physical, biological, and human worlds. He points out that over most of history, mathematics has been driven by the problems posed by practical, everyday living, ultimately addressing the question of whether mathematics is reducible to mechanical operations on sets of objects. He concludes that if this is so, then there is no possibility for a complete mathematicization of the human world.

An illustration of the way in which ideas from pure mathematics can be used to develop a language for speaking about human processes is provided in the chapter by Johnson. In his report, classical ideas from algebraic topology are used to associate a simplicial complex with a given human situation. The connective structure of the complex is then employed to infer various "deep structures" associated with the social situation. During the course of developing this structure, Johnson shows how the introduction of numerous concepts ***not*** entering into classical algebraic topology can play a significant role in teasing out the structure of the human situation.

The chapter by Grenander approaches the role of mathematics in human affairs from a different angle. Grenander's idea is that in order to speak about pictures and patterns, it is first necessary to develop an entire ***algebra*** of patterns. This algebra is then employed to mathematically characterize complex visual phenomena, and forms the basis of numerous pattern formation and recognition procedures. As Grenander points out, such an approach is really "mathematical engineering"; we build logical structures using the algebra of patterns, just as engineers build mechanical, electrical, or other physical structures.

CHAPTER 1

Allowing, Forbidding, but not Requiring: A Mathematic for a Human World

Peter Gould

In Morris Kline's extraordinarily thoughtful *Mathematics: The Loss of Certainty* (Kline, 1980), one aspect emerges clearly: over by far the greatest part of its history, mathematics has been driven by the need to describe the physical world of things. The distinction between pure and applied mathematics did not emerge until comparatively recently; no sharp distinction was made between mathematics and science in the seventeenth and eighteenth centuries, and *all* the great names contributed to the vast overlapping area between the two. In brief "there was some pure mathematics but no pure mathematicians" (Kline, 1980, p 281). It is unfortunate that, just as the biological and human sciences started to hive off from philosophy as separate fields of inquiry, so mathematics began to detach itself from the physical sciences. First, this means that few mathematicians today have ever been truly challenged by the biological and human sciences, with the result that old and inappropriate forms of mathemetics have been borrowed from the realm of the physical world to distort descriptions of the biological and human worlds. Second, the possibilities for creating new and appropriate qualitative mathematics have been diminished as mathematicians look increasingly to mathematics itself, rather than to the challenges beyond their distressingly private realm of discourse.

But like war and generals, mathematics is too important to leave to the mathematicians. Such a statement, unless it is read correctly, will not please many mathematicians (nor did it please the generals), but it may still be considered seriously by the few who have acquired a deep and intimate knowledge of the history of their own field. Like any one of many human endeavors, mathematics cannot reflect upon itself from the inside, but requires that creative sense of tension that the philosophical stance has always provided at its best. The human meaning of mathematics does not reside in mathematics, but in the larger arena of human discourse that places this particular, and quite peculiar, form of inquiry in relation to other aspects of our intention to know.

A Backcloth for Thinking

I would like to consider some of the problems of mathematization in the physical, biological, and human worlds, and I hope it will be considered appropriate that we start with the human – the world of ourselves. This is a curious, self-reflective world, and its capacity for self-reflection, for *thinking* about itself, is surely its most outstanding characteristic. Atoms and animals do not conceive of laws, nor do they generate the functional equations commonly used to describe them. In contrast, we are distinguished by that very ability. For this reason, it is perhaps appropriate to sketch a broad framework of concern within which we can discuss mathematically more specific questions and requirements. With the responsibility to represent the human sciences in this book, I hope you agree that it is entirely appropriate that we first take a step backwards – a sort of intellectual deep breath – and *think* about this world that is us; *think* about the creation of mathematics within and from out of this world; and then *think* about that sudden backward twist when mathematics is used to describe the world from which it originally arose. In doing so, it may seem that we are in a rather different realm of inquiry than some of the other contributions in this book, and yet perhaps we are always in this larger realm of thinking – even if we may not always realize it.

The mundane roots of mathematics

Not long ago, I read in *L'Express* (17–23 fevrier, 1984) a review of Giorello and Morini's *Paraboles et catastrophes: Entretiens avec René Thom* (1983), in which the reviewer wrote with great aplomb "Thus, mathematical structures have come before their use in physics, and not the reverse". Nowhere, perhaps, is the striking contrast between the French Cartesian approach to science, and the Anglo-Saxon empirical approach highlighted more vividly. It is the contrast between the creation of *a priori* structures into which the world is forced, and the creation of careful descriptions of the world which later suggest appropriate mathematical structures. The reviewer's mistake is a common one, although quite natural if the history of mathematics is not there to inform us about the facts. As an Anglo-Saxon, and one concerned with historical veracity, let me suggest, very generally and up to about one hundred years ago, that far from mathematical structures being created for their own sake, and *then* being applied to areas of scientific inquiry, the weight of evidence is for exactly the reverse. That is to say, over most of its history mathematics has been positively driven by the requirements and difficulties posed by the problems of practical, everyday living, as well as by those appearing in the world of physical science as we know it today. Such problems range from tallying sheep, building temples and pyramids with square corners, surveying field boundaries after floods, and other, literally mundane, tasks, to creating new algebras and topological structures in order to describe events on the quantum and cosmological scales.

Some mathematicians still know the story, but let me remind others of the infinitesimals of Isaac Newton, the extensions to the calculus by James Maxwell, the algebras and numbers of William Hamilton that were so outrageous in their time, and then point to the driving force behind Lagrange, Leverrier,

Legendre...and the literally scores of other distinguished mathematicians of the seventeenth to nineteenth centuries, who sought to describe faithfully the physical world. It is the intentional, and perhaps insatiable, curiosity to describe the world, that Greek legacy that pushes the questions to the limits (and may yet destroy us), that invariably comes *before* the mathematics, that often informs mathematicians where the deep problems lie, and that nourishes those who search for the mathematical solutions. John von Neumann knew this well, and stated so explicitly. And to accept this is not to deny in the least the desire of those who seek to develop mathematics for its own sake, and in giants like Friedrich Gauss and Henri Poincaré we see how both mathematics and science are illuminated by such penetrating thinking. Yet as a geographer, I cannot help noting, in a somewhat mischevious mood, that even the distribution that bears Gauss's name was devised to minimize the error in his instruments when he was making maps for the Duke of Hannover – a sound, descriptive enterprise going back to Babylon.

Now it is equally true that we can also point, in more recent times – the last eighty years? – to the reverse; namely, to the prior existence of mathematical structures which later became not just useful, but essential for the development of science. For example, Albert Einstein reaches for the tensor calculus of Ricci and Levi-Civita (created only a few years before), and Walter Heisenberg struggles to devise algebraic operations for rectangular tables of values, until Max Born tells him to go and study matrix algebra – already well-developed. In these particular cases, the appropriate mathematical structures were already in place, but even here the *thinking* about the physical world, from the cosmological to the quantum level, preceded the applications.

Why am I trying to direct thinking towards this brief, and necessarily superficial, historical review of mathematics? Because I want to point to the fact that historically, and even today, the physical sciences have, in general, set an impeccable example of thinking about the phenomena of interest, *and only then* devising and creating the mathematics that seems to be called forth by the descriptive requirements. The things themselves suggest the mathematical structures that must come into being to describe the physical phenomena that falls under the scrutiny of our curious gaze. What are the only other possibilities? First, to choose an already existing mathematical structure, and then run around the world desperately seeking something that will fit it. Rather like Diogenes searching for an honest man, our journals are full of reports of methodologists searching for an honest application. However, I think it is necessary to reflect whether most of the reports constitute science, in the sense that they genuinely illuminate a part of our world.

Second, we can borrow unthinkingly the mathematical structures devised to describe one aspect of our world, and use them, equally unthinkingly, to describe another aspect. If we take the mathematical structures devised to describe the worlds of celestial *mechanics*, statistical *mechanics*, quantum *mechanics*, continuum *mechanics*...and just plain *mechanics*, and then map the human world unthinkingly onto such structures, is it possible that the human world so described can look anything but *mechanical*? In brief, does the mathematical "language" chosen allow the description, and allow our thinking, to appear as anything but *mechanistic*?

The meaning of mathematics

Now mechanism is a world of lawful statements, statements we make about things in essentially their *deterministic* relations. And, simply as a stage aside, I do not want to argue the deterministic *versus* the probabilistic here. Throwing some probability distributions into the arena of methodological discourse does not, for me, solve the fundamental problems in the human sciences. It merely sweeps them under the rug, so that the real questions of transcending both approaches recede into concealment from our thinking. As for the purely statistical approach, which was so popular in the human sciences until a few years ago, it condemns a human scientist to be a calculator of moments of distributions. Quite apart from the shallowness of such descriptions, and the shallowness of the questions they purport to address, I cannot think of anything more boring as a lifetime's work.

Whether we consider deterministic or probabilistic descriptions, as they have been traditionally expressed, both are essentially functional in form. This means that the cog-wheel variables on the right-hand side of the equation turn and grind out, mechanically, on the left-hand side either a single value, or a mean value smeared with a bit of variance. The mechanical coupling of the conventional binary operations used on the set of real numbers is essentially the same, whether the model is deterministic or probabilistic. In the human world, we need to move beyond this simplistic dichotomy that arises from the descriptive requirements of the physical and biological worlds (where they may be perfectly adequate), to the fundamental facts of consciousness, reflection, and informed choice – not simply conditioned behavior – in the human world. The mathematics must enable non-mechanical interpretation in *allowing*, *forbidding*, but not *requiring* geometries. This, as we shall see, may be a contradiction if claims are valid that *all* mathematics is ultimately mechanical by its reduction to logically consistent operations. This, for me, is a frontier question for which I seek your most penetrating thought and insight. It may be, in a very deep sense, that the human world is not mathematizable. Which is *not* to say that we do not take certain aspects of the human world, and map them with enormously severe many-to-one mappings onto mechanistic structures devised in the physical world. For example, entropy maximization models (Wilson, 1970), straight out of Boltzmann, take counts of people, and counts of costs between residences and workplaces, and after some heroic assumptions, and a series of computer iterations, find a most-likely distribution that best fits the numbers of the journey-to-work census for a particular city. The result is a piece of social physics whose Langrangians tell us that a residential area far from the work places is less accessible than those close to the work places. Not terribly illuminating, and not terribly helpful when it comes to planning changes in the structure of the city to create a more humane and equitable world. We have crushed away so much in the mapping that constitutes our mechanical analogy that we cannot do very much with a solution that represents the most general case of numbers distributed in a constrained box.

This frontier question – whether the tyranny of the conventional binary operation forces mathematics and, therefore, the parts of our world described by mathematics, to *be* mechanical – this question leads us to reflect upon the meaning of mathematics itself. And here, as elsewhere in this chapter, I rely heavily upon the thinking of Martin Heidegger, who has reflected so deeply upon the original Ur-meaning of so many words we use daily. Unfortunately, the words used now have made a long journey through time from the Greek world where they were first coined, used, and reflected their original human meaning. To recapture one particular meaning, let us read Heidegger carefully for a moment (1977, p 118):

> Modern physics is called mathematical because, in a remarkable way, it makes use of a quite specific mathematics. But it can proceed mathematically in this way only because, in a deeper sense, it is already itself mathematical. ***Ta mathemata*** means for the Greeks that which man knows in advance in his observation of whatever is and in his intercourse with things...

He then goes on to elaborate on the characteristic of exactitude in physics, through the use of measurement, number, and calculation, where physics is "the self-contained system of motion of units of mass related spatiotemporally". It is obvious that we are extraordinarily close to Arthur Eddington (1935) here, with physics as a closed, self-contained system – in a sense, Heidegger's "object sphere". But then Heidegger continues (1977, p 120):

> The humanistic sciences, in contrast, indeed all the sciences concerned with life, must necessarily be inexact just in order to remain rigorous... The inexactitude of the historical sciences is not a deficiency, but is only the fulfillment of a demand essential to this type of research.

Now if the fundamental meaning of exactitude is grounded upon number and binary operation, and inexactitude is a necessary condition for the historical or reconstitutive sciences to be rigorous, we cannot approach and illuminate the human sphere through the mathematics (the ***ta mathemata***) of the physical sciences. And let us recall that this has traditionally been a mathematics of binary operations on sets of numbers – usually the reals – for which the continuum is required as a ***mathematical*** definition. In this classical analytical world, inexactitude may suggest the probabilistic smearings of the statistician, but these only represent a loosening up, a fuzzying operation after the creation of deterministic operations devised during the classical phase of the physical sciences. Instead of trying to tidy things up by the contradictory use of probabilistic smears, perhaps we should go back to the beginning and see what the human world, with real human beings center stage, is trying to say to us. This return to the clearing in the forest would be in the best traditions of classical science, even though the path dimly seen through the trees may not lead in the same direction as the one we are following now. Even to think of a mathematics that transcends the conventional deterministic–probabilistic dichotomy may constitute a challenge to modern mathematicians who are willing to leave classical analysis behind them.

In brief, we cannot employ conventional, physically inspired forms of mathematics in the human sciences, not if we wish to pay reverent heed to that world of conscious, sentient beings with the capacity to reflect upon any statement or description we make of them. And I use the rather poetic phrase "to pay reverent heed" with Heidegger, because it is here that our word ***theory*** is grounded in the Greek ***theōria*** (Heidegger, 1977, p 163). Its two parts are ***thea***, meaning outward appearance, the outward appearance we hear in our own word ***theatre***; and ***horaō***, which means to look attentively, to view closely. ***Theōria*** –

theory – is to look attentively at outward appearance. But there are perhaps even deeper roots, because with a slightly different stress the Greeks could also hear in *theōria* both *theá* and *óra*. Theá is goddess, and for Parmenides *alētheia* appeared as a goddess. And *a-lētheia* is *lētheia*, or concealment, negated in the Greek by the *a*, to create *un*concealment that for the Greeks was the *truth*. *Óra* is reverence, respect, and honor, so *theōria* now becomes the "reverent paying heed to the unconcealment of what presences" (Heidegger, 1977, p 164).

And now the long journey from the Greek world to us begins, a journey of successive translations, each of which constitutes a many-to-one mapping in which the original meaning is crushed out and lost. The Romans translate *theōria* as *contemplatio*, and the *templum* of *contemplatio* has in it "to cut", so that now we hear our own word, *template*. For what is a template but something created beforehand into which something later *must* fit – and if it does not, we cut it down, and chop it up, and force it until it does fit – usually with inappropriate applications of least-squares, or a myriad of other forcing acts that we euphemistically call "estimation procedures".

So what do we regard as our fundamental task as biological and human scientists? To pay reverent heed with the Greeks to allow that which *is* to come out of concealment? Or shall we cut up and shape and force into our preexisting template that which *is* in order that it shall become that which we want it to be? And I cannot help commenting here on the contrast between the scientific approaches of Rosalind Franklin (Sayre, 1975), Barbara McClintock (Keller, 1983), and Janet Rowley (Vines, 1984), and those *a priori* Roman templaters we call the model builders. The first, Rosalind Franklin, spent seven years paying reverent heed to hundreds of X-ray crystallography photographs, and the double-helix diagram was found in her notes after her early death. The second, Barbara McClintock, was ridiculed for years because she paid reverent heed to transposable genetic fragments, until she was awarded the Nobel prize much later (she actually used the phrase "you have to listen to the material"). The third, Janet Rowley, spent 25 years looking at the translocations of chromosomes, and so opened, almost single-handedly, an important and growing area of contemporary medical and cancer research. Is it possible that some women in our Western culture have a gentler Greek mode of questioning than many of the arrogant Roman templaters?

Let me suggest that after years of Roman arrogance, we try once more the gentler Greek mode, particularly as we approach the difficult task of thinking about the requirements for a mathematics that will describe, and allow that which *is* to come forth, without too much of the severe, *a priori* templating. Without, that is, so much forcing by severe many-to-one mappings of the human materials onto constrained structures that we know in advance (*ta mathemata*) the social physics that must be the conclusion. Let us also see the process of mathematical description in the larger context that constitutes human interest and inquiry. As scientists, we must always see mathematics as part of, as a contributor to, a much larger endeavor. For this task, I want to use the three perspectives of Jürgen Habermas (1971) as temporary pegs on which to hang our thinking, rather than as exclusive categories with which to fragment our thinking still further. As we shall see, his perspectives are actually intertwined and connected viewpoints that inform and shape each other.

Technical, hermeneutic, and emancipatory perspectives

At the start of any inquiry, we have to choose to observe some things and not others, and so we face the severe responsibility of thinking about what constitutes useful and fruitful definitions. We can never, of course, be sure about these until they have been tried, either to succeed or to fail in illuminating an aspect of our world. In their own sphere, mathematicians know all about these, often facing the same problems of trial and error, and therefore take such initial responsibilities seriously. Many in the human sciences do not, or they appear to think that the matter of definitions – of sets and operations and relations – is so obvious, even banal, that it is trivial and naive even to mention this first, but always crucial, step. Always crucial, in the literal sense of a crossing point, because it is here that we take our first step along one of a number of possible paths, so that virtually everything we can say thereafter is founded upon the choices we make here. We then have the further obligation of thinking about how we shall observe and record the relations between the sets, what operations we shall allow, how we shall notate the elements and operations, and even how we might express our thoughts symbolically or graphically. All these are essentially methodological questions that lie in the perspective that Habermas has termed the *technical*.

Even here, and not just simply as a stage aside, I think we should pause and remind ourselves, with Heidegger once again, of the deep and original meaning lying within *technē*. It is true that the word is the Greek for *art*, but it is also much more, and the deeper meaning only appears when we contrast it with *phusis* – which we translate as "Nature", but which really means that which "resides in itself" (Heidegger, 1979, p 81). *Technē* stands in contrast to that which resides in itself, it is the *knowledge* of beings, "that knowledge which supports and conducts every human irruption into the midst of beings". There is nothing mystical here: like all of Heidegger's thinking it is rock hard. The human irruption into the midst of beings is simply the field zoologist trapping lemmings, the glaciologist boring into the layered ice cap, the radar beam sweeping the thunderstorm, the earth satellite gathering its harvest of pixels with its electronic scythe. *Technē* is the mode of human irrupting into the *phusis*. So we can see *technē* as art, but as a broad conception of art, as a human capacity to bring forth. After all, what else *is* art and science but an irrupting into, and an adding of illumination to, the *phusis* that *is*? Thus, this irrupting, technical perspective imposes an enormous responsibility, for it determines what is brought forth in opposition to that which resides in itself, to that which is. That human irruption means something else is brought forth.

What, in scientific inquiry, is brought forth from the technical perspective? It is, I would aver, a *text*: not necessarily a text of words, but a text that may be offered in numerical, symbolic, geometric, graphical, or pictorial form. But whether as ordinary language, algebraic equation, tensor, Galois lattice, geometrical construction, systems diagram, bubble chamber photograph, or even computer output (the essential evidence for the four-color theorem), it constitutes an addition to Nature that was not present prior to the human act of irruption with Nature. And, as human scientists, perhaps we should reflect more deeply on the fact that in the human realm our irruptions are often re-irruptions, for we inquire into the human world constituted from both *phusis* and past human irruptions.

The efficient management of a large irrigation system, for example, requires a new irruption into a prior irruption into *phusis*.

But the real question is what does the text *mean*? The ten nonlinear equations of Einstein describing inertia, and centifugal and Coriolis forces have stood as a symbolic text for 70 years: only in 1984 did Jeffrey Cohen, after 25 years of trying with Eli Cartan's method, solve them. Now they can be given an extended meaning. So it is here that we have entered the ***hermeneutic*** perspective, for it is we who have to interpret and give meaning to the text. Sometimes, of course, the text may mean nothing; we can give no interpretation to the things we have created out of our technical definitions. I remember Hans Panofsky, the distinguished theoretical meteorologist, once telling me with complete candor that sometimes spectral decompositions of turbulent wind records mean something – and sometimes they do not! Thanks to Joseph Fourier, we know we can always decompose any continuously differentiable function into linearly additive pieces. But it is *we* who place and impose the structure with simple linear mathematics, and this may, or may not, illuminate the complexity of the turbulence of *phusis*.

The hermeneutic act, the act of interpretation of text, requires that we bring every scrap of knowledge, imagination, and insight to the task, and there is nothing to help us here – no books, no machines, only ourselves. But suppose we fail to interpret? To what can we ascribe our failure? Clearly, there are only two possibilities: we have either created a text from the technical perspective that is meaning*less*, or we have failed in the act of imagination. So back we circle to redefine and restructure our text, or we try to augment, heighten, and sharpen our imagination to bring to light that which lies still concealed. But suppose we succeed in our interpretation, suppose we suddenly "see" the meaning – and notice how we use the visual metaphor of "Oh, I *see*!", of "Voilá!", to describe that flash of illumination when understanding first breaks through. Even now our job is incomplete, for science is shared and verifiable knowledge, and we still have to persuade others to understand as we do. This may be no easy task: in all the sciences, physics included, we find case after case where the same text has been given different interpretations, and the advocates of different views had to persuade others that *their* interpretation was the....*true* one. But persuasion means *rhetoric* (Sugiura, 1983), and I use the word in its original meaning, without the perjorative connotations that it has gathered today. The art of rhetoric is an old and honorable one, and it is employed constantly in science. Of what use is your sudden seeing if you cannot persuade others to share it with you?

It is here that we find two crucial distinctions between the physical and human sciences – and I must let more knowledgeable people determine whether, or in what way, these distinctions characterize the biological sciences. First, the differences in interpretation in the physical sciences may be decided by the critical experiment. This is available because of the intrinsic mathematical nature that allows a looking ahead - the *pre*diction of the physical sciences, in contrast to the *post*diction of the historical sciences – using the adjective historical once again in the sense of the recreative or reconstitutive sciences. Prediction (in the absolute sense with which it is employed in the physical sciences) is seldom available to the human sciences for two reasons. First, the ethical stance does not, or should not, allow us to treat people as objects to be experimented with. Things do not care, the atom is indifferent to its radioactive decay, the rock is unconscious of the geologist's hammer. We do care: it is in our nature to care. Second, as human beings, either individually or as a social collectivity, we have the capacity to reflect upon the algebraic and geometric descriptions, and either change the geometries, or deny the algebraic expressions through our capacity as self-reflective, conscious, and *thinking* beings. Such a changing and denying of the geometry is unthinkable in the simpler and closed world of space–time, where our descriptions are underpinned by the assumption that unalterable laws of mass and motion hold eternally.

But already we have slid over into the third perspective of Habermas, the *emancipatory*, for ethical questions and feelings of caring do not arise directly in the technical and hermeneutic perspectives (although they may be deeply informed and shaped by them), and the idea that we can change the structures of our human world in the light of such values appears meaningless in the physical world. Again, where the biologists lie I must leave for them to decide. So we see that in all the sciences, the technical, hermeneutic, and emancipatory perspectives are intertwined, shaping and informing each other. The technical perspective shapes the text to which the hermeneutic responds; for example, we interpret today computer-shaped texts unthinkable 30 years ago. But it also shapes the emancipatory perspective wherein lie our value structures. Who can deny that our values have not been changed by the technical world? When do we tell someone of an incurable genetic disease just diagnosed by advanced technology (Connor, 1983)? When do we detach someone from a life support system and let them die in dignity? Moreover, these are not one-way streets: the hermeneutic perspective informs the technical and emancipatory – we interpret the values of ourselves and others – and the emancipatory informs both the technical and the hermeneutic. Let me provide a somewhat more extended, but quite concrete, example.

In a recent study of international television (Gould *et al.*, 1984), we had to create sets of words at different levels of generalized meaning to describe both the content and treatment of television programs. For example, a program like *Man and Woman* might be described by two sets of words; the first describing the content of the program as {physical health, individual relations, sexual relations, procreation, birth control, individual health maintenance, social health maintenance}, and the second the way that this connected structure of subjects was treated by {serious talk, social adjustment, ethical concern, documentary, northwest European culture}. Such treatments are very different from those such as {TV movie comedy, light performance}, which might turn the program into an amusing farce about the sexual adventures of young doctors.

Now, given a set of TV programs, a set of descriptive words, and two people coding (one a young TV executive from a major network in the US, the other a young Marxist from a university in Latin America), would we expect the description of the programs in the set to be the same? In our wildest dreams, I do not think the answer would be yes: at the technical level, the choice of descriptive words might be different, perhaps with the Latin American coder requiring words not even in the sets (these would be allowed to be added). Moreover, who can doubt that the interpretations of the derived structural texts would also differ? Why? Because both perspectives are informed and shaped by the underlying ideologies that express the values within the emancipatory perspective. So what price "shared and verifiable knowledge" now? Perhaps in the human sciences we can only have knowledge *modulo the ideology*? Perhaps the very phrase "human sciences" is an oxymoron – a phrase containing within itself a contradiction?

Some Traffic on the Backcloth

Within this broad, reflective framework, I now consider what an appropriate mathematics for the human sciences might look like. I shall ***assume*** that we must try to incorporate into our thinking three broad requirements. First, that any mathematical language we devise for our descriptive task shall make well-defined and operational the intuitive notions we have that ***structure*** is always central to our concern. If we talk (as we so frequently do), of the ***structure*** of an ecological system, a choral mass, a molecule, a family, a society, heart tissue, a university, a game of chess, a ballet, a conference of people concerned with the structure and evolution of systems...and the thousands of things that form the objects of human inquiry, we have to translate such a fundamental concept into concrete descriptive and operational terms. Second, we must allow our thinking to move out of the deterministic–probabilistic dichotomy toward structures that allow, forbid, but do not require. This, it seems to me, allows the most fundamental aspect of being human, namely an acknowledgment of consciousness itself, and its self-reflective capacity, to enter our structural descriptions. People are often parts of structures, or live in them. Finally, and in keeping with the empirical Anglo-Saxon spirit, we must do our best to start with the things themselves, and try to think what they require to describe them in their structural complexity.

Sets and hierarchies

To inquire is to make a choice, to choose to bring to our attention and observe some things and not others. Of this act, it has been said that it is essentially theoretical – theoretical in the ***a priori***, Roman templating sense. Of this I have some doubts, for it dresses simple and naked curiosity in something akin to the Emperor's clothes. Much of our inquiry is founded upon sheer curiosity or practical necessity, and it frequently involves bringing to our attention things that we, or others, have not thought about very much before. In such situations, I do not know how ***theory*** – in any well-developed, or even highly embryonic form – enters at this stage. But no matter: choosing to observe some things and not others means that we choose to observe, and therefore to define, ***sets***. That sets also form one of the fundamental building blocks in certain areas of mathematics should give us encouragement. Perhaps we are starting in the right place.

Not that sets are always easy to define (Couclelis, 1983), and sometimes the attempts lead to ambiguities, inconsistencies, paradoxes, and sheer nonsense. No one with actual experience in empirical research ever claimed that set definition was easy, quite the contrary, but if our sets are not well-defined then clearly this is our first problem, or all else is built on sand. However, and despite ingeniously constructed examples, it has been my empirical experience that set definition in actual research practice often appears fairly straightforward, although it may be time-consuming and tedious. In empirical research we define our sets extensionally, and I have the suspicion that such extensional definitions are close to simple naming propositions, such as "John is a man" (Kline, 1980, p 186). Difficulties seem to arise when intentional definitions are employed, perhaps closer to propositional functions, such as "x is a man". Set definitions for empirical research, rather than simply logical speculation, may also be an aspect of a research program that comes under vigorous reappraisal if we produce an uninterpretable text from our sets on the first analysis, and we have to circle back to think again. Even at this point, it should be noted that such a circling back is not a vicious circle, nor necessarily a sign of initial stupidity. It is a hermeneutic circle, or perhaps we should say spiral, because the circling back due to initial failure takes us to a starting point we were unable to reach before. If we are ignorant to begin with (as we must be, otherwise are we genuinely inquiring?), we should be capable of learning from failure. Indeed, the history of science is essentially a history of failures – some of them magnificent failures – and a history of renewed attempts to understand.

But ingenious paradoxes of set definitions do point to one thing: our words of everyday language, and the concepts they ennunciate, are often on different levels of generalization. For example, we feel instinctively that there is something awkward about the set M defined extensionally as {Algebra, Geometry, Topology, Mathematics}, for the element Mathematics is clearly at a higher level of generalization, say $N + 1$, and it covers Algebra and Geometry at level N. Where Topology lies in this hierarchical scheme is anybody's guess, but if you are going to talk meaningfully of the structure of mathematics, you had better decide. Similarly, a set of rooms at $N - 1$ (the base level N is arbitrary), aggregate up to a set of houses at N, which aggregate into a set of neighborhoods at $N + 1$, which aggregate into a set of towns at $N + 2$... and so on. Nor do our hierarchies of *cover* sets have to be formed from the usual tree-splitting partitions produced by equivalence relations: dandelion at $N - 2$ can aggregate by well-defined and empirically given relations to the $N - 1$ level sets Vegetables, Flowers, and Weeds. In medical diagnosis, the N-level symptom mouth ulceration may aggregate, with other diagnostic elements, to many different diseases at $N + 1$. It is conceptually important that we define very carefully the hierarchical structure of cover sets *before* we undertake further inquiry, or we shall end up confusing elements of sets at one level with members of their power sets at the next (Atkin, 1974).

Backcloth for traffic

But there is a further distinction to clarify: however we eventually define the structures that are of interest to us, it is clear that they exist *for* some purpose. However we create a structural text, and represent it algebraically as a polynomial, or even as a physical model constructed from a chemist's beads and connecting springs, that structure must *mean* something for something else. The reason it has importance for us is because it supports something, it provides a home for something, it carries something, or it is associated with something. In brief, the structure matters other than for itself. It is here that we arrive at the crucial difference between *backcloth* and *traffic* – two technical terms whose definitions we must grasp carefully in their specific meanings. Backcloth is structure, a multidimensional structure that allows and supports traffic. Backcloth structures, as multidimensional spaces, can exist without traffic: technically traffic, as a graded pattern, can consist of all zeros (if we happen to choose a number system to represent it). The reverse is not true: traffic needs a backcloth, a structural geometry, to exist, to support it.

Even though it means anticipating some of the points we discuss later, it is useful to have a concrete example here. Suppose we want to speak of the structure of a conference – say one on Structure and Evolution of Systems. We might be able to operationalize this seemingly valid, but initially quite intuitive notion, by considering the connections between the participants (a well-defined set), and the set of intellectual interests, carefully sorted out into their hierarchical levels of generality, and perhaps evaluated according to degrees of interest or competence. If the shared intellectual interests connect the participants, and so define *a* structure of the conference (perhaps one of many), what might be traffic on such a multidimensional structure? Clearly, ideas could be one sort of traffic, especially those ideas requiring certain combinations of intellectual interests to exist. A participant with limited professional competence in many interests that others have as mathematicians, physicists, biologists, zoologists, archaeologists, etc., could not have some of the ideas supported on other parts of the structure (the multidimensional geometry) of the conference.

Second, and perhaps as a result of the conference, some of the participants may collaborate in the future on research programs and papers. Those papers, as traffic, live on the pieces of the geometry that constitute the shared interests, or *faces*, that connect them. Or, perhaps one person brings some interests from those that define her, and another brings some of those that define him, and they create a new piece of geometry (perhaps a Leftschetz prism), that can support a paper of collaborative traffic that simply could not exist before – the geometry was not there to support it. Notice that such a structure *allows* (that is to say, ideas and research papers can exist if sufficient connective tissue is available in the structure); it *forbids* (papers on the effect of environmental change on the sexual habits of the Australian wombat will not exist, because the supporting geometry happens not to be there); but it does not *require*. Why? Because sentient, conscious, self-reflective human beings form an essential part of the geometry, and they can choose to think and collaborate or not. And who could predict whether they do or not? Also, of course, they are parts of other structures, and these geometries may also allow or forbid, but do not require in any absolute, law-like sense.

Or take a mathematical curriculum in a university, one of whose structures may be formed from a set of courses and how these are connected by the elements they share in common. The traffic that is supported by such a structure might be the students, and those that try to exist on a piece of the geometry defined by very advanced elements may not exist for very long. But this suggests (indeed, the very meaning of the word *curriculum* happens to be chariot race), that students may have to start on one part of the structure, and then move along or through it as they acquire the vertices that allow them to be transmitted to more advanced parts farther along the course of study. Such movement is referred to as traffic *transmission* (Johnson, 1982), and students will tell you that the *structure*, the way the pieces are connected together, affects the transmission. Is it a long chain, representing the sort of teaching that Herbert Simon called the "recapitulation of the field" method? Or have concerned and thoughtful mathematicians created an introductory course in abstract algebra, a high-dimensional, and perhaps well-connected, piece of structure that allows students to branch quickly into other areas of modern mathematics?

How are such multidimensional geometries defined in an operational sense? Clearly, they can only be defined by connections between and on sets, but here we face a number of choices, and I must make my definitions precise. These definitions are not traditional, although they were originally devised by mathematicians to make an important distinction that has turned out, quite fortuitously, to be useful in empirical work. Mathematicians accustomed to traditional ways may object, but then they are the first to insist upon clarity in this realm, and are often condescending when nonmathematicians object to *their* definitions. Thus, this observation constitutes an appeal for mutual tolerance and forbearance.

Functions, mappings, and relations

In science, almost universally, elements of sets are connected by *functions*, which I shall define as injective, surjective, or bijective mappings, usually the last because often the inverses exist and have empirical meaning. All of these are one-to-one or many-to-one. In the physical sciences, the function is used almost exclusively as a description of connections between the elements of sets. It has been highly successful – it seems to describe with fidelity many aspects of the physical world – and it has been borrowed by the biological and the human sciences, often constrained to linear form.

However, there is no reason why we should constrain and confine our descriptions to functions: we can enlarge the possibilities to mappings, where these allow one-to-many and many-to-many connections between elements of sets. Thus, all functions are mappings, but not vice-versa. But we can go further and relax the requirement that all elements in the domain must be assigned. For example, in the injective mapping or function, we are not required to employ all the elements of the co-domain, and there is no intrinsic reason, other than tradition, why we cannot relax this requirement for the domain. Indeed, it is the next logical step in the progression of relaxation and freeing the connective possibilities to allow more appropriate structural descriptions. Since we have no *a priori* knowledge of what these may require in any specific area of empirical inquiry, it seems prudent to provide for the most unconstrained description we can imagine. In this way, we can record what is there freely, without forcing our initial observations onto a constraining functional framework. Such an untraditional step may horrify classical analysts, because it takes them into an area which is no longer secured by convention, and where thinking has to start again. Others may find it more congenial, and even useful, and be prepared to ignore the idea that utility is somehow disreputable. As noted in the first part of this essay, historically the usefulness of a concept for empirical description was considered honorable. We might consider returning to this tradition. Thus, we define a relation, so that all functions are mappings are relations, but not vice-versa.

With this highly unconstrained or free definition of connecting elements of sets, we can represent a relation, say λ, between two sets, say P and I (perhaps people and interests), as an incidence matrix $\mathbf{\Lambda}$, where we might use 1 or 0 to indicate existence or nonexistence of a connection – although, in fact, any nonnumerical symbol, an asterisk or even a banana, could be used. Thus, $\lambda \subseteq P \times I$, and geometrically we represent each element in one set as a polyhedron, or simplex, whose vertices are in the other set. It is the union of the simplicial complex, notated $K_P(I : \lambda)$, made up of all the simplices, and its conjugate $K_I(P : \lambda^{-1})$, that is the backcloth. This is an important part of the text – essentially a geometric text – that we have to interpret, and to which we have to give meaning. In any empirical study, it is important to examine, think about, and interpret the complex and its conjugate, and normally this is undertaken by viewing the backcloth at all dimensional levels. This is equivalent to putting on spectacles with interchangeable lenses that can see only certain dimensional, or q, levels and above. For example, we can view the structure of a conference with such high-dimensional lenses or spectacles that no participant simplex comes into our view. As we gradually lower the q-value, or the dimensional level, the participant simplices of varying dimensionality appear and enter the complex, and gradually connect with others according to the interest vertices they share in common. The interests, of course, form the conjugate structure – polyhedra of interests whose vertices are defined by the participants.

Interestingly, the homology of the complex and conjugate are the same (Dowker, 1951), and this raises, in the particular context of empirical research, how such properties might be interpreted. For example, in a two-part invention of Johann Sebastian Bach, a q-hole appears in the conjugate structure, where the simplices are the notes, at $q = 2$ (and at $q = 3$ for the three-part invention), a homological characteristic of the music presumably governed by the rules of harmony and counterpoint of those days, which allowed (and perhaps forbade?) certain transitions, or connections, between the notes – but did not require them. Did not require them because a human being was writing the music and could make choices. A particular two-part invention of Bach is perhaps a form of traffic on an underlying musical backcloth. Arnold Schönberg changes the backcloth to allow traffic previously forbidden. Similarly, in a football (soccer) Cup Final (Gatrell and Gould, 1979), it was possible to see, in 1977, the way in which defensive players of Manchester inserted themselves into the Liverpool structure to alter the homology of the game.

Of course, the aim of a football team is to break up the structure of their opponents, and such characteristics of global structure can be captured by simply recording the number of pieces into which the backcloth falls at various q-levels, and the numbers of gaps that offer varying amounts of obstruction to the transmission of traffic. Needless to say, local structure, structure within a disconnected piece of the backcloth, may also be important, that is meaningful and capable of being given empirical interpretation. Furthermore, from the dimension of a simplex, and the dimension of the space where it first joins others, we can also derive simple measures of *eccentricity* that are in intuitive accord with the ordinary meaning of the word. For example, a highly eccentric person in a conference is going to be relatively disconnected from the rest. In a seminar with students, it is good to have a high-dimensional teacher to provide the initial glue to connect the somewhat eccentric students, and to help the student simplices in the complex to increase their dimensionality by the end of the seminar. Similarly, a highly eccentric football player may be disconnected from the rest of his or her team. Notice that what a defensive player tries to do is to increase the eccentricity of the offensive player by reducing the dimensionality of the face connecting him to his own team. Equally, in remote sensing, an eccentric pixel in the relation Pixels × Radiation Bands, or $\lambda \subseteq P \times R$, may indicate false transmission, or a land pixel forming an island in a large lake of water pixels.

Structural similarity

Such concern for global and local structure raises the question of what we mean by similar structures. In global terms, the mathematician is wholly involved here, for the fields of homology and homotopy theory are intimately concerned with such questions, and definitions are precise. In the finite, noncontinuous realm of empirical research, where we do not allow ourselves the luxury of things that we cannot observe (like the continuum of real numbers), but where we are nevertheless seeking appropriate mathematical descriptions, these rigorous views on structural similarity may have to be enlarged (Johnson, 1981). For example, two simplicial complexes describing two backcloth structures of empirical interest may have the same homotopic structure (or, as it has been called in this finite area, pseudohomotopic, or shomotopic, structure), but such general structural similarity may disguise vast differences in local connection that are of great empirical importance. The meaning lies in the substantive interpretation, not the mathematics, which is the language of the text, although even these sorts of questions may result in great technical difficulties. We still do not have an algorithm for the process of combinatorial search that tells us where the holes, the homotopic objects, are, and what simplices form their boundaries. Thus, we may have to be content, in an empirical sense, with simpler definitions of structural similarity – such as set-preserving properties.

For example, suppose we consider two backcloth structures composed of (a) a set of people in a small organization defined on a set of characteristic attributes or responsibilities, or $\lambda \subseteq P \times A$; and (b) the same set of people in which they are considered both as givers and seekers of advice, or $\mu \subseteq G \times S$. An interesting question is whether there is any similarity between the structure of responsibility and the structure of interaction of the people within the organization. Such a question requires a long and careful search to answer it: first, because the initial relations may be recorded as weighted, perhaps integer values, and a number of binary matrices may be derived by slicing (see below). Second, because set-preserving mappings may be sought at all dimensional levels. Now, it happened in the research I am referring to in this example (Gould, 1984), that no set-preserving mappings (except totally trivial ones at low slicing levels), could be found, implying that one structure of the organization (the structure of the people and their shared responsibility), was not reflected in the structure of interaction. The question then was why? – a question that was unlikely to be even raised outside of this particular and careful structural approach.

As we have seen, relations, and therefore structures, are described by binary matrices $\mathbf{\Lambda}$, and these may be derived by slicing weighted matrices, either by choosing a single slicing value Θ, a slicing row or column vector Θ_i and Θ_j, or even a slicing matrix Θ_{ij}. In essence, these are mappings of the form $\Theta : \mathbf{\Lambda} \to \{1,0\}$. Such mappings, when used to derive a series of geometric texts, often cause considerable discomfort to those approaching empirical problems from more conventional directions. Many practitioners appear to want a method, usually in the form of some sort of simplifying computer algorithm, that makes the research decisions for them, and gives them a strained and highly simplified text to interpret. But the human and biological worlds are complex, and many aspects require slow, careful, patient, and meticulous search to find meaningful (i.e. interpretable) structures. If we have integer-weighted relations, we may have to explore the structures with many slicing mappings, but this is actually in the best tradition of science, and has the advantage of keeping us very close to the original data. In these days of computer simplification, this is a positive asset: otherwise, we are back in *a priori* social physics, where everything is crushed down by least-squares to simple, usually linear, functional forms. Indeed, patient structural search, by trying to find the right mappings – right in the sense that they define a structural text that is rich in interpretable possibilities – can often disclose important structural change that is simply obscured by conventional approaches. For example, time series of Portugal's international trade show a small dip after 1974, and then a continuation of upward trends (Gould and Straussfogel, 1984). A careful examination of a series of international trade matrices, by finding the right slicing mappings, discloses the enormous structural change that Portugal actually experienced immediately after the 1974 Revolution, by being crushed down from a 21 to a 5 dimensional simplex. Only when Portugal was seen to go "left... but not too left" was she allowed to reconnect into the international trade structure and regain her former dimensionality.

Once again, this empirical example raises the important conceptual distinction of backcloth and traffic. International trade, used as a crude surrogate to define one aspect of international relations (and, therefore, international structure), is actually traffic being transmitted on a backcloth, a deeper structure of international connections that allows, forbids, but does not require. What are the elements of the set that form the deeper structure of international trade? And notice that most trade flows fairly freely today, implying that the geometry is relatively unconstraining, and that many country simplices share a common face of specific vertices that allows traffic transmission. But simplices that share a face form a *star* – an important part of a structure that is required for the transmission of traffic on the *backcloth* (Johnson, 1983; and Chapter 2 herein).

The association of backcloth simplices with traffic (often in the form of integer numbers), implies a mapping of the set of simplices in the complex to the positive integers, or $\Pi: K \to Z_0^+$, which implies that traffic can be considered as a pattern on the backcloth which can be resolved as $\Pi = \pi^0 \oplus \pi^1 \oplus \pi^2 \oplus \cdots \pi^t \oplus \cdots \pi^N$; $N = \dim K$. Such thinking leads to a rich body of concepts in algebraic topology, and helps us to recognize that the holistic concept of change may consist of change in the traffic on a relatively stable backcloth, in which case we ascribe change to a force (a *t*-force, since the pattern is graded, i.e. intimately associated with the dimensionality of the simplex on which it lives); or the change may result when the backcloth or geometry itself changes.

Some Things to Think About

In this chapter, I have only had space to outline some rather broad and general aspects of current explorations toward an appropriate mathematics for empirical structural research and description. However, I would like to end with a series of questions that I feel are certainly on the conceptual frontiers in the human sciences, and perhaps some of them are provocative enough to engage the attention of mathematicians.

First, if traffic exists on a simplicial complex, what happens to that traffic in the conjugate structure? Suppose we have a relation between a set of towns T = {Abisko, Kiruna...} and a set of amenities A = {post office, country store, garage...}, with people living as traffic on what a geographer would call a central place structure. For example, Mats-Olof Karlqvist exists, let us say, on the Abisko part of the simplicial complex, and contributes to the graded pattern π^2, either as a count of 1, or perhaps as a count of the number of kronor that he spends. But what happens to Mats-Olof in the conjugate structure? This is the structure where the amenities are the simplices, and we have to consider how a graded pattern on a complex is smeared over the conjugate, and what the empirical meaning might be.

Second, considerable attention is being paid to the old aggregation problem, that meso gap between the micro- and macro-levels that so many of us feel in our own areas of concern and research, not the least because algebraic hierarchies of cover sets make such questions explicit (Couclelis, 1977, 1982). It has been proposed, in principle, that we can characterize any thing or any person by a series of binary answers to a string of questions pertinent to our inquiries. Thus, at the lowest relevant level, say $N-3$, we start with a series of binary strings, actually a relation between the elements in one set (say the people in a town), and the set of pertinent characteristics (presumably carefully sorted out). To the degree that responses overlap, we may wish to aggregate to $N-2$ by recording the integer numbers of women and men, the children in the barndaghem (kindergarten), the police, and so on. Since we are often filtering away information, we may also wish to consider how the detailed properties of the geographic space are also discarded, from reality at one end (whatever we might mean by that), to a totally abstract geometry at the other. I use the word filter purposely, because mathematicians will recognize that we have been talking about filtrations at many points: for example, when we change our dimensional lenses and when we change our slicing parameters. But the question I want to pose is this: when we move from $N-3$ to $N+4$, from enormous and, by definition, incomprehensible detail, to total and utterly banal aggregation and geometric abstraction, are we, in a sense, moving from the relation to the mapping to the function? Is this why models of social physics work (to the degree that they do), because they are applied at a very high level of aggregation and abstraction in which all the multidimensional structure has been crushed down and filtered away?

Third, there are important questions to examine concerning relationships ***between*** hierarchical structures, where often one is a structure of physical phenomena and the other is a structure of human control. Electrical power systems, for example (Gould, 1982), start at the $N-3$ level of wires, bearings, circuit breakers, washing machines, and electrical eggbeaters, and aggregate eventually to the North American Power Control System at $N+5$. One more eggbeater whipping up a soufflé in New York City, and the aggregating relations produce a brownout, requiring reserve spin power in Northern Quebec to be switched in. But parallel to this is a whole hierarchy of human control, starting with line repair people, and even ordinary citizens telephoning in reports of a power break caused by lightning, to international control centers switching power across frontiers. What are the relations between these? Similarly, in a big irrigation system in India (Chapman, 1983), we have hierarchies of physical structures, from high dams storing water, to field channels and ***naka*** (the little gates that a farmer opens or closes to flood part of a field). Parallel to this physical structure is the hierarchy of control engineers who decide how much water to release (and even whether it shall be power for the cities today or water for the farmers tomorrow), right down through the local irrigation officer (a valuable post, often "bought" because it is so rich in bribery potential), to the individual farmer. To understand an irrigation system requires, first, a sound and appropriate structural description of the parallel, but intimately related, hierarchical structures.

Fourth, backcloth and traffic transmission can interact. For example, in a third world country overloaded lorries (i.e. transmitted traffic), may pound the laterite roads (the backcloth) to pieces. Moreover, what is a fire but the transmission of traffic on an inflammable, hierarchical backcloth? Furnishings, like curtains at $N-3$, aggregate to rooms at $N-2$, to houses at $N-1$, to neighborhoods at N...and so on. Yet the transmission of this traffic destroys the backcloth, and fires are often stopped by increasing the obstruction to transmission by changing the structure – as every firefighter knows.

Fifth, some attention is being given today to a variety of algebras (Heyting, Free Boolean, and so on), and their lattice representations – for example, a Galois lattice of a Heyting algebra (Ho, 1982). These are generated essentially from the use of the AND $\wedge$ and the OR $\vee$ binary operations, and the NEGATION $\sim$ on a set, and the claim has been made that because the simplices of a complex are created from vertices joined together, they are actually propositions formed from $\wedge$ AND. For example, a farmer simplex is a Tractor AND a Field AND an Irrigation pump AND...so on. Thus, an actual, empirically verifiable simplex is a point on a Galois lattice, or perhaps a lattice characterizing a Free Boolean algebra (Couclelis, 1983). I personally find this true, but unhelpful, and not particularly illuminating, and I wonder if the love of logic and mathematical formalism for its own sake (in mathematics perfectly legitimate and even desirable), has swamped thinking about what is required for the descriptive task of the substantive scientist? I have great difficulty with that OR $\vee$ operation, since I cannot interpret it when I am trying to build, and make operational, structural texts. Structures are composed of connections: this AND this AND this AND.... In marked contrast, the OR operation says this OR this OR this OR..., and I feel myself as traffic, perhaps decision traffic, bouncing around from one part of a structure to another. Furthermore, if empirically defined, simplicial complexes are large and difficult to comprehend, and these are only a very small subset of the lattice of a Free Boolean algebra, then how are we to grasp, and ultimately interpret, the combinatorially explosive possibilities of the lattice vertices? In aggregating up an algebraic hierarchy of cover sets, we may wish to use definitions employing $\vee$ OR – in the television research, for example, the employment of any descriptive term at the $N+1$ level, such as dance OR sculpture OR poetry..., may be sufficient to invoke the $N+2$ level term Art. But that is a matter of formalizing what we want to obtain as a meaningful, that is, interpretable, text or description, rather than being subject to the tyranny of a binary operation because it happens to characterize a particular propositional logic and its lattice representation.

And this leads, finally, to my last series of questions: is it true that all mathematics can be reduced to binary operations that have a few fundamentally logical counterparts, and does this mean that all mathematics, ***in principle***, can be reduced to ***mechanical*** operations? For example, is the binary operation of path composition in homotopy theory just another version of the logical AND $\wedge$

operation in the propositional calculus? In brief, can the computer, by turning on and off mechanical switches, do everything in mathematics? Are mathematicians ultimately Turing's men? Are laws in the physical sciences actually ***contained*** in the algebraic operations which are applied to the definitions made, and the number systems chosen (Gould, 1983)?

My intentions should not be misunderstood here: when I ask "Is mathematics ultimately mechanical, reducible to mechanical operations on sets of things", I do not mean to imply in the least that creative mathematicians are machines, that brilliant discoveries of proof form are somehow denigrated and dismissed. But once an area of mathematics is broken open, developed and extended, could those developments ***in principle*** be mechanized by Turing's machine? If so, I cannot see how the human world is mathematizable, since all the structures, ultimately grounded on some sort of binary operation, would have to be ***requiring***. When you do m binary opertion c^2 it requires E. But I know that the geometries of the human world are not requiring, and the proof of a theorem meaningful to the human world would imply the lack of that very capacity of consciousness that would allow self-reflection resulting in a choice to deny the consequences of the logical chain. Human beings have choices: farmer polyhedra in a small valley in Portugal (Gaspar and Gould, 1981) can, but need not, produce large traffic values of maize, apples, olives, and animals simply because they are defined as large multidimensional pieces of the structure of agriculture in the region. The capacity to reflect, to think, to alter the geometry ***consciously*** is ultimately ***us***. Can mathematics contain within itself allowing, forbidding, but not requiring, and, therefore, not mechanical, geometries? Or is our natural language, in which we think, and so have our being, also reducible to the software mechanisms of artificial intelligence and Turing's machine that requires us to say: I (binary operation) love (binary operation) you?

References

Atkin, R. (1974) ***Mathematical Structure in Human Affairs*** (London: Heinemann Educational Books).

Chapman, G. (1983) Underperformance in Indian irrigation systems: the problems of diagnosis and prescription. ***Geoforum*** 6: 23–32.

Connor, S. (1983) Gene probes find Huntington's Disease. ***New Scientist*** 99: 399.

Couclelis, H. (1977) ***Urban Development Models: Towards a General Theory***, PhD Dissertation, University of Cambridge.

Couclelis, H. (1982) Philosophy in the construction of geographic reality, in P. Gould and G. Olsson (Eds) ***A Search for Common Ground*** (London: Pion) pp 105–38.

Couclelis, H. (1983) On some problems of defining sets for Q-analysis. ***Environment and Planning B*** 10: 423–38.

Dowker, C. (1951) Homology groups of relations. ***Annals of Mathematics*** 56: 84–95.

Eddington, A. (1935) ***New Pathways in Science*** (Cambridge: Cambridge University Press).

Gaspar, J. and Gould, P. (1981) The Cova da Beira: an applied structural analysis in agriculture and communication, in A. Pred (Ed) ***Space and Time in Geography*** (Lund: G.W.K. Gleerup) pp 183–214.

Gatrell, A. and Gould, P. (1979) A structural analysis of a game: the Liverpool ***v*** Manchester United Cup Final of 1977. ***Social Networks*** 2: 253–73.

Giorello, J. and Morini, S. (1983) ***Paraboles et Catastrophes: Entretiens avec René Thom*** (Paris: Flammarion).

Gould, K. (1984) *Structural Approaches to Communications Research*, MSc Thesis, The Pennsylvania State University.

Gould, P. (1982) Electrical power failure: reflections on compatible descriptions of human and physical systems. *Environment and Planning B* 8: 405–17.

Gould, P. (1983) Things I do not understand very well. I: Are algebraic operations laws? *Environment and Planning A* 14: 1567–9.

Gould, P. and Straussfogel, D. (1984) Revolution and structural disconnection: a note on Portugal's international trade. *Economia* 7: 435–53.

Gould, P., Johnson, J., and Chapman, G. (1984) *The Structure of Television* (London: Pion).

Habermas, J. (1971) *Knowledge and Human Interests* (Boston: Beacon Press).

Ho, Y. (1982) The planning process: structure of verbal descriptions, *Environment and Planning B* 9: 397–420.

Heidegger, M. (1977) *The Question Concerning Technology and Other Essays* (New York: Harper and Row).

Heidegger, M. (1979) *Nietzsche, Vol. I: The Will to Power as Art* (New York: Harper and Row).

Johnson, J. (1981) The shomotopy bottle of Q-analysis, *Int. J. Man–Machine Studies* 15:457–60.

Johnson, J. (1982) Q-transmission in simplicial complexes, *Int. J. Man–Machine Studies* 16: 351–77.

Johnson, J. (1983) Q-analysis: a theory of stars, *Environment and Planning B* 10: 457–69.

Keller, E. (1983) *A Feeling for the Organism* (San Francisco: W. Freeman).

Kline, M. (1980) *Mathematics: The Loss of Certainty* (Oxford: Oxford University Press).

Sayre, A. (1975) *Rosalind Franklin and DNA* (New York: W. Norton).

Sugiura, N. (1983) *Rhetoric and Geographers Worlds: The Case of Spatial Analysis*, PhD Dissertation, The Pennsylvania State University.

Vines, G. (1984) Clues to cancer from a backwater of science. *New Scientist* 100: 10.

Wilson, A. (1970) *Entropy in Urban and Regional Modelling* (London: Pion).

CHAPTER 2

A Theory of Stars in Complex Systems

> Heavier-than-air flying machines are impossible.
> *Lord Kelvin*

Jeffrey Johnson

Introduction

A complex system such as a company, an institution, or a nation, can be thought of as being made up of very many interrelated parts experiencing local or global change through time. The practical need to control social institutions through planning, management, and government underlies the need to find scientific methods to describe and understand complex social systems, in the same way that the requirements of engineering underlie the need to find scientific methods to describe complex physical systems.

Let us assume that one can discriminate two kinds of data, the first concerns observing ***relations*** between things and the second involves assigning ***numbers*** to things. Somewhat simplistically, relational data lead to ***structure*** in an algebraic sense while numerical data lead to ***functions*** in a statistical or analytic sense. In other words, one has to combine both algebraic and functional methods to describe complex systems and the ways they change.

In his methodology of Q-analysis Atkin (1974, 1977, 1981) considers systems to consist of a relatively static backcloth which supports a relatively dynamic traffic of system activity. For example, an electrical network provides a backcloth for a traffic of electrical currents, a road system provides a backcloth for a traffic of vehicles, a university provides a backcloth for a traffic of ideas, and so on. It is argued that combinatorial structure in the backcloth constrains the behavior of the traffic, this combinatorial structure being determined by the notion of connectivity. Atkin's work begins with Dowker's observation that one can construct two sets of polyhedra from a binary relation between two sets A and B (Dowker, 1951); for example, the polyhedron determined by any particular a in A has as vertices those b in B which are related to a. The notion of connectivity is then defined in terms of the faces that polyhedra share; in other words the connectivity between two members of A is determined by the elements of B related to both. This is explained in more detail on pp 23–29 and is illustrated with the visual structure of a Tudor village and with the structure of the rectangular grid used in image processing.

The theory of Galois stars, developed on pp 29–35 and illustrated with the structure of an image and stars in the Bedford road system, generalizes Atkin's idea of connectivity by considering the intersection of many polyhedra in a star-like configuration. This intersection is identified as a hub polyhedron, where stars and hubs have a kind of mathematical duality related to the structure of the Galois lattice (Johnson, 1983b). It is argued that this star configuration is fundamental in analyzing the structure of a hierarchical backcloth of polyhedra defined by relations between hierarchically aggregated sets. Traffic is defined on this hierarchical backcloth as functions mapping the polyhedra to numbers, where the traffic functions themselves have hierarchical structure induced by the backcloth. It is a fundamental tenet of this methodology that the star structure of the backcloth constrains the values the traffic functions can take on the polyhedra – it allows and forbids, but does not require (see Chapter 1).

Within this context, in this chapter we define a general ***Q*-system** (Johnson, 1982b) which is a hierarchical backcloth supporting hierarchical traffic. It is suggested that the new star structure provides a superior method of analyzing the backcloth of Q-systems in order to understand their static, kinematic, and dynamic properties.

On pp 35–38 an introduction is given to the ideas of traffic and change being transmitted through the backcloth, this being illustrated by the structure of road system designs constraining vehicle flows. This is followed by discussion of a view of hierarchical structure developed during exceedingly difficult research into classifying television programs (Gould *et al.*, 1984), in which the nature of hierarchies and how they are defined to study complex systems is briefly described. The cone construction is introduced and used to illustrate the Non-partition Principle, The Nested Base Rule, the connectivity between words labeling elements and structures in the hierarchy, and The Principle of Usefulness. The star structure of Tudor Lavenham is briefly discussed to illustrate the idea of hierarchical set definition based on combinatorial structure.

On p 46 the idea of the Q-system is introduced and the hierarchical aggregation of system traffic is considered in some detail on pp 46–50. In particular, it is argued that some descriptions of complex systems which ignore the nature of traffic aggregation over the backcloth may profoundly misrepresent the system under study. Thus, models of complex systems that are equations alone are unlikely to capture the complexity that *is* the system.

Despite fundamental questions about the nature of time having caused a revolution in the conception of physical dynamics, analysts of complex social systems often accept Newtonian clock time as an *a priori* reality: time is another numerical variable. This flies in the face of observation that social events unfold in historical social time. On pp 51–60 the nature of social time and its relationship to clock time are investigated. In this context, the final section investigates the concept of prediction in social time and clock time, and gives eight different kinds of predictions that can be made. The practical necessity to tie social time to clock time in an effective way is investigated through the idea of a clock-time window for favored events, or ***goals***. The section ends with some speculations on heuristics for planning and managing complex systems, where these involve a structure between planned system trajectories through time and the system states (ideally the goals) which make up the possible trajectories. In particular, it is suggested that stars in this structure with large numbers of simplices and high dimensional hubs may be strategically advantageous.

As the main purpose herein is to convey ideas rather than prove mathematical results which can be found elsewhere, a conscious effort is made to keep notation to a minimum. Likewise, some definitions are not presented as pedantically as they might be, preference being given to illustrating the ideas with diagrams. All relevant references to the details are included herein. In particular, Johnson (1981a) establishes a notation of Q-analysis, Johnson (1983a) gives a detailed discussion of hierarchies in Q-analysis, Johnson (1982b) gives the complete definition of the Q-system, Johnson (1976) and Johnson (1981b) contain theorems which show that the road system backcloth constrains vehicular traffic flow, Johnson (1982a) gives a discussion of time in the context of q-transmission, and Johnson (1983b) presents the theory of stars. Atkin's texts (1974, 1977, and 1981) are standard references for Q-analysis. Other references of interest are Earl and Johnson (1981), which discusses the relationship between Q-analysis and graph theory, Gould (1980) and Beaumont and Gatrell (1982), which give pedagogic accounts, and Griffiths (1983) and Seidman (1983), which give different mathematical perspectives on Q-analysis. Many of these papers appear in two special issues on Q-analysis (Johnson, 1981d, and Macgill, 1983).

Polyhedra and Their Connectivity

Let λ be a relation between two sets A and B; that is, there is a rule which enables one to decide that a is λ-related to b or a is not λ-related to b, for each a in A and each b in B. For example, let A be a set of squares and B be a set of eight crossline types, so that each square is crosshatched by four of the eight crossline types. This is illustrated in Figure 2.1, where two squares are crosshatched by the line types β, γ, and δ, but square 1 also has line type ε and square 2 has line type ϑ. Clearly, the visual effect of these lines in combination is different to the visual effect of the lines seperately.

Each of the squares has four line types and can be represented by a tetrahedron, a polyhedron with a vertex for each line type, as shown in Figure 2.2(a). Figure 2.2(b) shows the polyhedra drawn such that they have a common β–γ–δ polyhedron, a triangle. For notation, the square 1 polyhedron is written as $< \beta, \gamma, \delta, \varepsilon >$ and the square 2 polyhedron as $< \beta, \gamma, \delta, \vartheta >$. The ***shared face*** of the polyhedra is the triangle $< \beta, \gamma, \delta >$.

In general, a polyhedron with n vertices exists in an $(n-1)$-dimensional $[(n-1)$-D$]$ space: a point (one vertex) is 0-D, a line (two vertices) is 1-D, a triangle (three vertices) is 2-D, a tetrahedron (four vertices) is 3-D, a pentahedron (five vertices) is 4-D, and so on. All polyhedra of dimension greater than two have to be drawn on 2-D paper as stylized representations, see Figure 2.3.

Since squares 1 and 2 share a triangle [Figure 2.2(b)], they can be described as two-dimensionally near, or 2-near. This idea can be generalized: let two polyhedra be ***q-near*** if they share at least $(q+1)$ vertices (Figure 2.4); that is, they

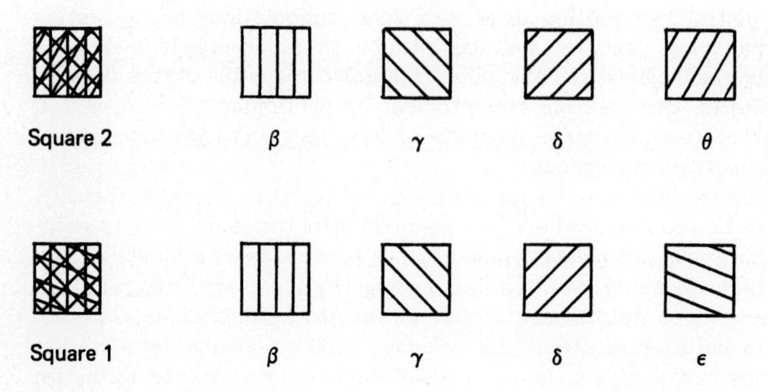

Figure 2.1 Squares crosshatched by different sets of line types.

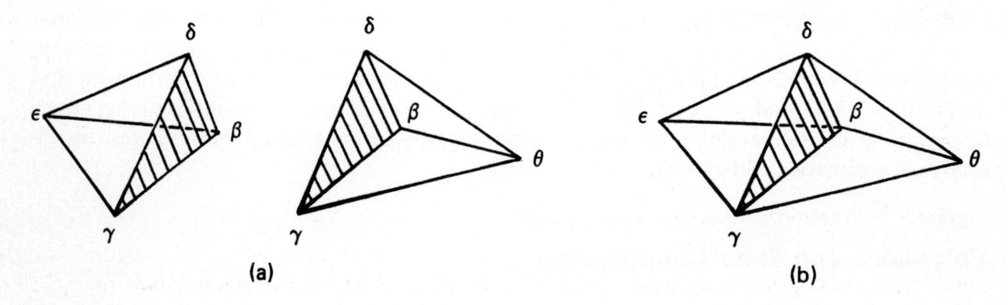

Figure 2.2 (a) Squares 1 and 2 of Figure 2.1 represented as polyhedra. (b) Squares 1 and 2 share a triangle.

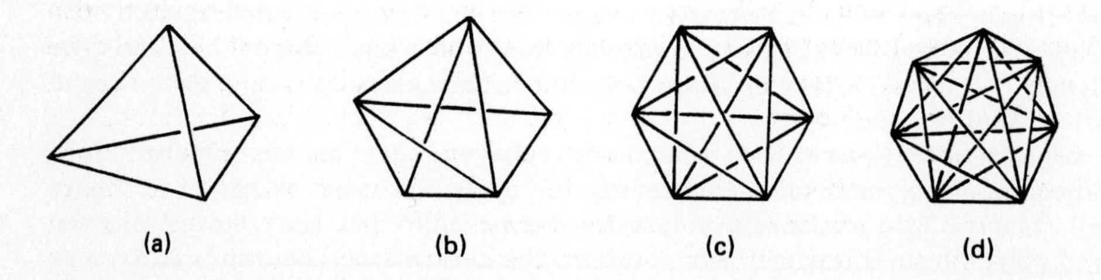

Figure 2.3 Multidimensional polyhedra represented on 2-D paper: (a) tetrahedron, (b) pentahedron, (c) hexahedron, (d) septahedron.

share a q-dimensional *face*. Figure 2.5 shows how this idea can be extended: let two polyhedra be ***q-connected*** if and only if there is a chain of pairwise, q-near polyhedra between them.

For any set of polyhedra of dimension q or greater, it can be shown that q-connectivity is an equivalence relation and so partitions the polyhedra into disjoint equivalence classes called ***q-connected components***. The Q-analysis

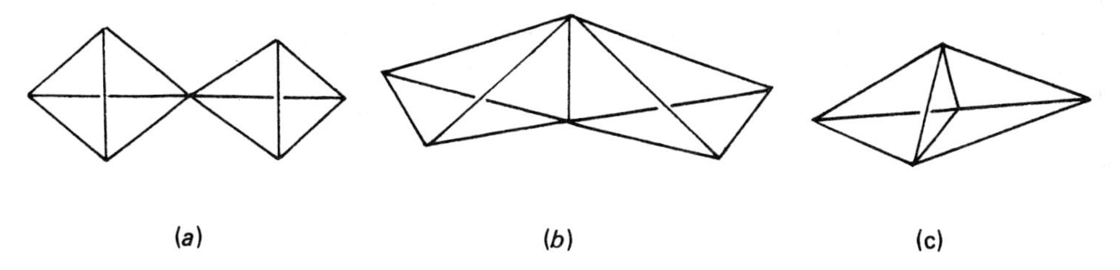

Figure 2.4 (a) 0-near, (b) 1-near, and (c) 2-near polyhedra.

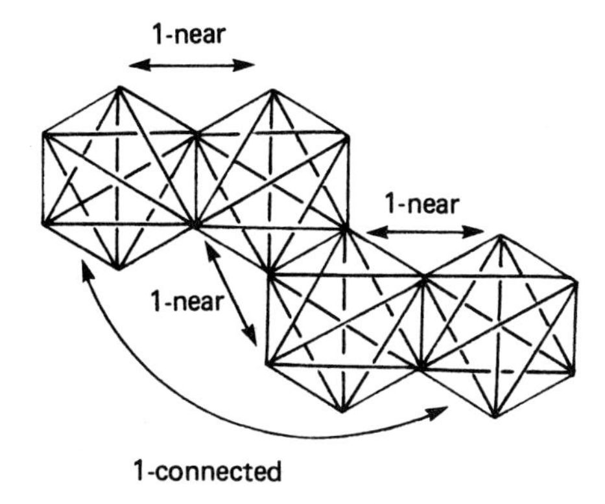

Figure 2.5 Polyhedra that are q-connected.

algorithm lists the q-connected components of a relation for each value of q, as illustrated in Figure 2.6.

Polyhedra are also called ***simplices***, an arbitrary set of simplices is called a ***simplicial family***, and a set of simplices which contains all the faces of its simplices is called a ***simplicial complex***. Let λ be a relation between $A = \{a_1, a_2, \ldots, a_m\}$ and $B = \{b_1, b_2, \ldots, b_n\}$, and let $\sigma(a_i)$ be a polyhedron with vertices all those members of B which are λ-related to a_i. The set of polyhedra $\{\sigma(a_i) | a_i \in A\}$ is a ***simplicial family*** denoted $F_A(B, \lambda)$, which, together with all its faces, is a simplicial complex, denoted $K_A(B, \lambda)$. Let the relation λ^{-1} between B and A be defined by the rule $b \; \lambda^{-1} a$ if and only if $a \; \lambda \; b$. Then the simplicial family $F_B(A, \lambda^{-1})$ is defined analogously to $F_A(B, \lambda)$ as a set of polyhedra $\{\sigma(b_j) \,|\, b_j \in B\}$ with vertices that are members of A. Similarly, $K_B(A, \lambda^{-1})$ is the complex containing these polyhedra with all their faces. $F_A(B, \lambda)$ and $F_B(A, \lambda^{-1})$ are said to be ***conjugate*** families, and the conjugate complexes $K_A(B, \lambda)$ and $K_B(A, \lambda^{-1})$ are sometimes called the ***Dowker complexes*** (Griffiths, 1983).

Example 2.1 The English village of Lavenham is highly regarded for its wealth of surviving Tudor buildings; those in the Market Square are shown in Figure 2.7. Atkin and his coworkers (Atkin, 1971) defined a relation between the buildings and the visual features listed in Table 2.1. The relation $\lambda \subseteq B \times V$ is defined by $B_i \; \lambda \; V_j$ if V_j is a visual feature of facade B_i, in which case building facades can be

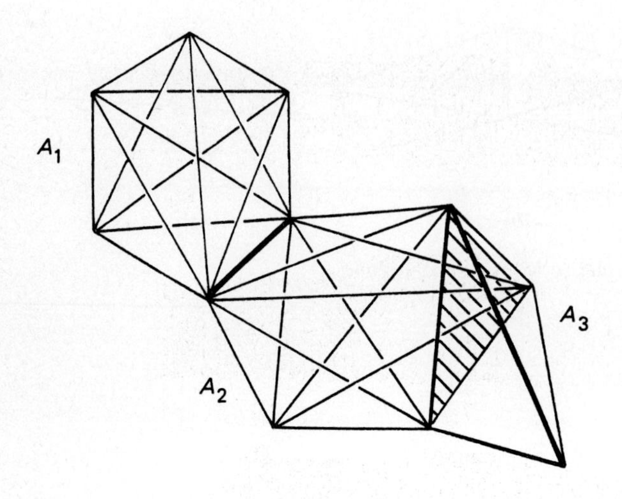

q	q-connected compoments		
5	$\{A_1\}$	$\{A_2\}$	
4	$\{A_1\}$	$\{A_2\}$	
3	$\{A_1\}$	$\{A_2\}$	$\{A_3\}$
2	$\{A_1\}$	$\{A_2,$	$A_3\}$
1	$\{A_1,$	$A_2,$	$A_3\}$
0	$\{A_1,$	$A_2,$	$A_3\}$

Figure 2.6 A Q-analysis of the polyhedra A_1, A_2, and A_3 of Figure 2.2.

Table 2.1 Visual features used to describe buildings in Lavenham Market Square.

Symbol	*Feature*	*Symbol*	*Feature*
V_1	Brick facade	V_{15}	Horizontal window
V_2	Rendered plaster	V_{16}	Square window
V_3	Exposed beams	V_{17}	Bay/Bow window
V_4	Brick and flint facade	V_{18}	Dormer window
V_5	Overhanging first floor	V_{19}	Shop window
V_6	Solid wooden door	V_{20}	Leaded window
V_7	Paneled wooden door	V_{21}	Elaborate chimney
V_8	Glass paneled door	V_{22}	Narrow plain chimney
V_9	Elaborately framed door	V_{23}	Wide plain chimney
V_{10}	Stone lintel	V_{24}	Clay tiled roof
V_{11}	Brick lintel	V_{25}	Slate tiled roof
V_{12}	Wooden lintel	V_{26}	Gable end
V_{13}	Molded lintel	V_{27}	Sloping gable
V_{14}	Vertical window	V_{28}	Passageway entrance

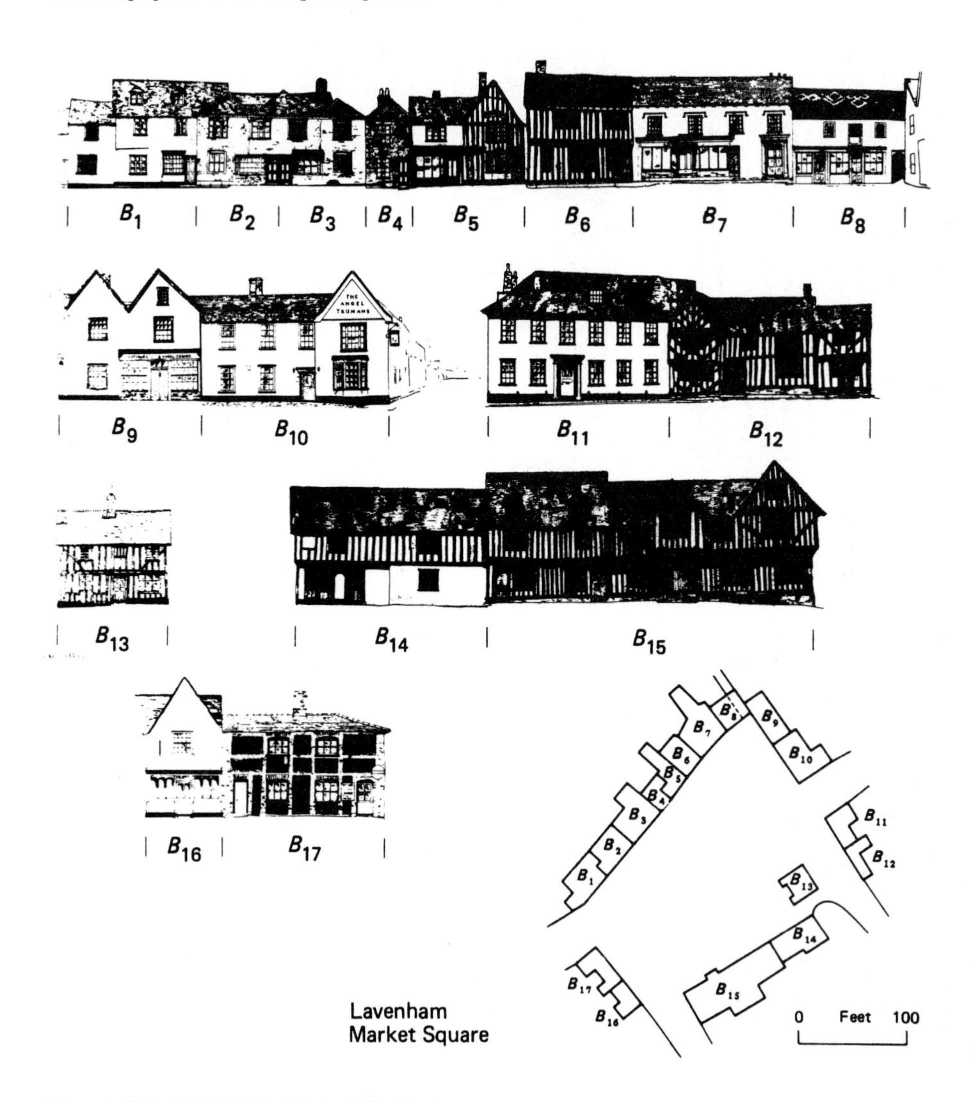

Figure 2.7 Lavenham Market Square.

Table 2.2 The visual polyhedra of some of Lavenham's buildings [the notation $\sigma(B_1)$ means building B_1 considered as a polyhedra].

Polyhedra	*Visual features*
$\sigma(B_1)$	$\langle V_2, V_5, V_7, V_{12}, V_{14}, V_{15}, V_{16}, V_{17}, V_{18}, V_{20}, V_{24}\rangle$
$\sigma(B_2)$	$\langle V_1, V_7, V_{11}, V_{15}, V_{16}, V_{17}, V_{24}, V_{27}\rangle$
$\sigma(B_3)$	$\langle V_1, V_7, V_{11}, V_{15}, V_{16}, V_{17}, V_{19}, V_{20}, V_{22}, V_{24}, V_{27}\rangle$
$\sigma(B_4)$	$\langle V_1, V_7, V_{10}, V_{14}, V_{23}, V_{25}, V_{27}\rangle$
$\sigma(B_5)$	$\langle V_2, V_3, V_6, V_8, V_{12}, V_{15}, V_{17}, V_{19}, V_{21}, V_{24}, V_{26}\rangle$

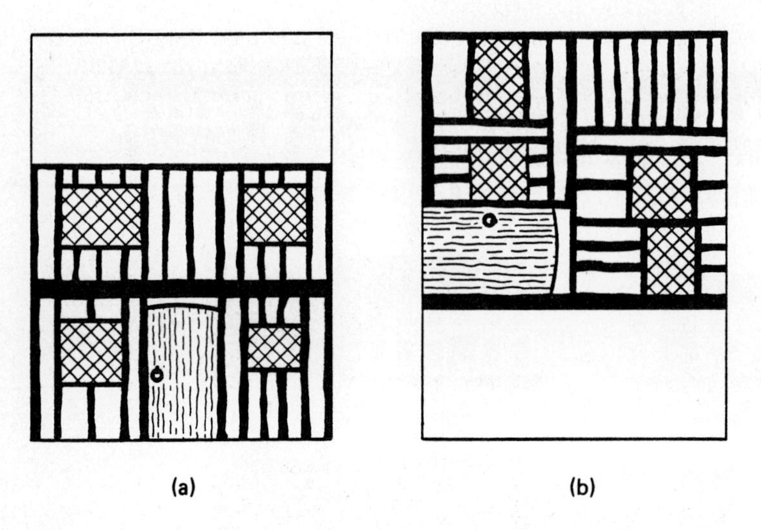

Figure 2.8 Structure *versus* cluster: (a) Tudor house and (b) Tudor house?

considered to be polyhedra made up of their visual features, as illustrated in Table 2.2. A Q-analysis of the buildings has a major component $\{B_{15}, B_6, B_{12}, B_1, B_{13}, B_{14}, B_5\}$ at $q = 6$, and $\{B_{15}, B_6, B_{12}, B_1, B_{13}, B_{14}, B_5, B_3, B_2, B_9, B_{11}, B_{16}, B_8, B_7\}$ at $q = 5$. The first of these contains buildings which conform to the popular concept of a Tudor house, while the Tudor features in the second component are somewhat more sparse. This example shows that the general concept of a Tudor house and what is actually a Tudor house can be different: in terms of appearance houses built in Tudor times have often been modified according to the needs and customs of later generations. Thus, at any time the visual structure of a town is evolving towards a new visual structure.

It would be misguided at this stage to consider Q-analysis as just another clustering algorithm. Figure 2.8 shows two identical sets of visual features, but clearly 2.8(b) could not be considered a Tudor facade or, indeed, any kind of facade. To resolve this problem more structure is necessary, namely a kind of grammar of shapes (*cf.*, Stiny, 1980) to combine the Tudor features properly.

Example 2.2 Here we consider the structure of tessellated space in remote sensing. In a square tessellation let square s be λ-related to square s' if they share an edge or a corner. Then every square is 5-near the squares to its right, left, above,

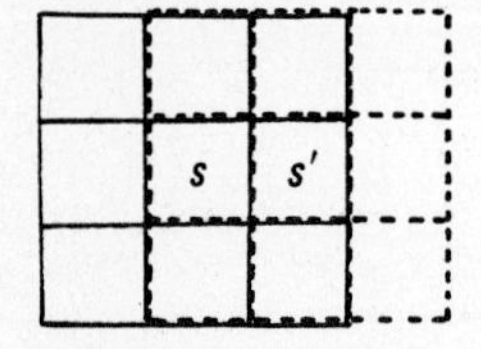

Figure 2.9 The squares s and s' are 5-near (that is, they share a 5-dimensional polyhedral face).

and below (Figure 2.9), and it is possible to show that the tessellated plane is everywhere 5-connected. However, Figure 2.10 shows that the tessellated plane is also everywhere full of 5-holes because diagonally adjacent squares are only 3-near each other.

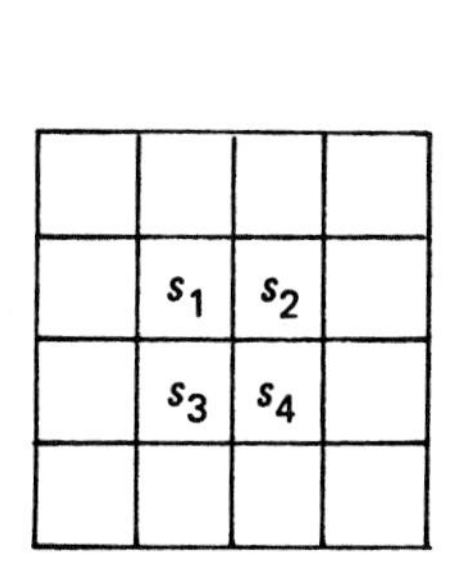

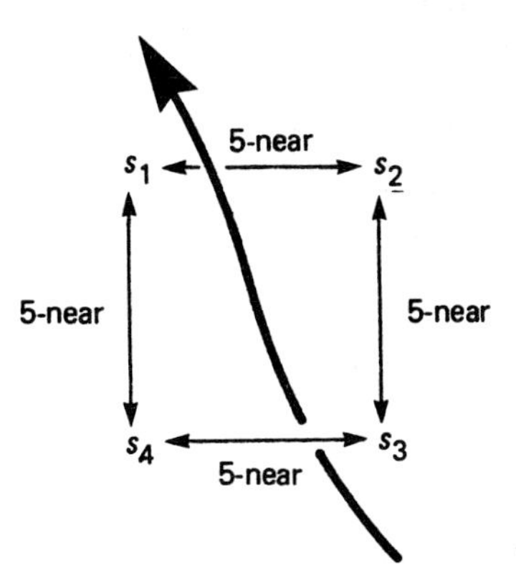

Figure 2.10 The tessellated plane is everywhere full of 5-holes (s_1 and s_3 are only 3-near, as are s_2 and s_4).

An Introduction to the Theory of Stars

Whereas q-nearness is defined by the intersection of *two* polyhedra, the concept can be extended to the intersection of *many* polyhedra. In Figure 2.11(a) there are six tetrahedra which share a common triangle (shaded). These tetrahedra can be brought together at the triangle to form a star structure, as shown in Figure 2.11(b). The common triangle is called the ***hub*** of the star.

Example 2.3 Recall from Figure 2.1 the crosshatching of squares by lines of different slopes. Figure 2.12 shows one hundred and seventeen squares, each related to four of the eight crosshatch line types. Close inspection of the squares shows that each of their polyhedra has the face $< \beta, \gamma, \delta >$ or the face $< \varepsilon, \zeta, \eta >$, but no polyhedron has both these triangles as a face. Thus the squares, viewed as polyhedra, define two distinct stars, as shown in Figure 2.13. These stars can be used to partition the squares according to the star they belong to, which reveals the substructures shown in Figure 2.14. In this particular case the star structure is exactly relevant to the process of pattern recognition.

The concepts of star and hub can be made more precise as follows. Given any polyhedron σ in a simplicial complex $K_A(B,\lambda)$, let the *Galois star* of σ, denoted star(σ), be the set of polyhedra $\sigma(a_i)$ which have σ as a face. The Galois star thus defined is a special case of the concept of star in topology, which not only contains those polyhedra having σ as a face, but also those polyhedra sharing one

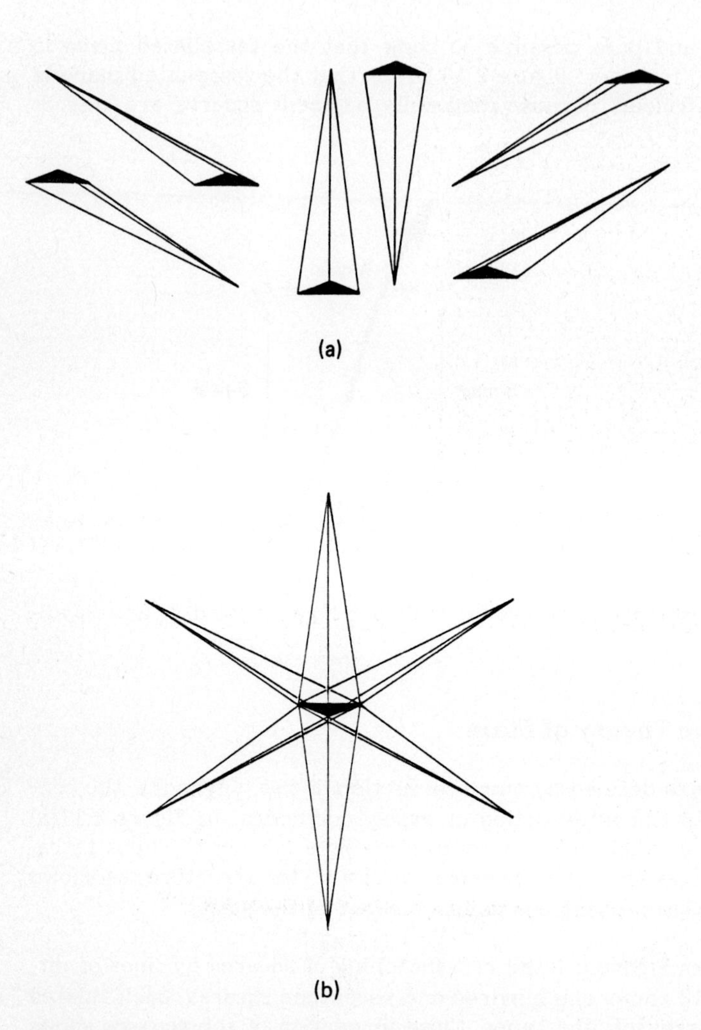

Figure 2.11 (a) Six tetrahedra share a common triangle and so (b) combine to form a star structure with the common triangle as a hub.

or more vertices with σ. In this chapter the term star means only the Galois star defined here. Given any set of polyhedra $\{\sigma(a_1), \sigma(a_2), \ldots\}$, let its ***hub*** be the polyhedron which is the intersection of the $\sigma(a_i)$. Then it can be shown that σ is a face of hub[star(σ)] and in the special case that σ = hub[star(σ)], the polyhedron σ is called a ***maximal hub***. It can also be shown that $\{\sigma(a_1), \sigma(a_2), \ldots\}$ is a subset of star[hub($\{\sigma(a_1), \sigma(a_2), \ldots\}$)], and in the special case that $\{\sigma(a_1), \sigma(a_2), \ldots\}$ = star[hub($\{\sigma(a_1), \sigma(a_2), \ldots\}$)], the set of polyhedra is called a *maximal star*.

Stars and hubs have a number of interesting mathematical properties, but the most important is that in **Proposition 2.1**.

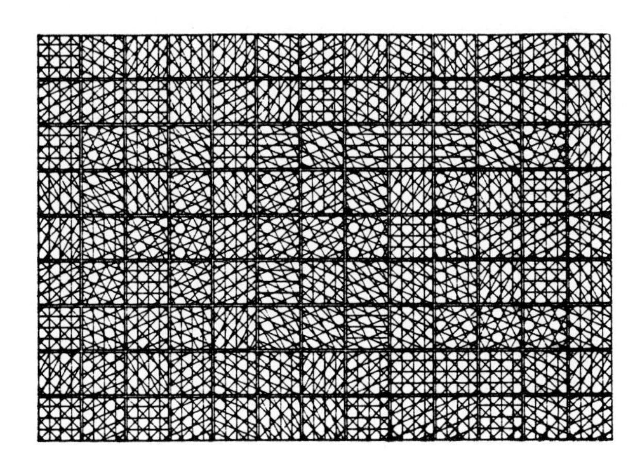

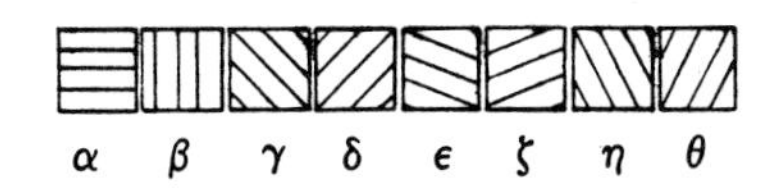

Figure 2.12 Squares crosshatched by four of the eight lines with different slopes ($\alpha - \vartheta$).

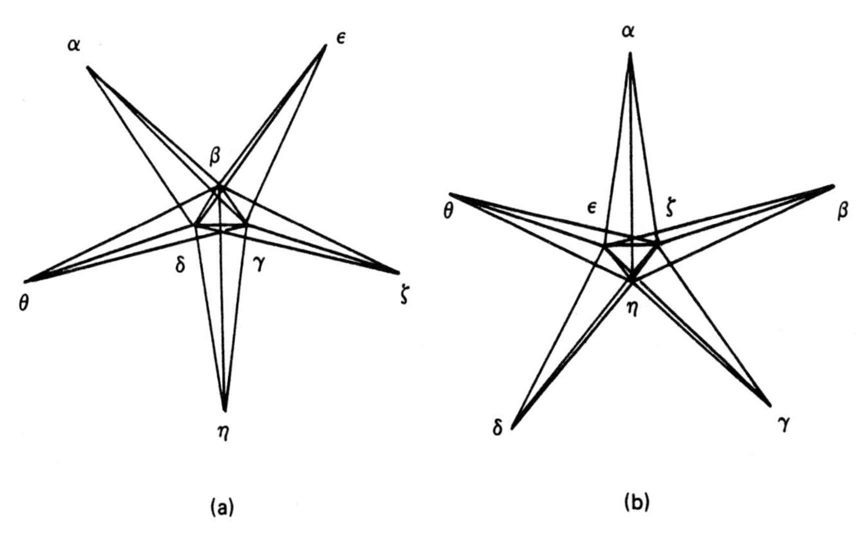

Figure 2.13 Stars formed by (a) $\beta - \gamma - \delta$ polyhedra and (b) $\varepsilon - \zeta - \eta$ polyhedra.

Proposition 2.1

$\{\sigma(a_1), \sigma(a_2), \ldots\}$ is a maximal star with hub $< b_1, b_2, \ldots >$ in $F_A(B, \lambda)$ if and only if $\{\sigma(b_1), \sigma(b_2), \ldots\}$ is a maximal star with hub $< a_1, a_2, \ldots >$ in $F_B(A, \lambda^{-1})$.

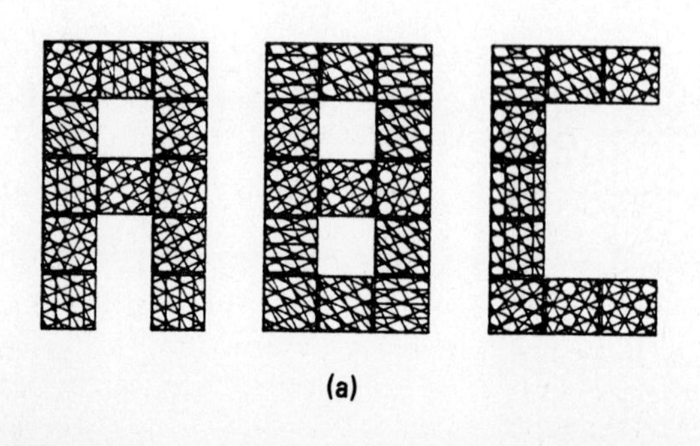

(a)

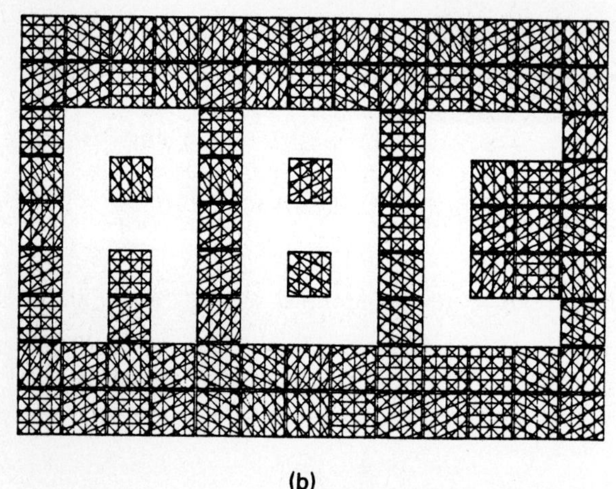

(b)

Figure 2.14 A pattern recognized by the star structure of the squares. (a) Squares belonging to the $\varepsilon - \zeta - \eta$ star and (b) squares belonging to the $\beta - \gamma - \delta$ star.

For the proof of **Proposition 2.1**, see Johnson (1983b, p 461).

In Q-analysis, two simplices are defined to be ***q-near*** if and only if they have a common q-face; that is, they share more than q vertices. Two simplices are *q-connected* if and only if there is a chain of pairwise, q-near simplices between them. Thus, q-connectivity is an equivalence relation on a complex and partitions its simplices into *q-connected components* and a *Q-analysis* lists these components for each value of q. Since the hub of a set of simplices is itself a simplex, let the ***dimension*** of a hub be its dimension as a simplex. To maintain numerical compatibility, let the *codimension* of a hub be defined as the number of named simplices in its star, minus one [in the case that two named simplices are identical, for example, $\sigma(a_1) = \sigma(a_2)$, then each counts separately]. Let a simplex be defined as an *n–q-simplex* if its codimension is n and its dimension is q, and let a set of simplices be *n–q-near* if they all share an n–q-simplex.

Proposition 2.2

q-nearness in Q-analysis is the special case of n–q-nearness, for $n = 1$.

Proposition 2.3

q-connectivity in Q-analysis is the special case of n–q-connectivity, for $n = 1$, defined as the transitive closure of n–q-nearness.

The literature on Q-analysis concentrates on these restrictions, but **Propositions 2.2** and **2.3** suggest that the Q-analysis algorithm can be extended to structures beyond those defined by q-nearness. However, they highlight a problem which is not apparent in the literature. Q-analysis claims to be a language of structure, but all *structure* analyzed through the Q-analysis algorithm is secondary to the q-nearness *function*. However, the Q-analysis algorithm is but a part of the methodology of Q-analysis and most applications of Q-analysis require the discussion to be *vertex-specific*. In other words, the specific vertices and simplices must be considered explicitly, with properties such as dimensions, q-nearness, eccentricity, etc., taking a secondary role as *indicators*.

The star structure developed in this chapter is not secondary to any measure; the stars and hubs simply exist as structures in their own right. We can impose mappings, such as dimension and codimension, on these structures, but these are secondary. The same is not true for the q-connected component structures of Q-analysis, which are secondary to the q-nearness mapping.

Example 2.4 Figure 2.15 shows a simplified version of the road system in the English country town of Bedford. As a simple experiment in Q-analysis, a route was defined for each through-traffic pair AB, AC,...,HA, IA. A route R_i is related to a link L_j if the route traverses that link. The roots are polyhedra connected to each other through shared links, and links are polyhedra connected to each

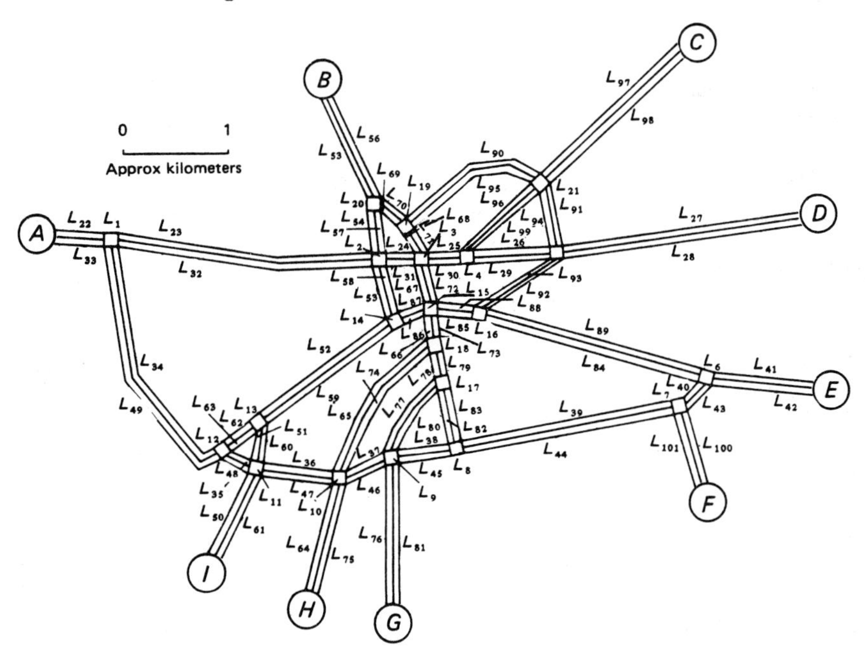

Figure 2.15 Major roads in the Bedford system.

Table 2.3 Major stars in the Bedford road system (through-traffic only).

Southern bypass west–east		
<AE,AF,GE,GF,HE,HF,IE,IF>	<—>	$<L_7,L_8,L_9,L_{38},L_{39}>$
<AE,AF,HE,HF,IE,IF>	<—>	$<L_7,L_8,L_9,L_{10},L_{37},L_{38},L_{39}>$
<AE,AF,AG,IE,IF,IG>	<—>	$<L_9,L_{10},L_{11},L_{36},L_{37}>$
Southern bypass east–west		
<EA,EG,EH,EI,FA,FG,FH,FI>	<—>	$<L_7,L_8,L_9,L_{44},L_{45}>$
<EA,EH,EI,FA,FH,FI>	<—>	$<L_7,L_8,L_9,L_{10},L_{44},L_{45},L_{46}>$
<EA,EI,FA,FI,GA,GI>	<—>	$<L_9,L_{10},L_{11},L_{46},L_{47}>$
Southern bypass interacting 2-way		
<AE,AF,EA,EI,FA,FI,IE,IF>	<—>	$<l_7,L_8,L_9,L_{10},L_{11}>$
<AE,AF,AG,EA,FA,GA>	<—>	$<L_1,L_9,L_{10},L_{11},L_{12}>$
A, origin		
<AE,AF,AG,AH,AI>	<—>	$<L_1,L_{11},L_{12},L_{22},L_{34},L_{35}>$
A, destination		
<EA,FA,GA,HA,IA>	<—>	$<L_1,L_{11},L_{12},L_{33},L_{48},L_{49}>$
E, origin and destination		
<AE,EA,EH,EI,HE,IE>	<—>	$<L_6,L_7,L_8,L_9,L_{10}>$
B, origin		
<BD,BE,BF,BG,BH>	<—>	$<L_3,L_{19},L_{20},L_{56},L_{70},L_{71}>$
B, origin and destination		
<BF,BG,BH,GB,HB>	<—>	$<L_3,L_{15},L_{18},L_{19},L_{20}>$
B and C, origin		
<BF,BG,BH,CG,CH>	<—>	$<L_3,L_{15},L_{18},L_{72},L_{73}>$

other through shared routes. Figure 2.16 shows those links in the system of dimension three or more (Johnson, 1984). These data were reanalyzed by a computer program searching for stars and the largest star–hub pairs are listed in Table 2.3.

As with the Q-analysis, these stars show the through-traffic structure to be dominated by the informal bypass south of the town, east to west and west to east, and northern origins B and C. Many British towns have a radial pattern similar to Bedford, with the problem that many through-traffic routes must pass *via* the town center, which causes severe congestion. It can be seen that Bedford has a significant substructure for the origins–destinations A, E, F, G, H, I through its links south of the center. Also, it is interesting to note how A and, to a lesser extent, E dominate this substructure, because they are at the ends of the bypass links. Hence many routes to and from A must share many links, to their mutual

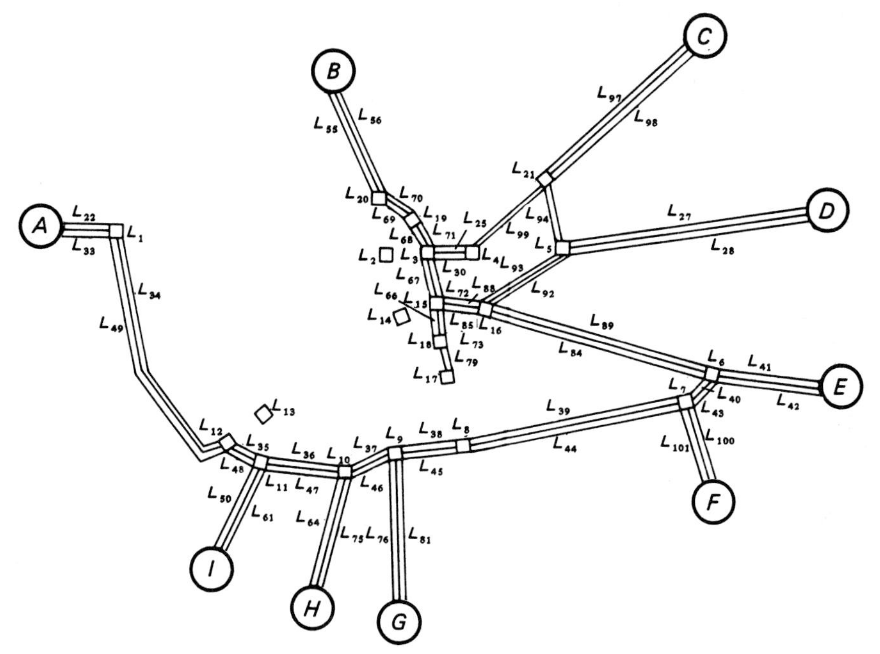

Figure 2.16 Bedford road system, through traffic links with dimension 3 or more.

detriment [disconnected routes perform better (Johnson, 1976)]. This is a property of the backcloth, which might not be significant if A generated little traffic. However, this is not the case; A connects Bedford with many towns to its west and with the M1 motorway, so carrying a large volume of traffic. It is interesting to note that since the original analysis a formal bypass has been opened, which extends the informal southern bypass in the west to avoid the difficult urban roads between I and A.

An Introduction to Traffic in Complex Systems

In complex systems one can make a distinction between *qualitative* data and *quantitative* data, the first being expressed as *relations* and the second being expressed as numerical *functions*. The methodology of Q-analysis makes a distinction between a relatively static *backcloth* structure (qualitative), which supports a relatively dynamic *traffic* of system activity (quantitative). Over and above this, the methodology claims that *the star structure of the backcloth constrains the behavior of the traffic* – it allows and forbids, but does not require (see Chapter 1). To illustrate this, the route-link structure supports a traffic of motor vehicles and constrains this traffic in the following ways. First, the vehicles cannot travel where there is no road: no backcloth implies no traffic. Second, a road may exist and allow vehicles to pass, but there may be none: the backcloth allows traffic, but does not require it.

The previous sections suggest that the structure of relations can be investigated through their families of polyhedra and the related connectivities. In the

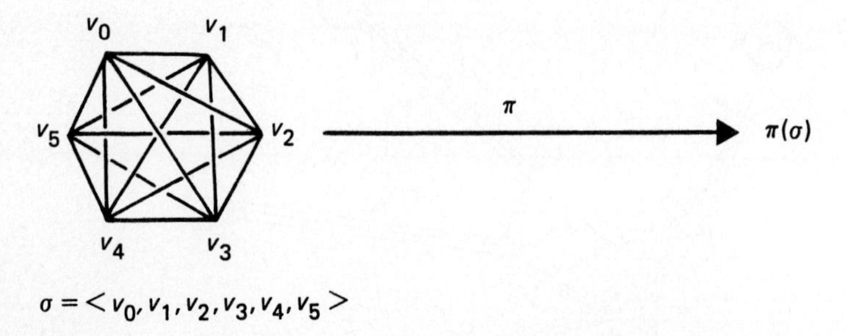

Figure 2.17 Traffic as a mapping from a polyhedron to a number.

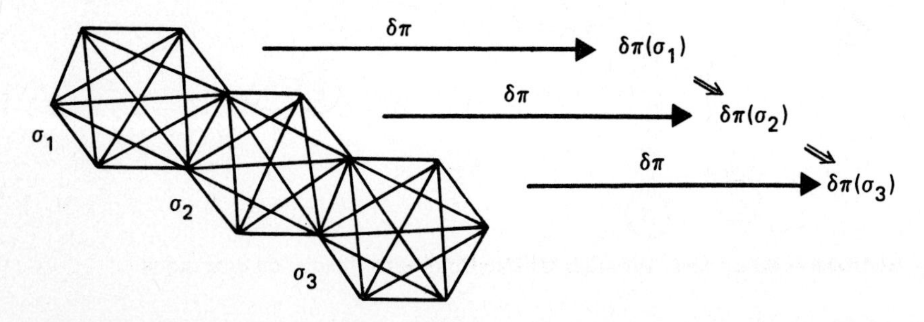

Figure 2.18 Illustration of q-transmission: a change $\delta\pi$ on σ_1 induces a change $\delta\pi$ on σ_2, which induces a change $\delta\pi$ on σ_3.

first instance, the representation of traffic can be thought of as an arrow mapping a polyhedron to a number (Figure 2.17). The theory of q-transmission (Johnson, 1982a) suggests one method in which backcloth structure can constrain traffic. In its simplest case, we can define *change* in traffic to be transmitted from one polyhedron to another through a shared q-face. For example, Figure 2.18 shows three polyhedra with σ_1 1-near σ_2 and σ_2 1-near σ_3. Thus, σ_1 and σ_3 are q-connected, but they do not share any vertices. Under some circumstances, it is possible that a change in traffic, $\delta\pi$, on σ induces a change $\delta\pi$ on σ_2 because they are 1-near. Similarly, the change on σ_2 may induce a change $\delta\pi$ on σ_3 because they are 1-near. In this way, changes in traffic may be transmitted from one polyhedron to another, with which it has no shared vertices, via intermediate connectivities. Figure 2.19 illustrates this for a simple road system with routes between A and C, B and C, and B and D. Suppose there is an accident on link L_2 which restricts its capacity. The resulting traffic flows on L_5 and L_6 decrease because fewer vehicles are able to travel AC in unit time. If the road system is congested it may be that vehicles on route BC are normally slowed by congestion on L_5 and L_6. A reduction in traffic on AB reduces congestion on links L_5 and L_6, and hence vehicles on route BC spend less time on L_{10}, L_9, L_7, and L_8. In turn, this reduction in congestion may reduce travel time on route BD. By this argument reduced flow on AC is 1-transmitted as reduced travel time on BC, which is 1-transmitted as reduced travel time on BD.

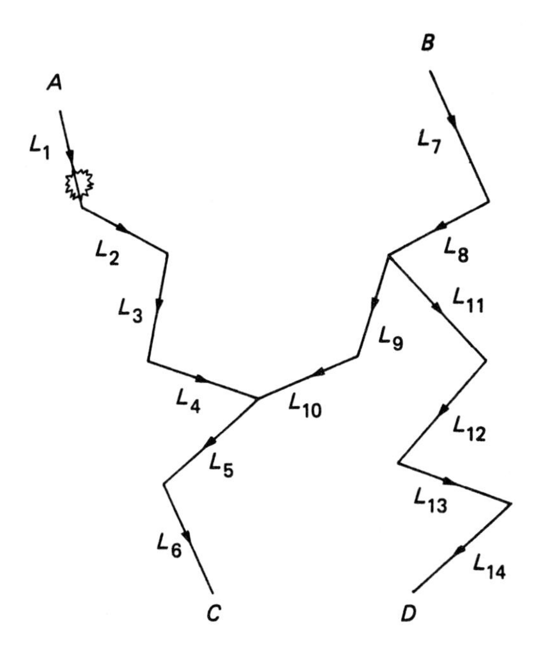

Figure 2.19 Changes transmitted through a single road system.

A number of theorems (Johnson, 1976) show that disconnecting the link and route structures enables road systems to perform better. Dramatic examples of this occur at complicated road intersections which, in Britain, are often roundabouts (called rotaries or traffic circles in North America). For instance, in 1973 one of the roundabouts in the town of Hemel Hempstead was causing severe congestion and an experiment was devised which implemented six mini-roundabouts and two-way flows (Figure 2.20). Intuitively, it is clear that the design before the experiment caused many routes to be highly connected and therefore caused many vehicles to interact with each other, to their mutual detriment. Similarly, it is clear that the experiment has caused at least some pairs of routes (e.g., *ab* and *if*) to become disconnected, to their mutual advantage. These intuitions become more difficult when considering the whole intersection. However, the computer analysis and theorems show the new design to be significantly disconnected and that the traffic flow must improve, as, indeed, it has:

> The current experiment being undertaken at the Plough has, in terms of capacity and reduction of delays, proved a success and a new lease of life has been given to the junction. (*The West Herts Transportation Study*, Hertfordshire Country Council, 1974.)

Vehicular traffic is an example of a much more general idea; for example, Johnson and Wanmali (1981) considered a traffic of retail goods in Indian periodic market systems, Gould (1981) considered a traffic of electricity on an unstable

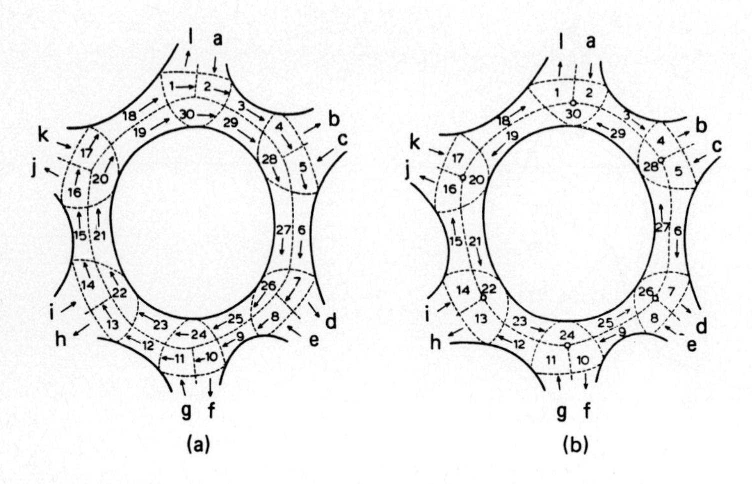

Figure 2.20 Hemel Hempstead system: (a) preexperiment, (b) postexperiment. The post-experiment design significantly disconnects the link and route structure, thereby dramatically increasing capacity.

physical backcloth in the US, Gould *et al.* (1984) considered a traffic of television programs across a world backcloth, Atkin (1977) studied many kinds of traffic associated with university life, and so on.

Hierarchical Structure

As is well known, large complex systems often have hierarchical descriptions in a hierarchical vocabulary. A common problem when studying such systems is that the description may be too coarse, with insufficient detail, or too fine, with overwhelming detail (Figure 2.21). The intermediate word problem involves finding useful words between too high and too low a hierarchical level. This is illustrated in Figure 2.22, where the purpose is to describe television programs. Many hierarchical schemes attempt to give exclusive classes, despite real systems having a natural cover structure with intersections (Figure 2.23). In practice, exclusive classes are not necessary and they violate natural structure in the observations.

An intermediate word like SPORT has many things aggregated into it, and this set can be represented in the usual Venn diagram. If this is joined to a point representing SPORT at a higher level, a cone is formed (Figure 2.24). The cone representation can be useful in a number of ways; for example, it gives a graphic illustration of the Non-partition Principle (Figure 2.25) and it illustrates the kind of structural requirements one would probably impose on a hierarchical scheme (Figure 2.26). These cones can also be thought of in a structural way, as illustrated in Figure 2.27. Here the intermediate words Tudor-1, Tudor-2, and Tudor-3 are connected in terms of their lower level structure. It is interesting to ask where archetypes like TUDOR come from, see Postulates 2.1 and 2.2.

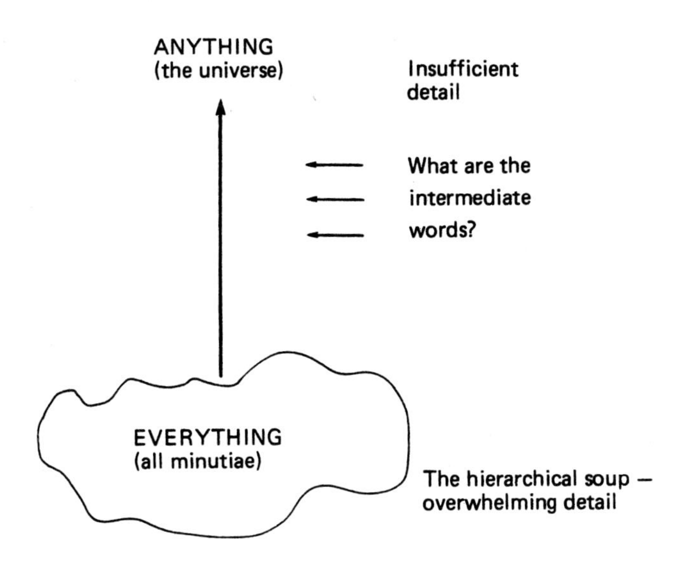

Figure 2.21 The intermediate word problem.

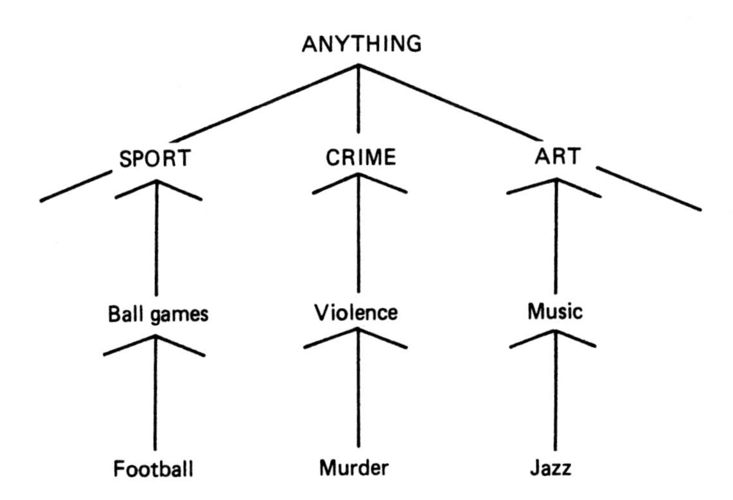

Figure 2.22 Part of a heirarchy for describing television programs.

POSTULATE 2.1 Out of the combinatorially many ways things combine, human beings select a relatively small subset to name.

POSTULATE 2.2 The principle of usefulness; that is, names are given to particular combinations because it is *useful* to do so.

For example, the archetype "Scrambled Tudor" did not exist before, but it does now because it is pedagogically useful (Figure 2.28).

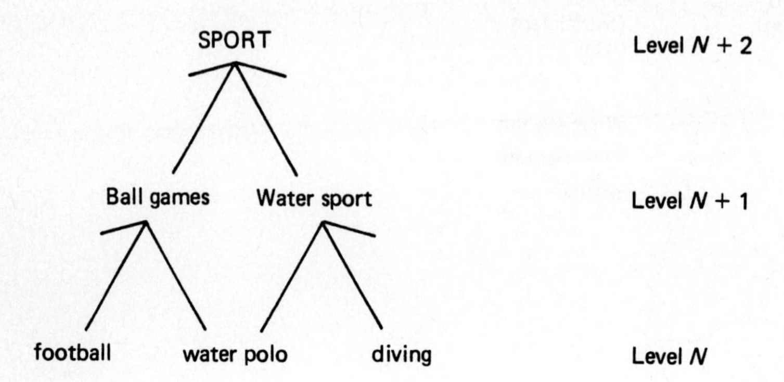

Figure 2.23 Level (N + 1) covers level N, but it ***does not*** partition it.

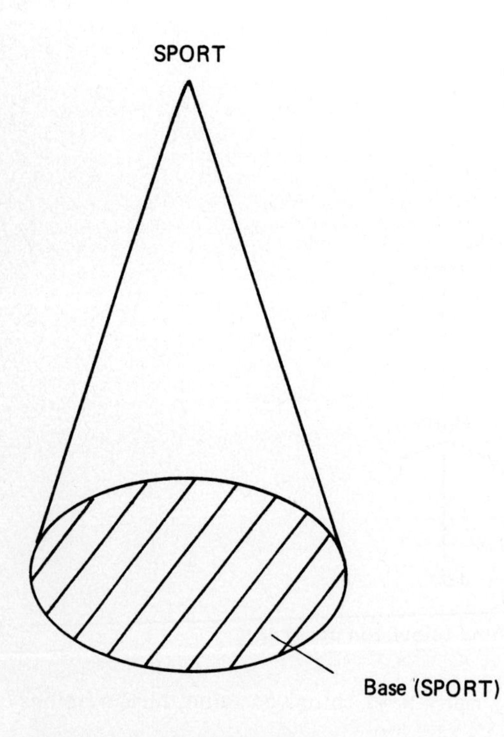

Figure 2.24 The hierarchical cone construction with apex SPORT and base the Venn diagram of the set of things aggregated to SPORT.

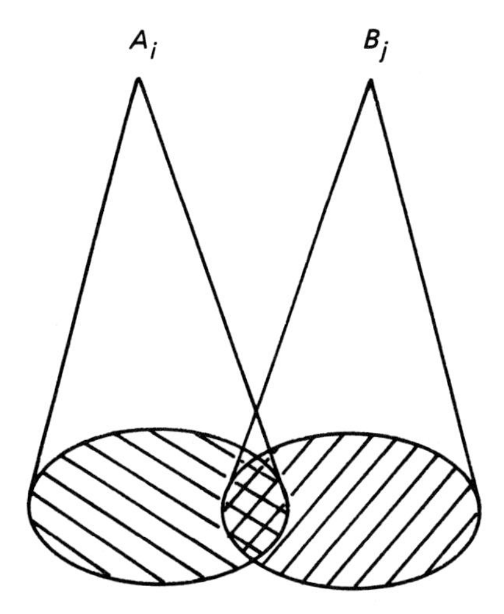

Figure 2.25 The Non-partition Principle.

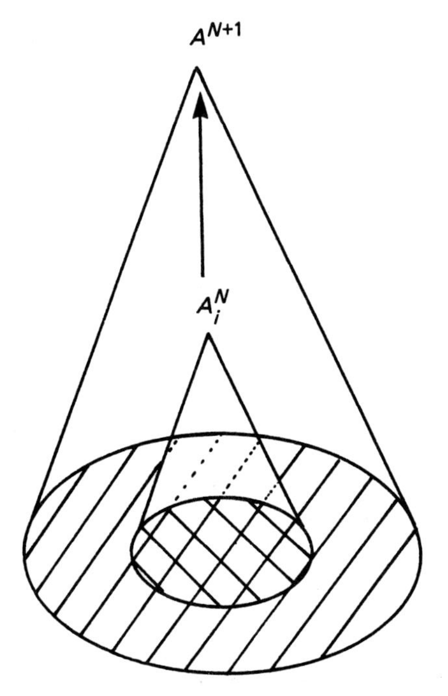

Figure 2.26 The Nested Base Rule [if A_i^{N+1} aggregates into A^N then base(A_i^N) $\subseteq$ base(A^{N+1})].

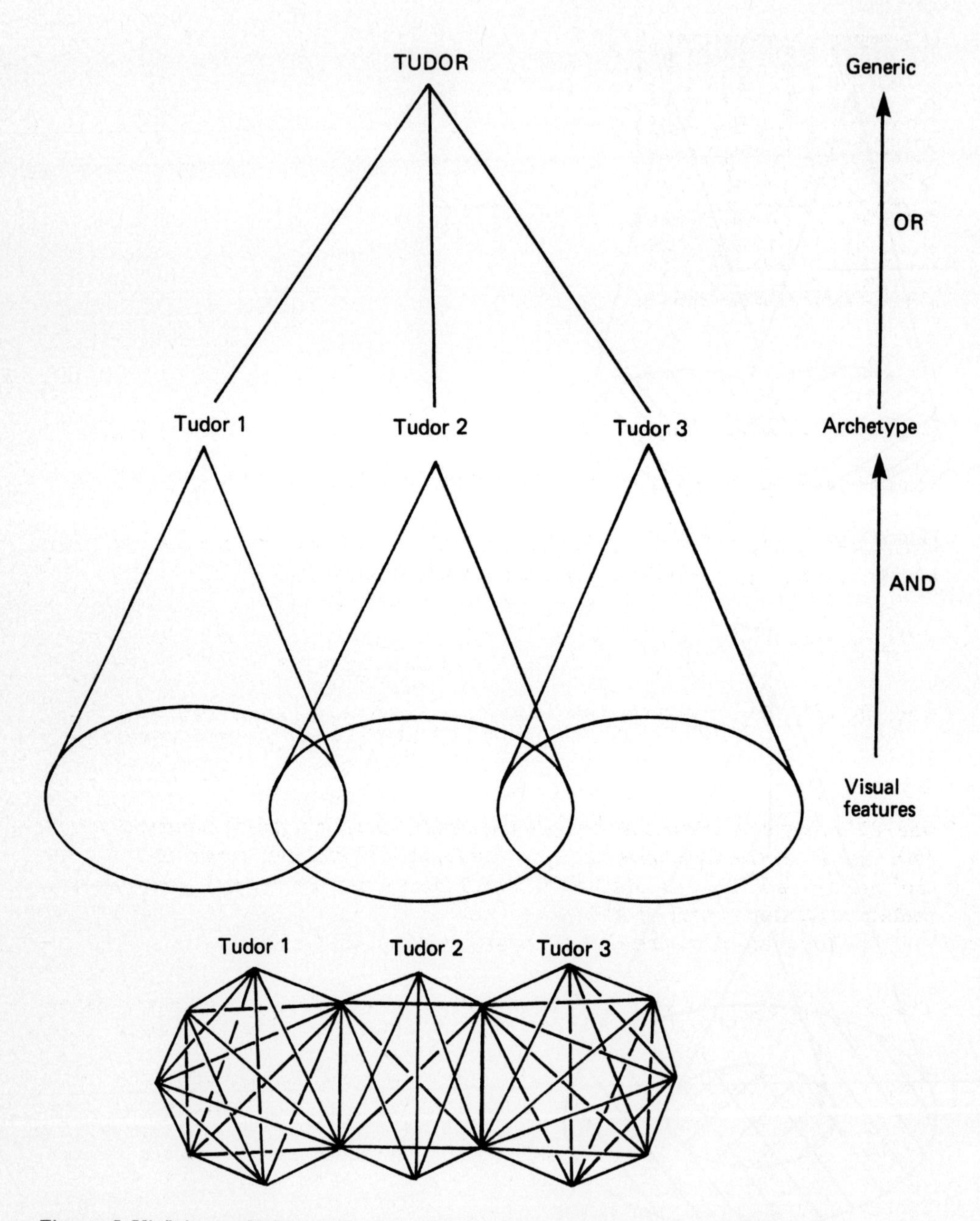

Figure 2.27 Intermediate words are connected through lower level words.

Figure 2.28 Scrambled Tudor house style (compare with Figure 2.8).

Example 2.5 Stars and the hierarchical definition of Tudor house style. The relation between Lavenham building facades and visual features contains the stars:

$< B_1, B_6, B_{12}, B_{13}, B_{15} >$	<—>	<wooden lintel, horizontal window, square window, leaded window, clay-tiled roof>
$< B_6, B_{12}, B_{13}, B_{14}, B_{15} >$	<—>	<exposed beams, wooden lintel, square window, leaded window, clay-tiled roof>

which identify sets of houses that are definitely Tudor in style. It is interesting to note that B_1 in the first does not have the feature of exposed beams in its facade; the beams are doubtless present in the Tudor construction, but they are concealed by plaster rendering and white paint.

The following stars are also associated with a Tudor-style house:

$< B_1, B_6, B_{12}, B_{13}, B_{15} >$	<—>	<wooden lintel, horizontal window, square window, leaded window, clay-tiled roof>
$< B_1, B_6, B_{12}, B_{14}, B_{15} >$	<—>	<first floor overhang, wooden lintel, square window, leaded window, clay-tiled roof>
$< B_1, B_6, B_{13}, B_{14}, B_{15} >$	<—>	<wooden lintel, vertical window, square window, leaded window, clay-tiled roof>
$< B_1, B_{12}, B_{13}, B_{14}, B_{15} >$	<—>	<wooden lintel, square window, leaded window, clay-tiled roof>
$< B_5, B_6, B_{12}, B_{13}, B_{15} >$	<—>	<exposed beams, wooden lintel, horizontal window, clay-tiled roof>

The following stars are interesting because they contain anomalies in terms of a definition of Tudor style:

$< B_3, B_6, B_{12}, B_{13}, B_{15} >$ <—> <horizontal window, square window, leaded window, clay-tiled roof>

$< B_1, B_2, B_3, B_6, B_{15} >$ <—> <paneled wooden door, horizontal window, square window, clay-tiled roof>

In the first of these, B_3 is included amoung the Tudor houses, although it appears Victorian at first sight. In this case the leaded windows are definitely of a later Victorian or Edwardian style which means they, at least, do not date back to Tudor times: perhaps the descriptor leaded window requires further clarification? Another anomaly with this house is its clay-tiled roof, while the construction of its sloping gables is very similar to the Victorian cottage next door. For economic reasons, most Victorian houses have slate roofs, but this one has a clay-tiled roof. Perhaps it has been reroofed since Victorian times with clay tiles, or was it built using old clay tiles from another building?

The second star contains a mixture of Victorian and Tudor house styles, so similar considerations to the previous case apply. Now we have Tudor buildings fitted with paneled doors more characteristic of later times; again, one can speculate that old doors have been replaced with new.

These anomalies show that house styles exist which are hybrid, neither acceptably Tudor nor obviously Victorian. As noted before, the visual structure of an area evolves over time as one generation after another effects repairs and improvements. It is interesting to note that the anomalies are seldom named and that people very often *restore* old properties, which involves removing visually incongruous features and replacing them with others more original. It may be that initially we name combinations of features somewhat arbitrarily, but it would seem that once something has been recognized and named there are strong pressures to make similar objects conform to the name.

The problem of hierarchical set definition, or classification, involves discriminating things which are different and aggregating things which are similar (Johnson, 1981c). Clearly, stars tell us which things are similar and are potentially very useful in this. However, a binary relation between, say, facades and features is not sufficient to discriminate Tudor from scrambled Tudor, because these styles involve relations *between* the elements in the facade: Georgian style windows are arranged in vertical and horizonal rows, roofs are on top, and so on. In other words, there is a complicated set of rules which fit all the parts together properly. Such rules can be thought of as compound propositions which take the value true when things fit together properly and false when they do not. In fact, the propositions can be subdivided in such a way that testing any of the subpropositions as true is sufficient to show that the compound proposition is true; the subpropositions are linked by the conjunction OR. Furthermore, each subproposition is a set of simple propositions involving only the conjunction AND. Any single one of the subpropositions is effectively a set of rules to establish whether a given set of features combines properly for aggregation. Clearly, a necessary condition for this is that the building possesses all the relevant features – that is, it has the correct star structure.

Propositions and relations are effectively the same things; one says $a_1, a_2,...,a_n$ are R-related if and only if $P(a_1, a_2,...,a_n)$ = true, where P defines R. This allows the definition of different polyhedra to those given by a binary relation: let $\langle a_1, a_2,...,a_n \rangle$ be a polyhedron if and only if $P(a_1, a_2,...,a_n)$ = true.

In the case of visual features and buildings it is possible to obtain two different sets of polyhedra: the first by observing the relation λ between building facades and features; the second by defining a set of propositions which test whether a given set of features aggregates to a recognized type. By definition, the aggregation propositions can only be applied if the necessary set of vertices is present; that is, is a face of the λ-polyhedron. Thus, the polyhedral analysis becomes a necessary part of the aggregation procedure and also has considerable heuristic value in eliciting appropriate aggregation propositions. The relational part of an aggregation applies to what has been called the AND aggregation; archetypes are aggregated by a simple OR (Johnson, 1983a). This is illustrated in Figure 2.29.

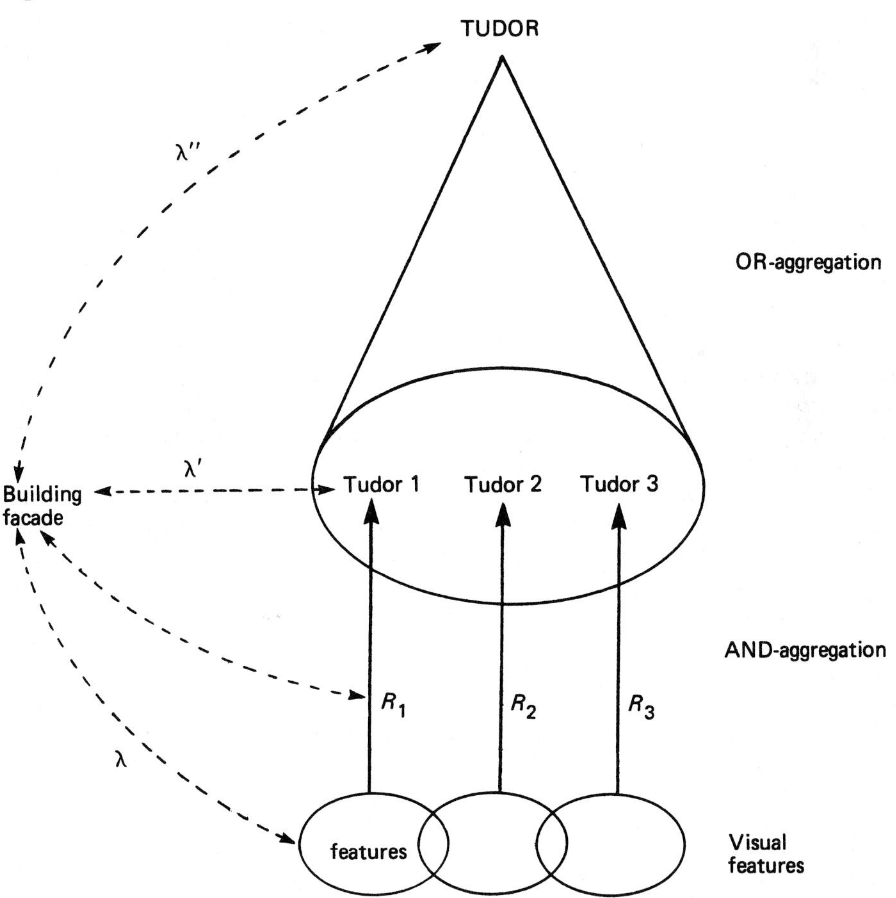

Figure 2.29 Hierarchical aggregation through AND and OR aggregations.

Q-Systems

Let H be a hierarchical scheme of sets; that is, various sequences of cones as defined in the previous section. Let $\{\lambda\}$ be a class of relations between those sets. The simplicial complexes formed from all the relations between the cones are then called a ***hierarchical backcloth***, denoted $(H, \{\lambda\})$. Typically, a simplex in one of these complexes combines a number of vertices from different cones, as illustrated in Figure 2.30. In this diagram traffic is represented by an arrow mapping the simplex into a number; for example, the number of vehicles traversing a link, the amount of grain grown on a piece of land, and so on.

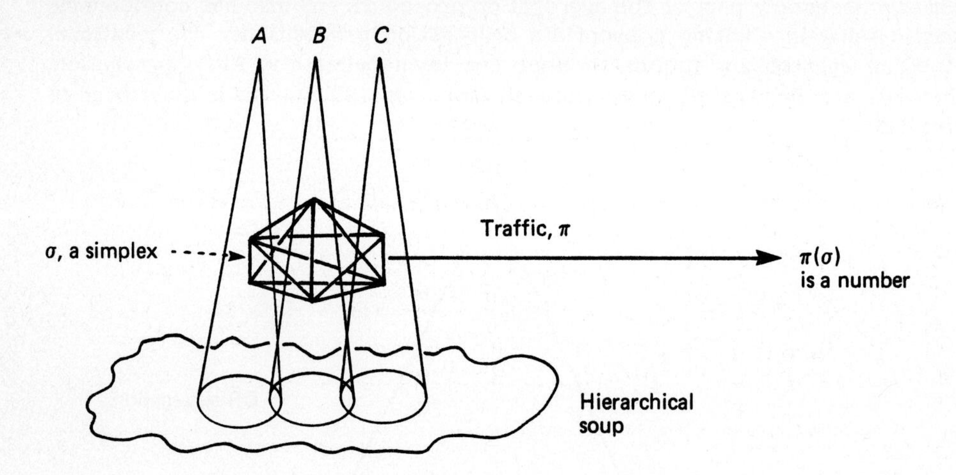

Figure 2.30 Schematic representation of a simple Q-system.

As a simplification of the original definition (Johnson, 1982b), let us say that a ***Q-system*** consists of a hierarchical backcloth, $B = (H, \{\lambda\})$, and traffic on the hierarchical backcloth, T. Given a hierarchical scheme H of sets, any particular class of relations $\{\lambda\}$ between those sets defines a ***state of the backcloth***. Given a backcloth in a particular state, any particular values for a given set of traffic patterns define a ***state of the traffic*** relative to the state of the backcloth. ***A state of the Q-system*** is a particular state of traffic relative to a particular state of backcloth.

Hierarchical Aggregation of Traffic in Q-Systems

Manufacturing companies provide examples of complex systems which are subjected to detailed analysis at many hierarchical levels. Figure 2.31 shows a simple company with two factories and a head office. In Figure 2.32 one of the factories is described by the structure of its workshops, stores, and so on. In turn, a workshop is described by the structure of the detailed manufacturing backcloth, the part of the system where particular things are actually made. Figure 2.31 suggests just a few of the many types of traffic that exist on this backcloth, traffic that is explicitly known in most well-run companies.

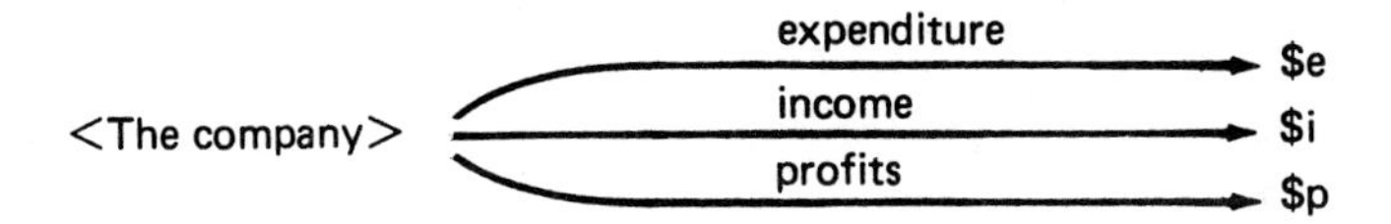

Profits(<The company>) = Income(<The company>) – Expenditure(<The company>)

(a)

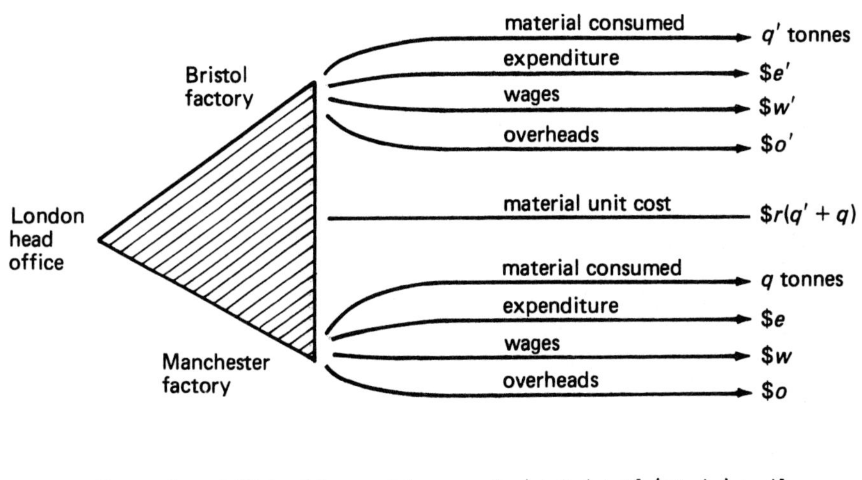

Expenditure(<Bristol factory>) $= \$w' + \$o' + \$[q' \times r(q' + q)]$
Expenditure(<Manchester factory>) $= \$w + \$o + \$[q \times r(q' + q)]$

(b)

Figure 2.31 Two relatively highly aggregate descriptions of a company, (a) more highly so than (b).

Without laboring the point, Figure 2.31(b) shows that some traffic must refer to more than one vertex of a polyhedron. It is clear that the unit cost of material to *each* factory depends on the quantities required by *both* factories, since in this kind of system unit costs usually decrease with quantity. Also, unit costs and quantities are usually step functions with discontinuous changes in price – even simple systems are unlikely to be linear, or even continuous, in their traffic.

Figure 2.32 shows how the company can be described in increasing detail down to the level of the machines used and specific articles manufactured. Although academics tend to dislike such a detailed level of description, it is interesting to note that those responsible for complex systems, such as the management of a company, do, indeed, have long lists of the parts used, do know precisely who is employed and how much they are paid, do know how many machines are owned and where they are, and so on. It is clear that without this information the company would be unmanageable.

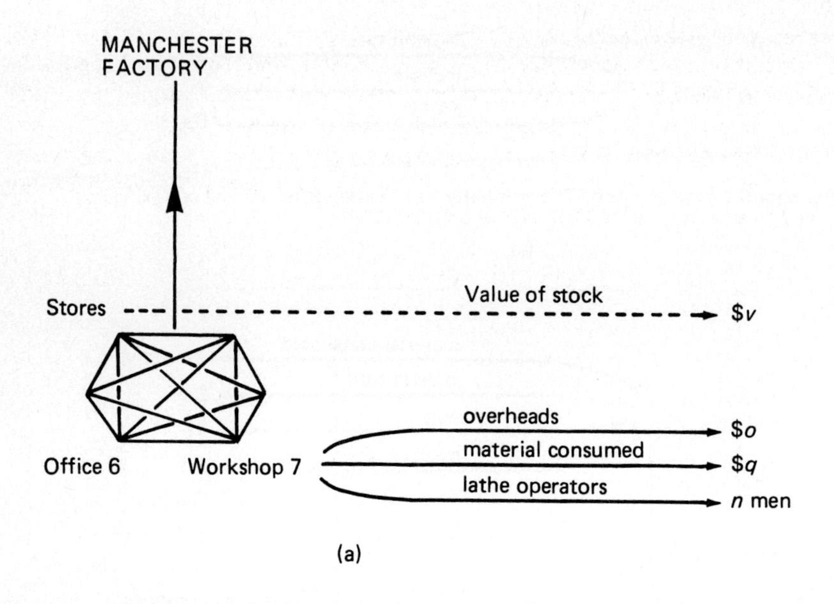

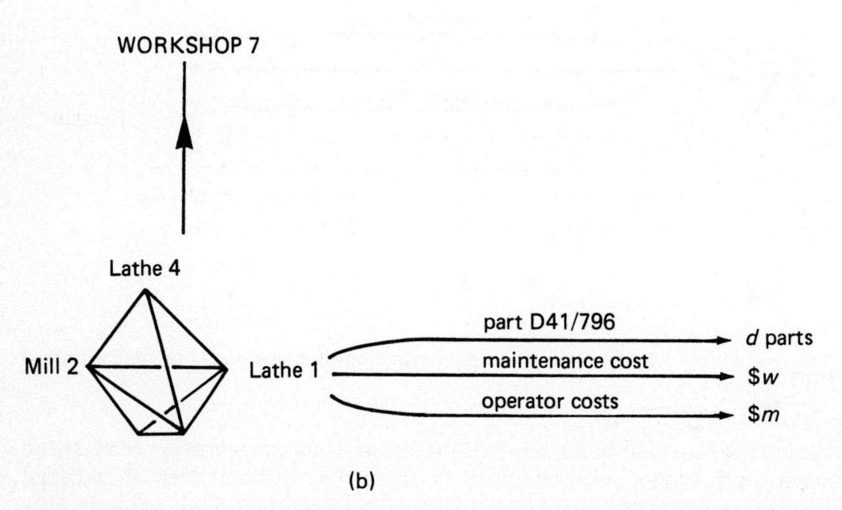

Figure 2.32 Two relatively disaggregate descriptions of parts of a company, (b) more so than (a).

This hierarchical view of complex systems highlights a number of aspects that many analysts prefer to overlook. The most important of these is the nature of the aggregation of traffic (Johnson, 1983c) and the implications of this for predicting future system states. Figure 2.33 is an abstraction of a system at, say, levels $N + 5$, $N + 4$, and $N + 3$, corresponding to a complicated organization. Let us assume the backcloth for the analysis and consider the traffic. Between level

$N + 4$ and level $N + 5$ the question of hierarchical aggregation may resolve itself in terms of a function, with $y = f(x_1, x_2, \ldots, x_n)$. In *real* systems this function is probably quite complicated, almost certainly piecewise discontinuous at many places, possibly multiple-valued, depending on the system's history, and so on. Note that the expression $y = f(x_1, x_2, \ldots, x_n)$ almost certainly does not mean $y = f_1(x_1) + f_2(x_2) + \cdots + f_n(x_n)$, except for the simplest systems, and that $f(x_1, x_2, \ldots, x_n)$ is a function of the *whole* polyhedron.

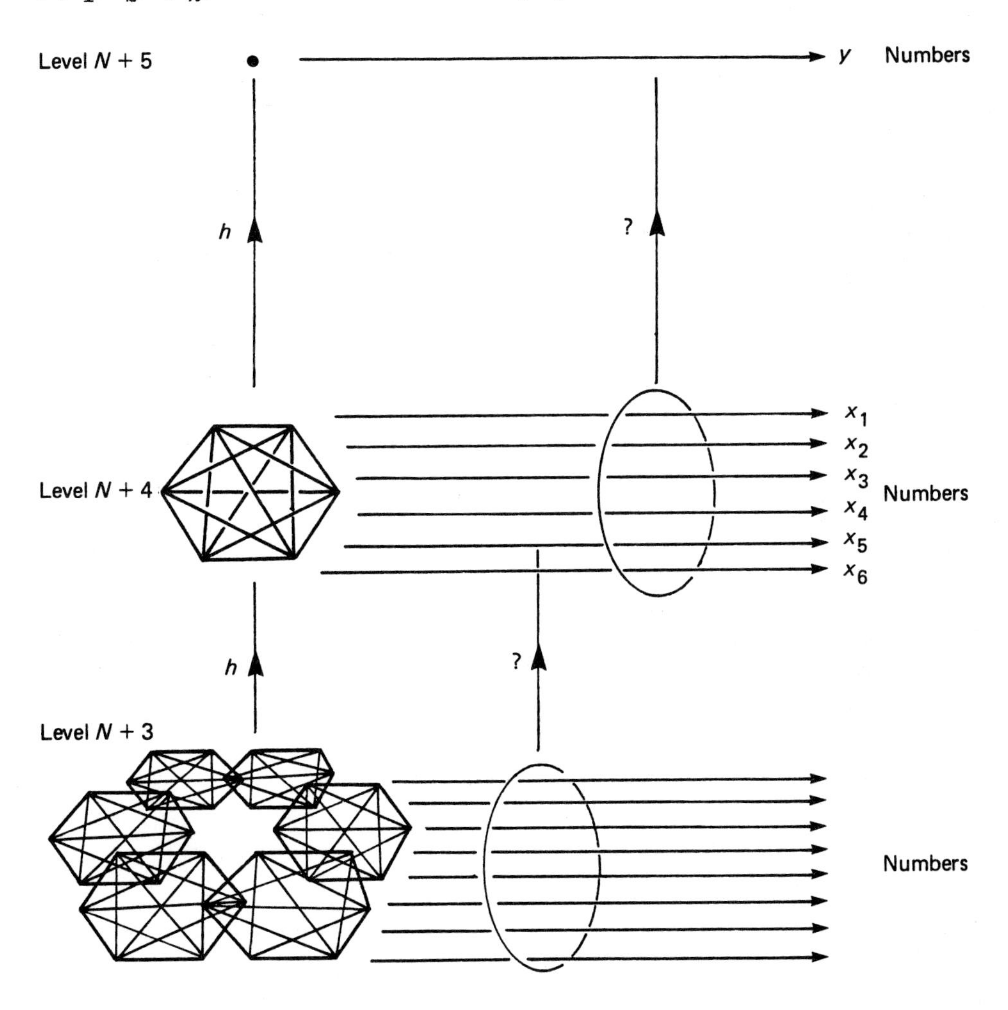

Figure 2.33 How does traffic aggregate relative to the backcloth?

Descending the hierarchy from level $N + 4$ to level $N + 3$ the backcloth itself is more complicated – the $N + 4$ vertices now become $N + 3$ polyhedra with associated discontinuities between some of the polyhedra. Recall that traffic behavior depends on the star structure of the backcloth in some systems: how are these more subtle properties of the traffic aggregated? As drawn in Figure 2.33 the variable x_5 may depend on a combination of traffic functions defined on a number of $N + 3$ level polyhedra. It is then likely that the aggregation is conditional on both the value of the $N + 3$ traffic and its supporting star structure.

Finally, one can ask how the traffic on a complex hierarchical system is sometimes required to sustain the system. In some cases the maintenance of relations may consume money; for example, the backcloth that supports telecommunications traffic also consumes income traffic – to the extent that the least expensive backcloth necessary to support the communication traffic is sought: telephone, telex, electronic mail, dispatch rider, etc. In other cases income traffic is invested in creating new backcloth structures – a new factory, a new machine, a new employee, for example. The decision to make such investment reflects an awareness of the managers at $N + 5$ that $\$p$ is, indeed, determined by structure at lower levels in the system. Since complex systems are complex these examples can only hint at the application of Q-system theory to real cases. However, an important point remains to be made in criticism of the kind of analysis frequently involved in the prediction of future system states of nations, international groupings, or even the world.

Figure 2.34 shows some typical variables involved in the equation of unemployment for Britain. Some analysts seek functional relationships between such variables and provide predictions on this basis. Despite such predictions being notoriously unreliable, there is little evidence of these analysts asking fundamental questions about the methods they use: indeed, some do not even admit that they deliberately shut their eyes to the real system because it is too complicated. Complex systems are undeniably complicated in the numbers of elements, relations, and functions they involve. This complexity involves the intimate relationship between traffic and stars (at least); and that complex systems are invariably hierarchical with complicated hierarchical relations on the backcloth *and* traffic. Given this, a main argument in this chapter is that one will not understand complex systems by ignoring everything but highly aggregate numerical relationships.

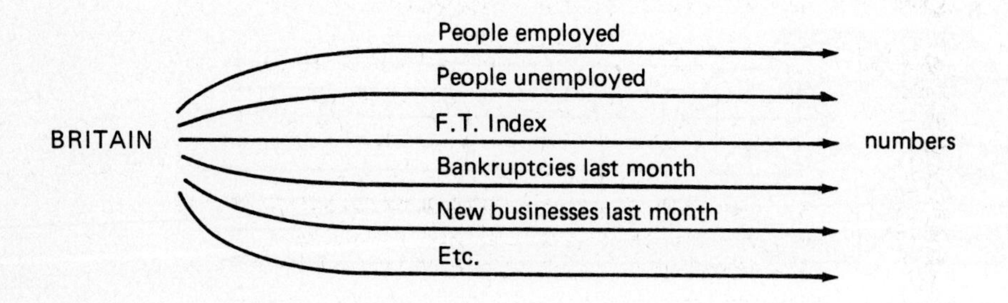

Mean income(<BRITAIN>) = f(People employed, . . .)

Figure 2.34 An adequate description of a complex system?

An Appropriate Concept of Time for Describing Change in Complex Systems

Time is implicit in our interest in systems, but too often an inappropriate Newtonian concept of time is used. Consider the pendulum in Figure 2.35 which is used to measure *clock time*. Suppose an observer can detect the presence or absence of the pendulum bob at point p, and counts aloud the number of times the *presence* of the bob changes to the *absence* of the bob at p. When the pendulum is swinging one would hear the observer saying "one, two, three,...", which can be used to construct a sequence of *now-moments*: now-1 means the observer says "one", now-2 means the observer says "two", and so on. This sequence of now-moments {now-1, now-2, now-3, ... } is totally ordered and has the usual properties of Newtonian time. Note that it is not the position of the bob which marks time, it is the *change in position*; the stationary pendulum bob cannot be used to mark time in this way.

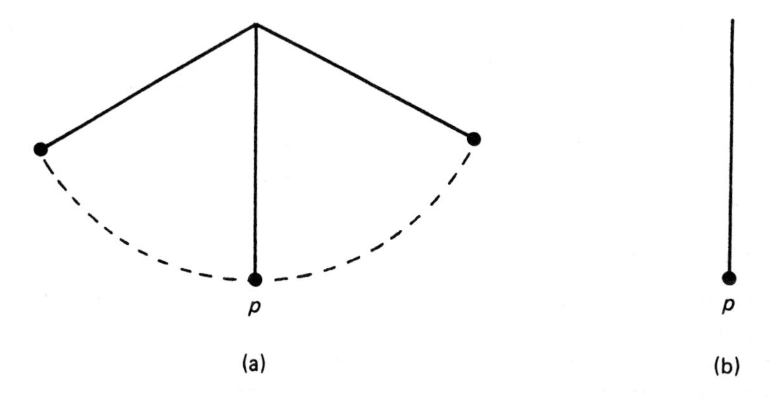

Figure 2.35 The pendulum used for measuring clock time: (a) moving and (b) stationary.

In physics the now-moments are mapped to numbers according to their sequential occurrence and, by a further process of abstraction, it is assumed that subtracting one of these numbers from another gives an appropriate measure of *duration*. This view of clock time is supported by the observation that different physical systems give consistent durations when compared. For example, pendulum clocks, spring clocks, water clocks, candle clocks, quartz clocks, etc. all give more-or-less coincident now-moments signifying the passing of, for example, an hour. However, these clocks have to be calibrated against each other, there being no absolute duration. Similarly, some clocks are internally more consistent than others, which we interpret as being the ability to measure time more accurately. However, it can be argued that clocks do not measure time, they *define* time relative to their now-moments. In classical physics, motion and change in motion are used to observe (pendulum) clock time in the first place; then time is subsequently used to define and observe motion as velocity, acceleration, etc. In other words, motion is tautologically defined by motion, or time is tautologically defined by time, in physical systems.

The now-moments of the pendulum clock are particularly simple, with only two observable *states*; namely, the bob is at p, and the bob is not at p. Thus *the*

now-moments of a physical system can only be observed as changes in state of the system.

The discrepancy between physical time and social time is well known (e.g., Sorokin and Merton, 1937); time flies, time drags, and so on. This is sometimes interpreted that the social system is in some way out of step with true (clock) time. However, it is equally possible to say that clock time may not be appropriate for measuring time in social systems. Certainly, social events do not appear to be totally independent of clock time, since many human activities are defined in terms of days, weeks, months, years, etc. On the other hand, historians measure social time by social events such as wars, treaties, births, deaths, marriages, inventions, and so on. Although they find it useful to associate such events with dates in clock time, there is seldom the suggestion that the cycles of history can be related to exact clock-time durations.

Social events are characterized by structure; they are made up of many necessary parts which must be combined in the correct way for the event to have happened. Events have the property that the whole is more than the sum of the parts. For example, consider the event Workshop on Structure and Evolution of Systems. Somewhat simplistically, suppose it is made up of the activities <travel from home to Abisko>, <listen to papers>, <academic discussion>, <social discussion>, <contemplate nature>, and <skiing>. The academic event is characterized by <travel from home to Abisko, listen to papers, academic discussion>, presumably to be experienced by all participants, with all parts necessary for the event to have happened. However, most experience a richer event: <travel to Abisko, listen to papers, academic discussion, social discussion, skiing> for the gregarious sportsperson; <travel to Abisko, listen to papers, academic discussion, contemplate nature> for the introvert, and so on. For each of the participants, the parts of the event form by themselves or with others (evolve?) until the event is recognized to have happened. Some have a feeling that the event develops over a clock time week, others may feel everything gels on the last day. In either case, it is likely that the event of participating in the workshop marks in their memory when May 1984 occurred, rather than vice-versa.

Atkin has described such processes as ***p-events***, where the number p reflects the number of parts of the event (Atkin, 1974, 1977, 1981). These p-events reflect the observation that something has changed in the system. In just the same way that changes in the physical system of the pendulum seem to be appropriate for defining physical clock time, it does not seem unreasonable to let changes in social systems define social time. In other words, it is suggested that the time appropriate to a Q-system is defined by changes in state of the Q-system. In a Q-system the ***past*** can be defined to be all states of the system which have been observed, excluding the state of the system as it is being observed now in the ***present***. The ***future*** can be defined to be all possible states of the Q-system and, of the many possible states in the future, only some will be observed. A discussion of past, present, and future in logic can be found in Prior (1967). If one accepts that a Q-system can have a single state, s, at a given now-moment (effectively a definition of now-moment), it is possible to produce a kind of stochastic lattice (Figure 2.36) which corresponds to what Atkin (1979) has called the NOW-horizon. Let a path through this lattice be called a Q-system ***trajectory*** through social time. In other words, a Q-system trajectory is a set of Q-system states $\{s^1,\ldots,s^n\}$, where s^{i+1} can follow s^i.

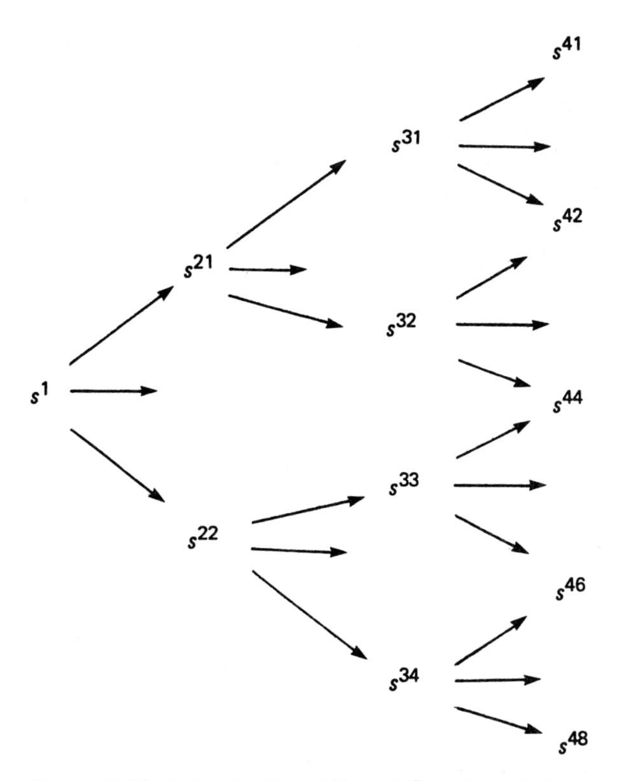

Figure 2.36 A stochastic lattice of Q-system states.

It is possible to consider Q-system trajectories as p-events which grow as superevents of previous events [Figure 2.37(a)], where the Q-system does not "forget" its past. However, there are many Q-systems in which earlier p-events are lost by accident or design [Figure 2.37(b)]; for example, the erection of scaffolding in the construction of a building is an event which is subsequently removed as a means to the event "the building is finished". Thus, some system trajectories are developments of events from events which have ceased to exist.

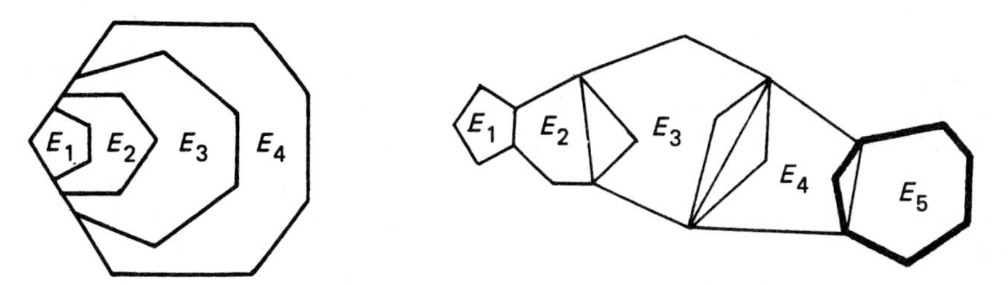

Figure 2.37 The development of events from previous events (p-simplices are represented by polygons): (a) p-events as superevents of previous events which still exist; (b) the development of events from previous events which have ceased to exist (events E_1, E_2, E_3, and E_4 cease to exist as E_5 evolves).

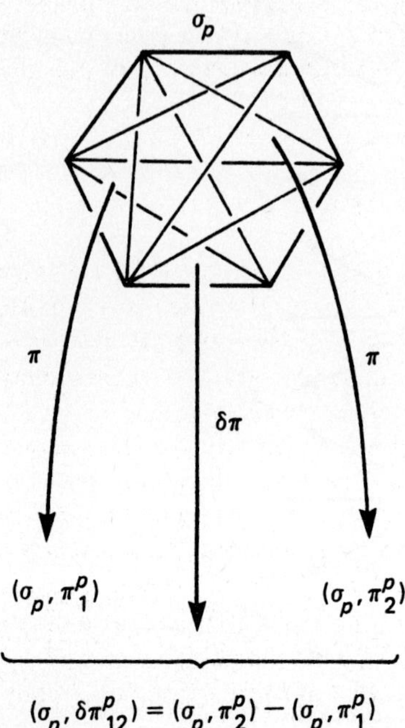

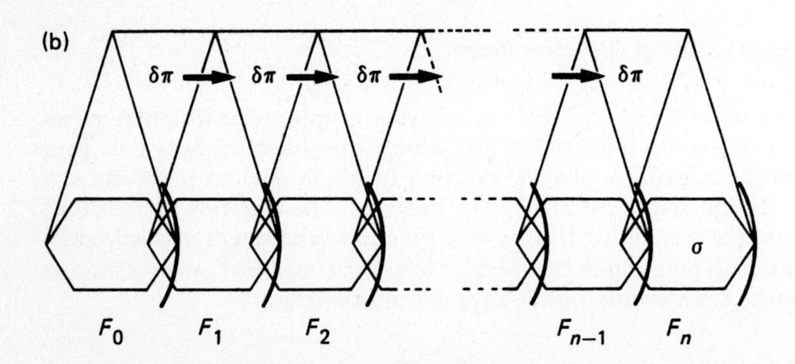

Figure 2.38 Changes in state determined by changes in traffic: (a) p-events as changes in traffic (p-forces); (b) changes $\delta\pi$ can be predicted to arrive at σ at now-moment n (changes are q-transmitted through the fronts F_0, F_1,...,F_n).

In some cases, the trajectories are characterized by changes in the traffic state without any change in the backcloth state (Figure 2.38). For example, the backcloth to the pendulum does not change with counting the passing of now-moments and their related intervals. The theory of q-transmission (Johnson, 1982a) examines the change of traffic values over a backcloth, the general idea being that q-dimensional changes (q-forces) are transmitted down q-connectivities. It transpires that simplicial complexes have a *transmission front* structure which relates to the time it takes for changes to occur in the traffic. Given a q-connectivity, one might ask when a q-force at one end is q-transmitted to the other. The answer can be given precisely as the number of transmission fronts along the chain, which gives the tautological prediction that a q-force (a change in q-dimensional traffic) arrives at a simplex precisely at the moment it arrives [Figure 2.38(b)].

Time and Prediction in Complex Systems

Suppose every social Q-system exists in a single state when observed at any given clock time and that Q-systems are not observed and recorded continuously in clock time. The history of a social Q-system is then composed of a series of clock-time defined states, s^t. In general, let the expression $\delta s = s^{t_2} - s^{t_1}$ represent the change of state between clock times t_1 and t_2.

The relationship between social and clock time can be schematically drawn as in Figure 2.39, in which the polygons represent simplices that are associated with change as events. Implicit in this picture is a large amount of omitted backcloth that does not change with these events and, for simplicity, it is assumed the events are not simultaneous. Thus, the transition from system state s^{t_1} to state s^{t_2} is schematically represented by a hexagon (5-simplex) changing to an octagon (7-simplex), this change happening between clock times t_1 and t_2. If it is now t_3, all the past events can be ordered in retrospective clock time (it is assumed),

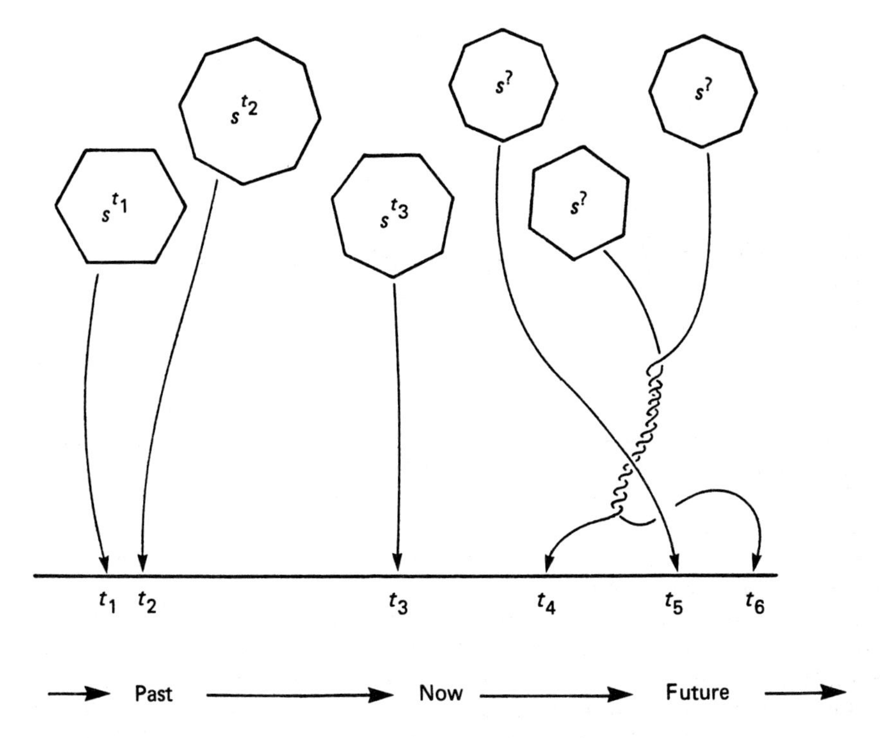

Figure 2.39 Mapping social time into clock time (Q-system events are represented as polygons).

but it is not always possible *a priori* to predict whether other events will occur before or after each other in future clock time. Thus, we can know what the possible system states are, but we may be unable to reliably predict which will occur first.

In his book *Multidimensional Man*, Atkin (1981) speculates on some very interesting relationships between the dimensions of events and clock time and uses the formula (Atkin, 1978, p 292)

$$\sum_{t=1}^{p+1} \binom{p+2}{t+1} \times \binom{t+1}{2}$$

as the time ratio of structural interval to Newtonian interval to derive the relationships below. Notice

$$\binom{p+2}{t+1}$$

is the number of t-faces of a $(p+1)$-simplex and the number of edges in each t-face is

$$\binom{t+1}{2};$$

Atkin also assumes hierarchically unbound time-intervals of successive 8-events at successive levels, with 24 h taken as a given fact. Although Atkin does not argue a precise relationship between social time and clock time, his point that high-hierarchical, high-dimensional social events take a relatively long clock time to occur is very convincing; for example he gives relationships between human events and clock time (Atkin 1981, p 196):

Quickest impression	*Breath*	*Waking/sleeping*	*Life*
1.34×10^{-4} s	3.4 s	24 h	70 y

There is no doubt that many human activities relate closely to clock time and the related physical events. Attempts to ignore clock time when traveling can result in the social–psychic–physical disorder called jet lag; the period of gestation for humans is about nine calendar months; it takes three years of study to obtain a bachelor's degree at some universities, and so on. Some of these correspondences reflect biophysical properties which genuinely occur in clock time and others are conventions which apparently depend on social considerations.

POSTULATE 2.3 Prediction hypotheses:

(1) Some social events, for example birthdays, can be tautologically predicted in clock time. These predictions are exact.

(2) Some social events, for example gestation and birth, can be predicted in clock time (inexactly), because physical processes are involved.

(3) Some social events, for example football matches and television watching, can be predicted in clock time because they have been planned to occur at

specific clock times. These (exact) predictions may become incorrect for other socially determined reasons and events.

(4) Some social events can be predicted in clock time on the basis of previous observations; for example, taking an annual holiday. These predictions may be inaccurate or incorrect.

(5) Some social events predictably must precede other events; for example, the road must be built before people can drive along it.

(6) Some social events can be seen retrospectively as predictable from previous events, but systematic collection of data about those previous events would be impractical, for example: the car broke down; I walked home; it was fine; I went the long way round; I heard a bird; I looked up; I tripped; I broke my leg; I am not at work, ... the car breaking down caused me to be absent from work.

(7) Some events can be tautologically predicted in social time; for example, when I am good enough I will play in the first team; when I feel better I will get out of bed.

(8) Some events seem chaotic with respect to clock time; for example, whether I win or lose when playing dominoes.

This can be summarized as follows: there are (at least) two types of time: they are measured by physical events (clock time) and social events (social time); these events are determined by changes in the state of physical or social Q-systems; social time can be precisely associated with clock time for past and present events; some social events can be precisely predicted in clock time; some social events can be roughly predicted in clock time; and some social events are not predictable in clock time beyond the possibility that they might happen.

The speculations which attempt to relate future states of Q-systems to clock time have one of the forms:

(1) The Q-system will have the state s^i at some unknown i, $i >$ now.

(2) The Q-system will have the states s^i and s^j, and the former precedes the latter in clock time.

(3) The Q-system will have the state s^i after some t_{min} in clock time, where $t_{min} < i$.

(4) The Q-system will have the state s^i before some t_{max} in clock time, where $t_{max} > i$.

If an event must occur before a system state s^j occurs in clock time and after a system state s^i occurs in clock time, then $[s^i, s^j]$ is defined to be a *Q-system window* for that event. If it is speculated that an event will occur before a clock time t_{max} and after a clock time t_{min}, then $[t_{min}, t_{max}]$ is defined to be a *clock-time window* for that event. For example [the universe at your birth, the universe at your death] is a Q-system window for the event of you reading this chapter. More speculatively, [1984 AD, 2084 AD] is a clock-time window for the event of you telling me what you think of it.

When planners and decision makers are attempting to locate social events in clock time, it will probably be helpful if (as far as possible) they arrange their *goals* (desired states of the Q-system) into:

(1) Those with time window [now, t_{max}].
(2) Those with time window [t_{min}, t_{max}].
(3) Those with time window [t_{min}, ?].
(4) Those with time window [now, ?].

Here t_{min} and t_{max} exist in clock time. The first three of these relate goals to clock time and the last states that the goals belong to the unknown clock-time future including, for practical purposes, never happening.

In principle, the time windows for a set of goals can be superimposed on each other, as illustrated in Figure 2.40. With any set of goals, the intersection of their time windows is either empty or a subwindow in clock time. Let W be the set of all subwindows that can be generated by the time windows for a set of goals G. Then it is an interesting research question to ask: Does the structure of the relation between a set of goals G and a set of time windows W have any (heuristic) value for strategic planning in clock time? Certainly the complex $K_G(W,\lambda)$ shows how the

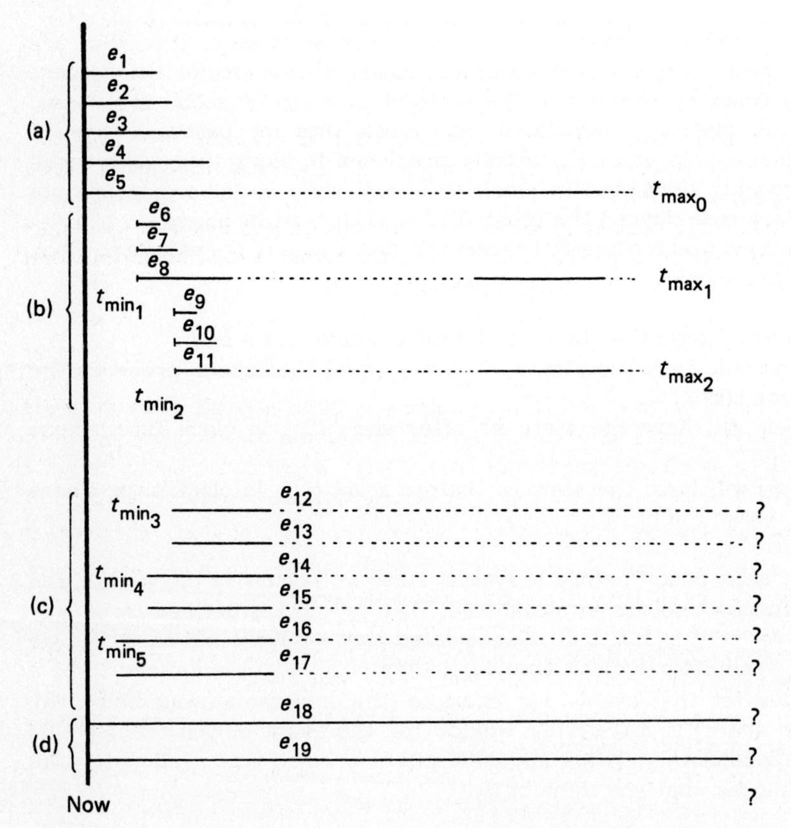

Figure 2.40 Juxtaposition of time windows for future events, e_i, where the time windows are (a) [now, t_{max}]; (b) [t_{min}, t_{max}], (c) [t_{min},?], and (d) [now, ?].

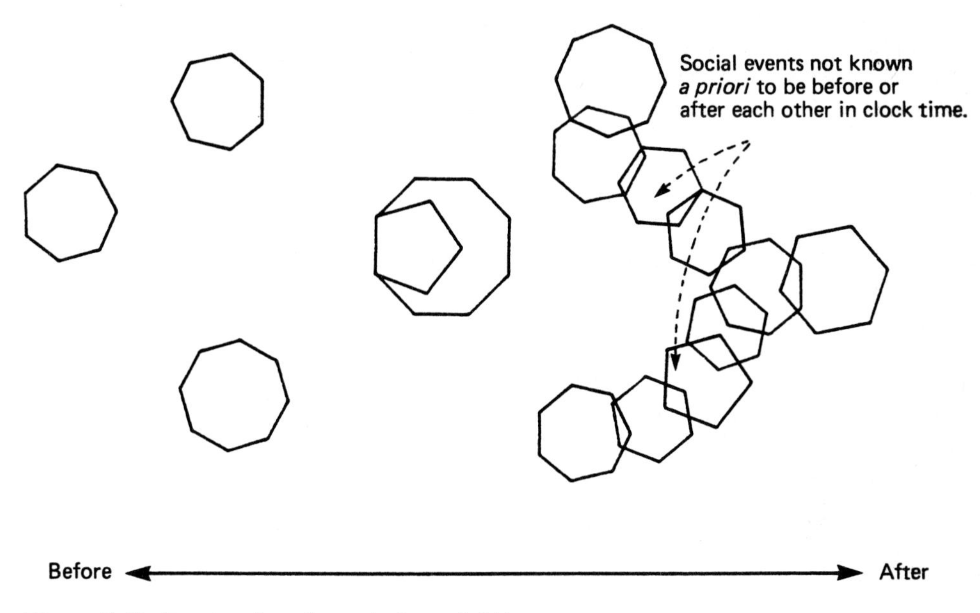

Figure 2.41 Quasi-ordered events in social time.

goals are related to each other through shared subwindows of clock time. Also, the complex $K_W(G,\lambda^{-1})$ shows how intervals of clock time are related to each other through shared goals.

The subintervals not containing other subintervals can be ordered by clock-time occurrence, w^i occurring before w^{i+1}. Let G^t be the set of goals that can occur in time window $w^t \in W$. Then the expression

$$\prod_{t=1}^{n} G^t ,$$

the Cartesian product of the sets G^t, contains all those Q-system trajectories which begin with a goal that can occur in w^1 and end with a goal that can occur in w^n, where $w^1, w^2, \ldots, w^n$ have no subintervals. These trajectories could be written $(G^1_{i_1}, G^2_{i_2}, \ldots, G^n_{i_n})$, where $G^j_{i_j} \in G^j$; that is, the Q-system is steered toward the goal $G^n_{i_n}$ via a set of desired system states which are, by definition, themselves goals. Theoretically, decision makers could devise tactical sets of actions to proceed from one goal to another in a Q-system trajectory.

However, those goals which will occur a long while into the clock-time future may be possible in very many clock-time subwindows. Thus, the further ahead that is planned, the more possible system states that have to be considered for tactical planning. Clearly, tactical planning should not be attempted beyond the ability to cope with the combinatorial possibilities.

Although events are totally ordered in historical clock time, there seems no good reason to insist that they are necessarily totally ordered in the future. For example, if two events share the same clock-time window, one does not know which will occur first. Certainly, there is no suggestion that the events will necessarily occur simultaneously. Thus, one can view future events as being quasi-ordered by their occurrence in future time (Figure 2.41). In turn, this means that the stochastic lattice of Figure 2.36 must be revised; its vertical columns of system states are implicitly alternatives for a single future time.

Suppose that decisions about a Q-system are made in a language which can have a single state of the system as a vertex. Then the political part of making decisions consists of 0-dimensional traffic on this metastructure, and choosing strategies (sets of system states) can be viewed as p-dimensional traffic on a strategy which is a system trajectory with $(p + 1)$ states. Considered thus the strategies are connected through shared system states and the system states (including the goals) are connected through shared strategies.

Speculation 2.1

The technical problem of deciding between strategies to achieve agreed goals may be assisted by the heuristics:

(1) High-dimensional goals in the system state *versus* trajectory complex may be more attainable (strategically superior) than low-dimensional goals in this structure.
(2) Sets of highly connected strategies (Q-system trajectories) in the trajectory *versus* system state complex may be more attainable and more manageable (strategically superior) than highly eccentric strategies.
(3) Large, high-dimensional stars will indicate significant commonality between many strategies and goals.

Conclusion

In this chapter I have directed a number of arguments at different readers. To the uncommitted, I have presented a view of complex systems which I believe to be along the correct lines, if not perfect. To those familiar with Q-analysis, I have presented the star structure as a generalization of q-connectivity, which I believe is necessary, if not sufficient, for understanding complex systems. To statisticians and economists I have presented here the argument that sometimes they may have ignored important backcloth structure while concentrating on functional relationships. To all I would say this: "heavier-than-air complex systems *are* possible" – for those able to go beyond conventional methods of thinking.

References

Atkin, R.H. (1971) *Research Report 1, Urban Structure Research Project* (Colchester, England: Dept. of Mathematics, University of Essex).
Atkin, R.H. (1974) *Mathematical Structure in Human Affairs* (London: Heinemann Educational Books).
Atkin, R.H. (1977) *Combinatorial Connectivities in Social Systems* (Basel: Birkhäuser).
Atkin, R.H. (1978) Time as a pattern on a multidimensional structure. *J. Biological Structure* 1: 281–95.

Atkin, R.H. (1979) *A Kinematics for Decision Making*. Mimeo (Colchester, England: Dept. of Mathematics, University of Essex).

Atkin, R.H. (1981) *Multidimensional Man* (Harmondsworth, England: Penguin Books).

Beaumont, J. and Gatrell, A. (1982) *An Introduction to Q-analysis*, CATMOG 34 (Norwich, England: Geo Abstracts).

Dowker, C.H. (1951) Homology groups of relations. *Annals of Math* 56: 85–95.

Earl, C.F. and Johnson, J.H. (1981) Graph theory and Q-analysis. *Environment and Planning B* 8: 367–91.

Gould, P. (1980) Q-analysis, or a language of structure: an introduction for social scientists, geographers and planners. *Int. J. Man–Machine Studies* 13: 169–99.

Gould, P. (1981) Electrical power system failure: reflections on compatible descriptions of human and physical systems. *Environment and Planning B* 8: 405–17.

Gould, P., Johnson, J.H., and Chapman, G. (1984) *The Structure of Television* (London: Pion/Methuen).

Griffiths, H.B. (1983) Using mathematics to simplify Q-analysis. *Environment and Planning B* 10: 403–22.

Johnson, J.H. (1976) The Q-analysis of road intersections. *Int. J. Man–Machine Studies* 8: 531–48.

Johnson, J.H. (1981a) Some structures and notation of Q-analysis. *Environment and Planning B* 8: 73–86.

Johnson, J.H. (1981b) The Q-analysis of road traffic systems. *Environment and Planning B* 8: 141–89.

Johnson, J.H. (1981c) Q-discrimination analysis. *Environment and Planning B* 8: 419–34.

Johnson, J.H. (Ed) (1981d) Special issue on Q-analysis, *Environment and Planning B* 8(4).

Johnson, J.H. (1982a) q-transmission in simplicial complexes. *Int. J. Man–Machine Studies* 16: 351–77.

Johnson, J.H. (1982b) The logic of speculative discourse: time, prediction and strategic planning. *Environment and Planning B* 9: 269–94.

Johnson, J.H. (1983a) Hierarchical set definition by Q-analysis: 1. The hierarchical backcloth. *Int. J. Man–Machine Studies* 18: 337–59.

Johnson, J.H. (1983b) Q-analysis: a theory of stars. *Environment and Planning B* 10: 457–69.

Johnson, J.H. (1983c) Hierarchical set definition by Q-analysis: 2. Traffic on the hierarchical backcloth. *Int. J. Man–Machine Studies* 18: 467–87.

Johnson, J.H. (1984) Latent structure in road systems. *Transportation Research B* 18: 87–100.

Johnson, J.H. and Wanmali, S. (1981) A Q-analysis of periodic market systems. *Geographical Analysis* 13: 262–75.

Macgill, S. (Ed) (1983) Special issue on Q-analysis, *Environment and Planning B* 10(4).

Prior, A. (1967) *Past, Present and Future* (London: Clarendon Press).

Seidman, S.B. (1983) Rethinking backcloth and traffic: perspectives from social network analysis and Q-analysis. *Environment and Planning B* 10: 439–56.

Sorokin, P.A. and Merton, R.K. (1937) Social time: a methodological and functional analysis. *Amer. J. Sociology* 42: 615–29.

Stiny, G. (1980) Introduction to shape and shape grammars. *Environment and Planning B* 7: 343–51.

CHAPTER 3

Pictures as Complex Systems

Ulf Grenander

Introduction

When we try to process images – restore, analyze, or understand them – we can approach the problem by viewing the image as a complex system of units, combined by certain rules of regularity. The choice of units is by no means straightforward. Initially, it may appear that a natural choice would be the picture elements (pixels), say black and white, or gray scale, or color, etc. A closer scrutiny, however, reveals this approach to be too superficial and that we must use more intrinsic, more informative, units that lead to the construction of ***random geometries***, several instances of which we study herein.

In attempting to build regular structures (random geometries) suitable for image analysis we use the concepts and techniques of the general theory of patterns that has emerged in recent years. This is really ***mathematical engineering***: to build logical structures just as the engineer builds mechanical, electronic, or other physical structures. In so doing, we attempt to express our prior knowledge of the world or microworld, for example, as seen by a digital camera, in a precise mathematical form. It is reminiscent of what those in artificial intelligence call ***knowledge engineering***.

Once this has been achieved for a particular system, so that we have created a picture ensemble together with a probability measure of it, we turn to the analysis of the ensemble, and to the problems of deriving algorithms, based on the regular structure used, for image analysis. Our knowledge of how to do this is far from complete, but the last few years have witnessed a rapid development of analytical results.

One circumstance that has contributed to the increased interest in applying theoretical ideas of patterns to image processing is the recent advance in computer architecture. Computers have not just become faster and cheaper – their fundamental architecture is also changing from purely sequential von Neumann machines to truly parallel ones. This fits exactly the mathematical properties of the regular structures discussed above and makes possible hardware implementations of the algorithms we are deriving.

It is not enough to prove optimality or other properties for our algorithms. We must also demonstrate that they are computationally feasible, which is done by systematic mathematical experiments on the computer. To date conventional machines, a VAX, a timeshared IMM3081, and various microcomputers have been used. These have restricted the size of the images that can be processed, but are sufficient to establish feasibility and give an indication of what computing power is required for larger images.

Before we examine how to implement the above in actual cases, let us first relate what follows to the general theme of the book. We attempt to show why pictures should be viewed as complex systems. Space does not allow more than an outline of this approach, but it is sufficient to illustrate the main ideas. Nevertheless, the discussion may give the erroneous impression that it concerns only picture processing; this is not true. On the contrary, many other complex systems are being analyzed in similar terms; for example, shape formation in embryology, medical diagnostic patterns, and large software systems. All of these applications of general pattern theory share one essential feature: ***the analysis employs as units certain concepts*** – the generators – ***that need not be directly observable.***

It has sometimes been argued that scientific theories should only involve observables. But this is wrong, as is obvious from even a cursory reading of the history of science. Instead, one could state, as René Thom has, that such theories typically use imaginary units – unobservable ones. When the theorist tries to invent suitable units he or she strives for simplicity: thus the units should interact as simply as possible. In the following, as well as for other complex systems, this is achieved in the sense that they interact ***locally***, where locality is expressed by the connections of the σ graph discussed below. It is this graph that determines the type of regularity that characterizes the complexity of the model. Herein, the type is given by the random geometrics that are introduced.

Bear in mind that the examples given are intended as illustrations of a methodology of general scope. For didactic reasons we concentrate on the main principles and do not present the computer programs that have been used to implement them.

Basic Concepts in General Pattern Theory

The combinatory approach

The starting point of spectral analysis is Fourier's theorem, which states that the set of all functions can be generated by trigonometric functions. The *generators*, $g = A \exp i(\lambda_1 x + \lambda_2 y + \varphi)$, where the trigonometric functions are written in complex form for convenience, span the set of all pictures that could conceivably be of interest.

Fourier analysis is mentioned because it is an example of how complicated phenomena can be described by combining simple units: the trigonometric functions. We know that other systems of orthogonal functions can also be used to build analytical methods, but we also know that only certain orthogonal systems are natural. What system to use depends on what sort of pattern ensemble we have at hand. The choice of generators, decisive for the success of the method of analysis that we arrive at, is difficult and requires both inventiveness and insight into the pattern structure.

Suppose that we have selected a set, G, of generators. The generic element of G (the *generator space*) is denoted by g with suitable subscripts when needed. To synthesize patterns we form combinations, $c = \{g_1, g_2, \ldots, g_n\}$, but it requires a good deal of thought to understand how such a c should be interpreted. Therefore, let us return to the Fourier case for guidance. If

$$c = \{A_1 \exp i(\lambda_1^1 x + \lambda_2^1 y + \varphi_1),\ A_2 \exp i(\lambda_1^2 x + \lambda_2^2 y + \varphi_2), \ldots\},$$

we interpret this by saying that it represents the image with the series expansion

$$I(x,y) = A_1 \exp i(\lambda_1^1 x + \lambda_2^1 y + \varphi_1) + A_2 \exp i(\lambda_1^2 x + \lambda_2^2 y + \varphi_2) + \cdots .$$

Hence, if we have two combinations,

$$\begin{cases} c' = (g_1', g_2', \ldots) \\ c'' = (g_1'', g_2'', \ldots) \end{cases},$$

they represent the same image if functional identity holds between the two series; that is

$$\sum_i g_i'(x,y) \equiv \sum_i g_i''(x,y) . \tag{3.1}$$

Thus, we identify two combinations, c' and c'', of generators if the *identification rule* R in equation (3.1) holds, or, symbolically if $c'\ \mathrm{R}\ c''$ is true. We encounter many other sorts of identification rules herein, usually not linear as rule (3.1), but the important thing to remember is that *one image* I *can be analyzed in several combinations* c. The image is what the observer can see, while c is an *analysis or explanation* of it.

So far, we have merely imitated Fourier analysis, but we now find a fundamental difference. In the Fourier case we are free to select the generators arbitrarily, but here we have to introduce *restrictions between them*. It is best to explain this with some examples.

Example 3.1 Consider a linear spline, a piecewise linear function for which the line segments have been joined continuously, as in Figure 3.1(a). The continuity condition is the restriction mentioned above. Here, the line segments play the role of generators and we can *parameterize* them, Figure 3.1(b), so that g is represented as the four-vector (a,b,k,l), where (a,b) is the interval on which the segment is defined and $y = l + kx$ is its equation.

To ensure continuity each generator emits two signals to its neighbors; one, β_1, to the left and the other, β_2, to the right. These signals, the *bond values*, are two-vector, in this example. The left one, β_1, tells the left neighbor that the left endpoint of the interval of definition is a and that the corresponding ordinate is $l + ka$. The bond value to the right says that the right endpoint is b and the corresponding ordinate is $l + kb$.

We now join generators as in Figure 3.1(c). At each juncture, where two bond values β' and β'' meet, we require that a *bond relation* ρ hold, $\beta'\ \rho\ \beta''$; in this example ρ means simply EQUAL, so that

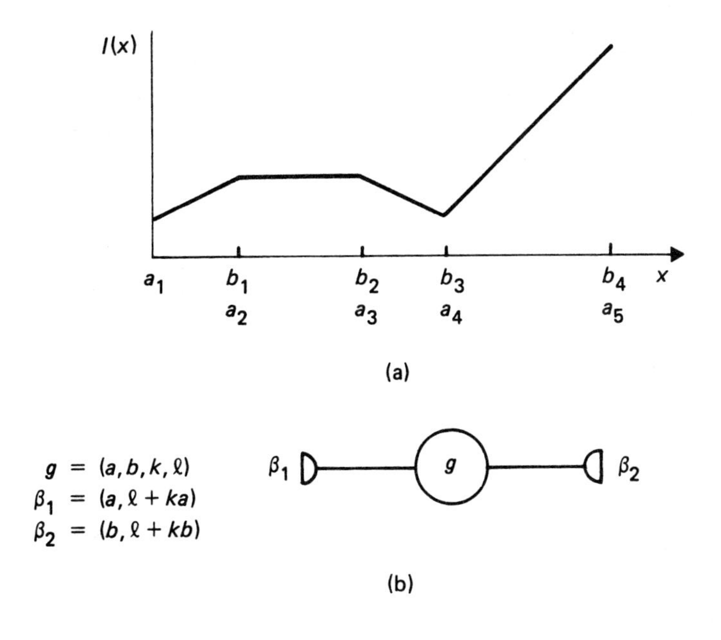

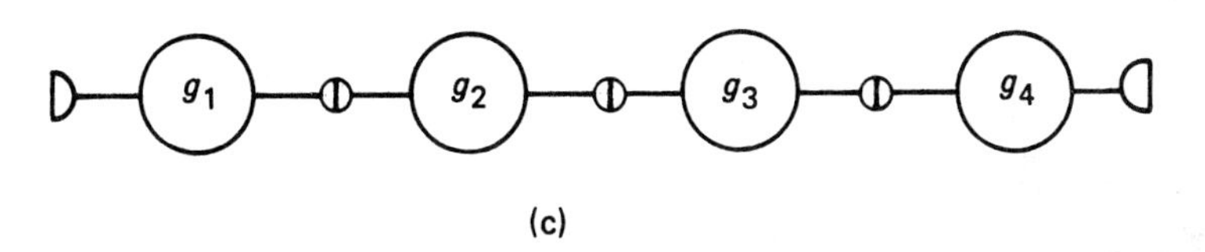

Figure 3.1 (a) A linear spline; (b) parameterized generators; and (c) linked generators.

$$\begin{cases} \beta' \, \rho \, \beta'' \iff b' = a'' \text{ and } l' + k'b' = l'' + k''a'' \\ \beta' = (b', l' + k'b') \\ \beta'' = (a'', l'' + k''a'') \end{cases}$$

Note that we have joined generators in a LINEAR chain; other topologies (or *connection types*) are used later. Formally, we write a *configuration*, $c = \sigma(g_1, g_2, g_3, \ldots, g_n)$, where g_i are the generators used from G, and σ, the *connector* or connection graph, states how they should be joined. If ρ holds for all bond couples occurring in σ, we say that the configuration is *regular*, and the set $C\ (\sigma,\rho)$ of all regular configurations over ρ is the *configuration space*.

Quite often one image (here = function) can be represented by two or more different regular configurations. In Figure 3.1(a), for example, if we split the fourth interval (a_4, b_4) into two parts and let each subinterval support one line segment, the image is then also represented by a five-generator configuration. Hence an image corresponds to a set of R-identified regular configurations and analysis consists of studying this correspondence.

Example 3.2 Now let us examine how this appears in a two-dimensional (2-D) version, Figure 3.2(a). Consider again linear splines, that is piecewise linear functions (defined on the triangles), continuously joined to each other. For simplicity, we assume that the triangle system is fixed to the regular one in the figure, although this is not really necessary.

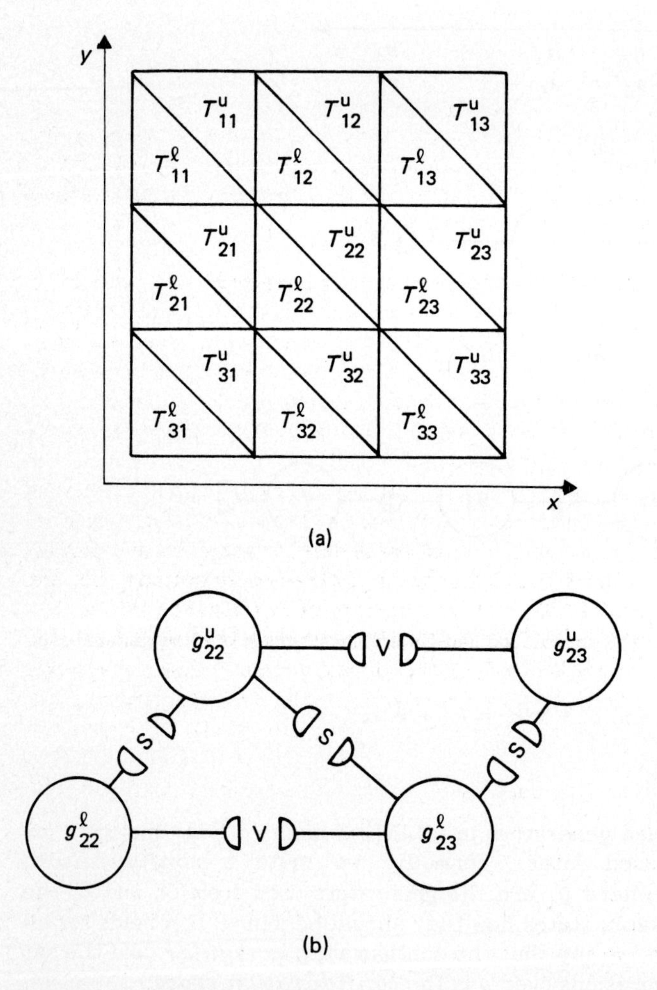

Figure 3.2 (a) Two-dimensional linear splines, defined on the triangles; (b) relationship of generators.

We can parameterize the generators as

$$g = (T, a, b, c)\,,$$

which means that the linear function is defined on the triangle T and has the equation $I = a + bx + cy$. Since we have two types of triangles, upper and lower, the generator space splits into the union $G = G^l \cup G^u$. Each triangle borders on 12 others: T^u_{22}, for example, has the neighbors T^l_{23}, T^u_{23}, T^l_{13}, T^u_{12}, T^l_{12}, T^u_{11}, T^l_{11}, T^u_{21}, T^l_{22}, T^u_{32}, T^l_{33}, T^u_{33}. Two neighbors can share a side, S, or a vertex, V, as in Figure 3.2(b), where g^l_{22} shares a side with g^u_{22} and a vertex with g^l_{23}. The bond values in a bond couple for a side consist of two real numbers, one for the ordinate at each endpoint of the side. If the bond couple corresponds to a vertex the bond values need only consist of a single, real number for the single ordinate. As before we let ρ stand for EQUAL:

$$\beta' \rho \beta'' \Longleftrightarrow \beta' = \beta'' .$$

In this example, the generators have an arity of 12 (number of bonds from the generator). Also the connector σ is a good deal more complicated than in **Example 3.1**. Actually, one could simplify the construction, but we do not attempt this here. We have not included any information about the location of a T in the bond values as it is not necessary, since the triangulation was specified in advance – with nonregular triangulation this would be needed.

Example 3.3 Let us turn to a rather different pattern ensemble. Suppose we operate on the background space $\mathbf{Z}^2_L$, the discrete 2-D torus, and use a binary background space $\{0, 1\}$, say 0 for white and 1 for black. We want to form (discrete) horizontal lines, not extending over the whole range $x = 0$ to $x = L - 1$, and not crowding each other vertically. This pattern is wholly artificial and not particularly interesting in itself. It does illustrate, however, some important principles in pattern synthesis and we return to it repeatedly.

How should we choose the generators? It is clear that we must let the 1-generators send signals to the 0-generators vertically: keep away! But not horizontally, where we want contiguous segments of ones. What is required are ***anisotropic*** generators sending out different bonds in different directions. So, we choose four generators and use a four-neighbor topology ($NJ = 4$), enumerated j = 0, 1, 2, 3: the first, generator 0, sends $\beta_j = 0$ in all directions [Figure 3.3(a)] and plays the role of white or zero; generator 1 is shown in Figure 3.3(b) and plays the role of the left endpoint of a line segment; generator 2 [Figure 3.3(c)] represents an internal point of a line segment; and generator 3 [Figure 3.3(d)] represents a right endpoint.

This is summarized in Table 3.1 (later usually referred to as the GE, for reasons to be discussed), which contains the respective bond values. Now we decide which bond couples, β_1, β_2, are regular and introduce the bond relation ρ by the truth value table (1 for true, 0 for false). Inspection of Table 3.2 together with Figure 3.3 shows that only such patterns as we initially decided to synthesize are,

Table 3.1 Relationship between bond coordinates and generators.

	Bond coordinate, j			
Generator	0	1	2	3
0	0	0	0	0
1	1	3	1	1
2	1	2	1	2
3	1	1	1	3

indeed, regular. Figure 3.4(b) shows one regular configuration with G = 0, 1, 2, 3, represented by the symbols 0, |–, × –|, respectively.

Example 3.4 The regular structure defined in **Example 3.3** generates patterns consisting of (discrete) line segments, all horizontal. If, instead, we wanted vertical line segments they could be obtained by rotating the generators through 90°. If we require both horizontal and vertical line segments to give a new generator space, G_1, we can use both types of generators together, $G' = G \cup G_1$.

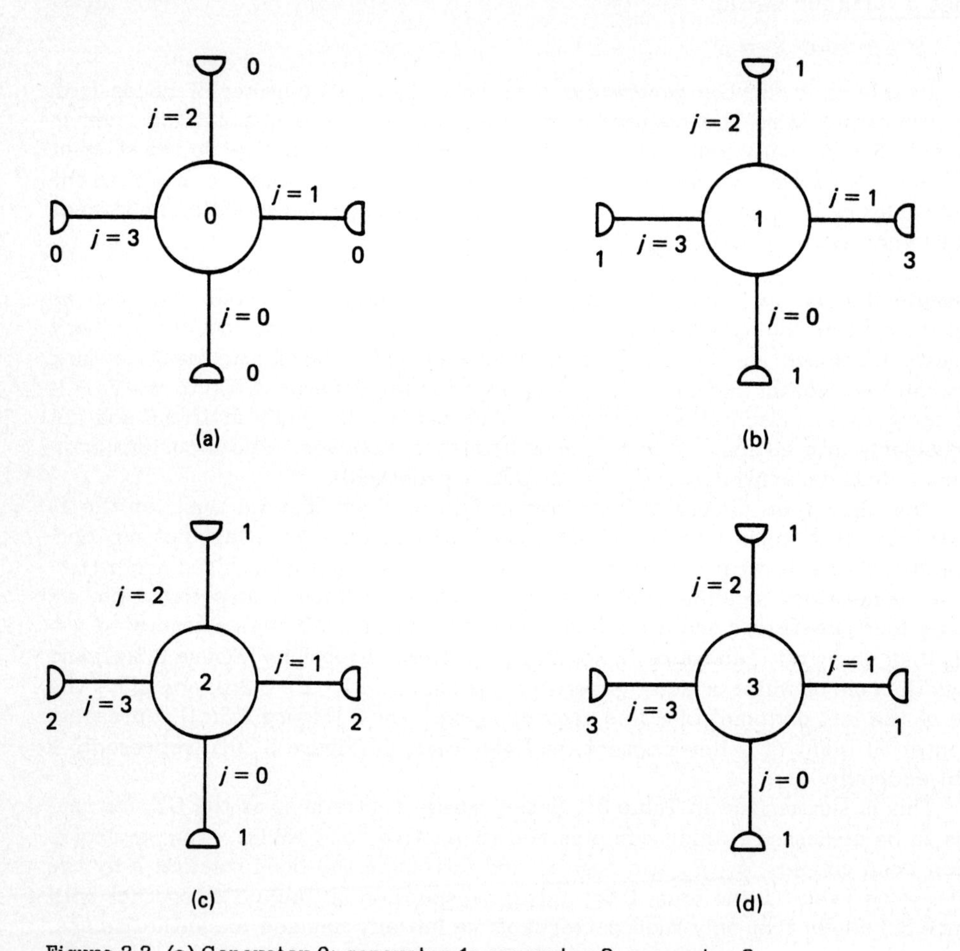

Figure 3.3 (a) Generator 0; generator 1; generator 2; generator 3.

Table 3.2 Truth value table (1 = true; 0 = false).

β_1 \ β_2	0	1	2	3
0	1	1	0	0
1	1	0	0	0
2	0	0	1	1
3	0	0	1	0

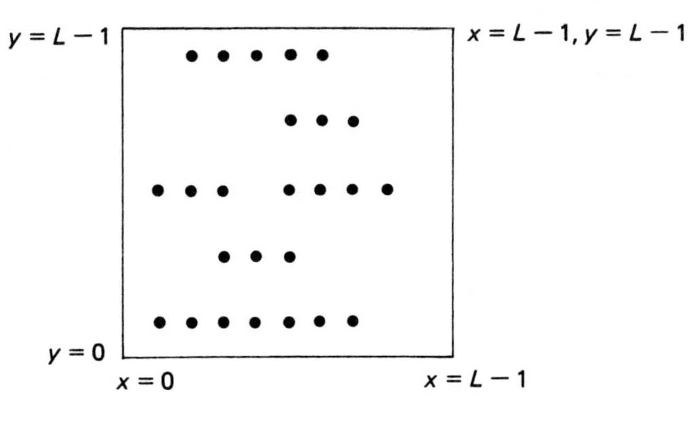

(a)

y = L − 1 x = L − 1, y = L − 1

y = 0

x = 0 x = L − 1

(b)

Figure 3.4 (a) Image produced by (b) configuration with $G = \{0, 1, 2, 3\} = \{0 \mid\!- \times -\!\mid\}$.

To make this general we introduce the concept of the ***bond structure*** group. Consider a generator spae GO with all its generators of arity NJ (sometimes denoted ω) and with the bonds enumerated by ***bond coordinates*** $j = 0, 1, 2, \ldots, NJ - 1$. The set J of NJ bonds is now subjected to permutations and we choose the set of permutations so that it forms a group (BSG, the bond structure group), but with no other general restriction.

For example, if $NJ = 4$ and we allow all rotations, then we can represent the bond structure group as

$$BSG = \begin{pmatrix} 0 & 1 & 2 & 3 \\ 1 & 2 & 3 & 0 \\ 2 & 3 & 0 & 1 \\ 3 & 0 & 1 & 2 \end{pmatrix},$$

where each row is a (cyclic) permutation of $J = (0, 1, 2, 3)$.

We now extend the ***original generator space*** GO to the extended generator space GE produced as the set of all generators that can be obtained from GO as the result of an arbitrary permutation. Returning to **Example 3.4**, we see that GE consists of 16 elements (with duplication) and, in general

$$|GE| = |GO| \times |BSG| \ .$$

Starting from a generator space G and a bond relation ρ we have formed the configuration space $\boldsymbol{C}(\boldsymbol{R})$ of all regular configurations over a connecting graph σ; $\boldsymbol{R} = < \rho, \sigma >$. In this chapter we concentrate on the case where σ is fixed in advance, but in other situations it is necessary to let it be variable, taking values in some set Σ, the *connection type*, of possible connectors. We speak of the regularity $\boldsymbol{R}$ as comprising the ***local regularity***, ρ, and the ***global regularity***, σ (or Σ).

It was mentioned earlier that the configuration is an abstraction used to synthesize the images. ***An image is what an ideal observer can see***, and we have identified some configurations in terms of an identification rule R. An image is an equivalence class

$$I = \{c \mid c \in \boldsymbol{C}(\boldsymbol{R}),\ c\,\mathrm{R}\,c_0\} \subset \boldsymbol{C}(\boldsymbol{R}) \tag{3.2}$$

of all the configurations that are R-equivalent to a fixed ***prototype*** c_0. The set of all such images forms the ***image algebra*** J. It is an algebra in the sense that given two images, I_1 and I_2, we may be able to combine them into a new one, $I_3 = \sigma_0(I_1, I_2)$, where σ_0 is some new connector. Note, however, that this is only possible if $\boldsymbol{R}$ is not violated for the new I_3. Therefore, the algebraic operation is not always defined (similar to not being allowed to divide by zero for real numbers). In **Example 3.1**, if we start from the two images I_1 and I_2 in Figure 3.5 and let the algebraic operation be the combination of I_1, moved to the right, with I_2, it is clear that local regularity holds for $\sigma_0(I_1, I_2)$. Indeed, the right bond value

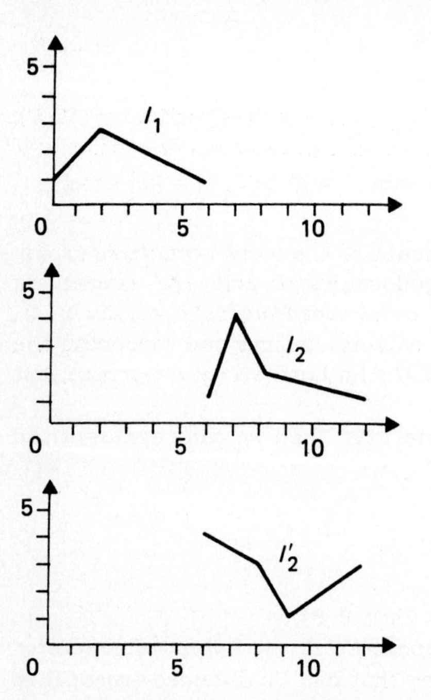

Figure 3.5 Images used to illustrate **Example 3.1**.

of I_1 is 1 and the left bond value of I_2 is also 1 so that ρ, meaning EQUAL, holds. On the other hand, I'_2 has a left bond value of 4 so that ρ is false and $\sigma_0(I_1, I'_2)$ is not defined.

The image algebra J expresses the ***combinatory*** properties of images and describes ***a rigid pattern*** (or pattern ensemble), while an $I \in J$ is ***a realization*** of the pattern.

Metric pattern theory

If our patterns were perfectly regular the image algebra would be the appropriate reference space for the ensuing pattern analysis. Unfortunately, most real patterns are not that rigid – they exhibit much variability – which must be taken into account in any realistic analysis. We do this by replacing the strict dichotomy of regular and nonregular with a graded evaluation in which a probability is attributed to each configuration. The purpose of ***metric pattern theory*** is to introduce and analyze such probability measures, for both configurations and images. This randomness is conceptually different from the observational noise that deforms images – deformations are discussed later.

Consider, for a fixed connecator σ, the configuration space

$$C(\sigma) = \{\sigma(g_1, g_2, \ldots, g_n),\ \forall g_i \in G\} \ . \tag{3.3}$$

A given configuration $c = \sigma(g_1, \ldots, g_i, \ldots, g_n)$ is parameterized by:

(1) Generator coordinates enumerating sites in σ, $i = 1, 2, \ldots, n$.
(2) Bond coordinates, $j = 1, 2, \ldots, \omega_i$.

Hence a bond couple in σ can be written as (k, k') with $K = (i, j)$ and $k' = (i', j')$. In order that c be regular we must have, as discussed in the previous section, the proposition

$$\bigwedge_{(k,k') \in \sigma} \beta_j(g_i)\ \sigma\ \beta_{j'}(g_{i'}) = \text{TRUE} \ , \tag{3.4}$$

where the conjunction is taken over all bond couples appearing in σ.

In order to relax condition (3.4), we replace ρ, defined on $B \times B$ (B = bond value space), by a function A on $B \times B$, taking as values nonnegative real numbers. The values of A, the ***acceptor function***, need not be probabilities, but serve as defining probabilities. In $\boldsymbol{C}(\sigma)$ we then introduce the probability densities

$$p(c) = \frac{1}{Z} \prod_{(k,k') \in \sigma} A[\beta_j(g_i), \beta_{j'}(g_{i'})] \ . \tag{3.5}$$

In order to make p a legitimate probability density, with $P[C(\sigma)] = 1$, we must choose the normalizing constant Z so that the sum of $p(c)$ over all $c \in C(\sigma)$, if $\boldsymbol{C}(\sigma)$ is discrete, or the integral with respect to some fixed measure, if $C(\sigma)$ is a continuum, is equal to one. Expression (3.5) is related to probability models studied intensively in statistical physics, where the constant Z is given the name partition function. Note the similarity between equations (3.4) and (3.5). The first, rigid regularity, is a special case of the second, relaxed regularity, if we make Z = 1, let A take only the values 0 and 1, and interpret these as truth values.

It is convenient to write equation (3.5) in a slightly different form, mathematically equivalent, but easier to manipulate during the design of inference machines, discussed later; namely,

$$p_T(c) = \frac{1}{Z} \prod_{i=1}^{n} Q(g_i) \prod_{(k,k') \in \sigma} A_0^{1/T}[\beta_j(g_i), \beta_{j'}(g_{i'})] \ . \qquad (3.6)$$

Here Q is a density in the generator space G, A_0 is a fixed acceptor function, and T a positive parameter called the *temperature* of the pattern, in analogy with statistical physics. If we write

$$A_0(\beta, \beta') = \exp a_0(\beta, \beta') \ ,$$

where a_0 should be thought of as the *affinities* between bond values, we obtain

$$p(c) = \frac{1}{Z} \prod_{i=1}^{N} Q(g_i) \exp\left\{\frac{1}{T} \sum_{(k,k') \in \sigma} a_0[\beta_j(g_i), \beta_{j'}(g_{i'})]\right\} \qquad (3.7)$$

The role of Q is to give different emphasis to different generators. It is not the same as the marginal distribution of all gs in the random configuration c, but is useful for easy manipulation of the probability measure when we want to increase or decrease the occurrence of a particular generator.

What happens when $T \to 0$? Well, if $a_0(\beta', \beta'')$ is positive, then $(1/T)$-$a_0(\beta', \beta'')$ becomes very large, and if it is negative, then $(1/T)\, a_0(\beta', \beta'')$ becomes very small. Hence a small T accentuates any difference in affinities between bond values. Certain configurations, where the bond values fit well, have large probabilities, and so all others are highly unlikely. Therefore, *cold patterns will, with high probability, become regular* in the sense of the bond relation

$$\beta' \rho \beta'' \Longleftrightarrow a_0(\beta', \beta'') = \max_{\beta_1, \beta_2} a_0(\beta_1, \beta_2) \ .$$

The opposite case, $T \to \infty$, renders the exponents in equation (3.7) close to zero, which means that the generators in c behave, almost, as an i.i.d. sample from G with the marginal distribution equal to (actually proportional to) Q. Hence *hot patterns are purely random* and very chaotic. It is only in this special case that Q represents the marginal distribution.

The probability density p in equation (3.7) on C induces a probability measure on $\mathbf{J}$ equal to the set of all, R-equivalence classes of C, so that an image I, or a set of R-equivalent configurations $I \in \mathbf{J}, I \in C$, has the probability

$$P(I) = \sum_{c \in I} p(c) \ .$$

For any fixed T, P in equation (3.7) *represents a flexible pattern*. It is a relaxed form of the regularity described by an image algebra.

Example 3.5 First a simple example: suppose that $G = \{0, 1\}$ with $\beta_j(g) \equiv g$, and that we use the trivial identification rule R = (0, 1), so that configuration here means the same thing as image, and suppose the acceptor matrix is

$$A = \begin{pmatrix} 1 & .1 \\ .1 & 1 \end{pmatrix} = (a_{kl};\, k,l = 0,1) \ .$$

A typical image is given in Figure 3.6, and shows little geometric activity.

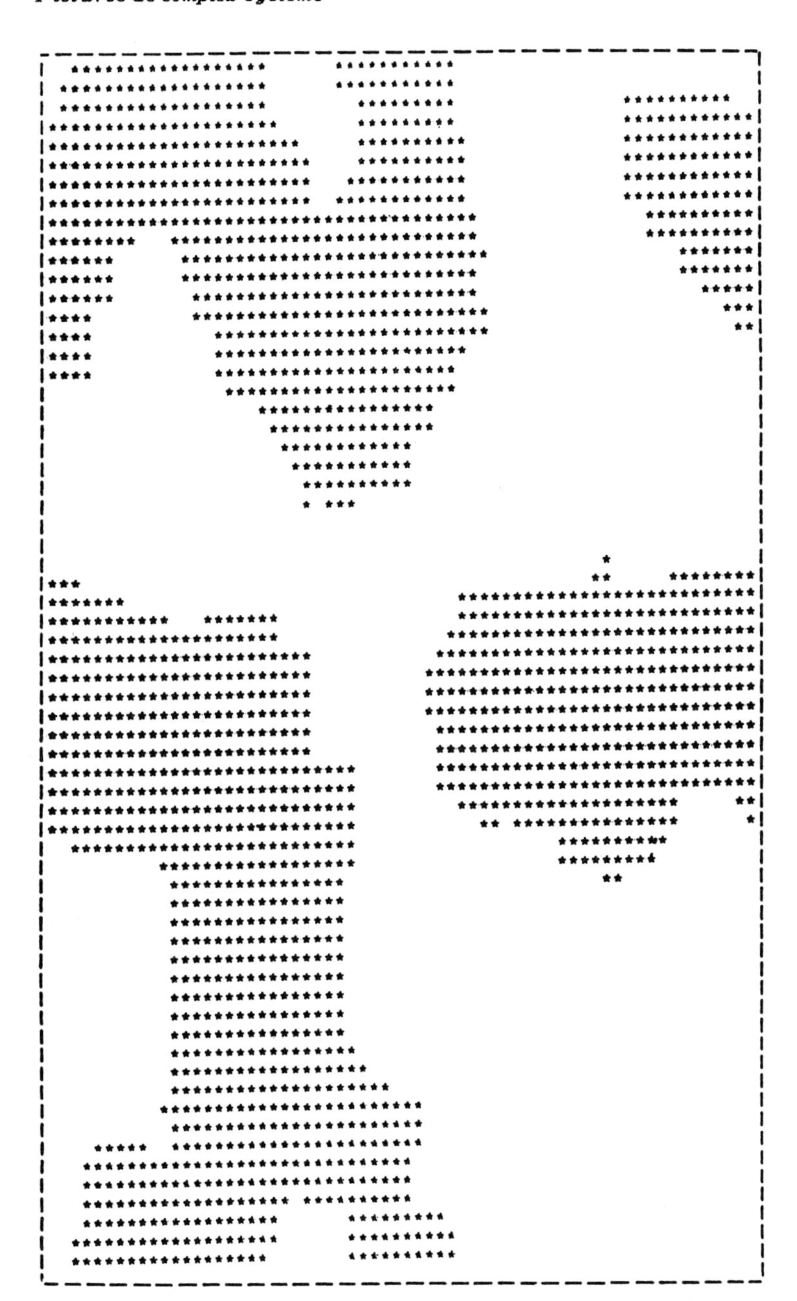

Figure 3.6 Image typical of **Example 3.5**.

If we increase the size of the entry $a_{01} = a_{10} = .1$, the picture becomes more chaotic; if we lower it the picture contains larger clumps of zeros or ones. This can, of course, also be achieved by replacing A with

$$A_T = \{a_{kl}^{1/T}; k, l = 0, 1\}$$

Example 3.6. If we use the generators in **Example 3.6** with the acceptor matrix

$$A = \begin{pmatrix} 1.01 & 1.01 & .01 \\ 1.01 & .01 & .01 \\ .01 & .01 & 1.00 \end{pmatrix},$$

and if we identify 0 with a blank and 1, 2, or 3 with a star we obtain, in one particular case, the configuration shown in Figure 3.7. Note that horizontal segments such as 12223 are surrounded by zeros. This configuration corresponds to the image in Figure 3.8.

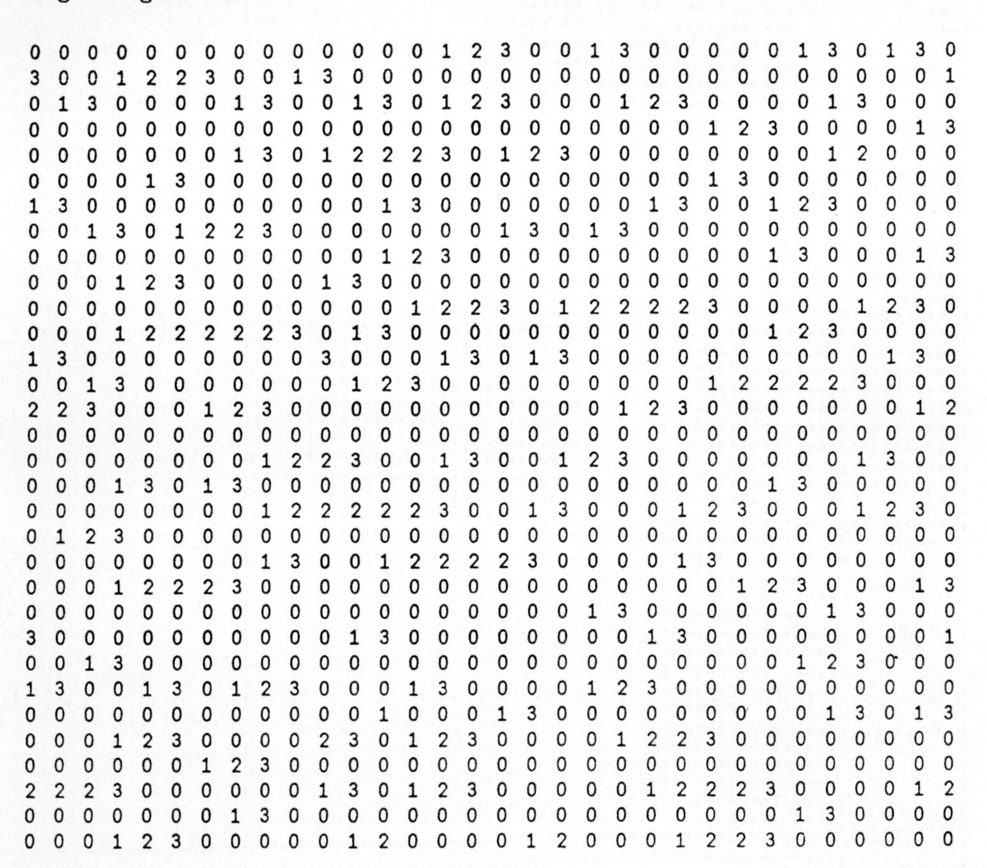

```
0 0 0 0 0 0 0 0 0 0 0 0 0 0 1 2 3 0 0 1 3 0 0 0 0 0 1 3 0 1 3 0
3 0 0 1 2 2 3 0 0 1 3 0 0 0 0 0 0 0 0 0 0 0 0 0 0 0 0 0 0 0 0 1
0 1 3 0 0 0 0 1 3 0 0 1 3 0 1 2 3 0 0 0 1 2 3 0 0 0 0 1 3 0 0 0
0 0 0 0 0 0 0 0 0 0 0 0 0 0 0 0 0 0 0 0 0 0 0 1 2 3 0 0 0 0 1 3
0 0 0 0 0 0 0 1 3 0 1 2 2 2 3 0 1 2 3 0 0 0 0 0 0 0 0 1 2 0 0 0
0 0 0 0 1 3 0 0 0 0 0 0 0 0 0 0 0 0 0 0 0 0 0 1 3 0 0 0 0 0 0 0
1 3 0 0 0 0 0 0 0 0 0 0 1 3 0 0 0 0 0 0 0 1 3 0 0 1 2 3 0 0 0 0
0 0 1 3 0 1 2 2 3 0 0 0 0 0 0 0 1 3 0 1 3 0 0 0 0 0 0 0 0 0 0 0
0 0 0 0 0 0 0 0 0 0 0 0 1 2 3 0 0 0 0 0 0 0 0 0 0 1 3 0 0 0 1 3
0 0 0 1 2 3 0 0 0 0 1 3 0 0 0 0 0 0 0 0 0 0 0 0 0 0 0 0 0 0 0 0
0 0 0 0 0 0 0 0 0 0 0 0 0 1 2 2 3 0 1 2 2 2 2 3 0 0 0 0 1 2 3 0
0 0 0 1 2 2 2 2 2 3 0 1 3 0 0 0 0 0 0 0 0 0 0 0 0 1 2 3 0 0 0 0
1 3 0 0 0 0 0 0 0 0 3 0 0 0 1 3 0 1 3 0 0 0 0 0 0 0 0 0 0 1 3 0
0 0 1 3 0 0 0 0 0 0 0 1 2 3 0 0 0 0 0 0 0 0 0 1 2 2 2 2 3 0 0 0
2 2 3 0 0 0 1 2 3 0 0 0 0 0 0 0 0 0 0 0 1 2 3 0 0 0 0 0 0 0 1 2
0 0 0 0 0 0 0 0 0 0 0 0 0 0 0 0 0 0 0 0 0 0 0 0 0 0 0 0 0 0 0 0
0 0 0 0 0 0 0 0 1 2 2 3 0 0 1 3 0 0 1 2 3 0 0 0 0 0 0 0 1 3 0 0
0 0 0 1 3 0 1 3 0 0 0 0 0 0 0 0 0 0 0 0 0 0 0 0 0 1 3 0 0 0 0 0
0 0 0 0 0 0 0 0 1 2 2 2 2 2 3 0 0 1 3 0 0 0 1 2 3 0 0 0 1 2 3 0
0 1 2 3 0 0 0 0 0 0 0 0 0 0 0 0 0 0 0 0 0 0 0 0 0 0 0 0 0 0 0 0
0 0 0 0 0 0 0 0 1 3 0 0 1 2 2 2 2 3 0 0 0 0 1 3 0 0 0 0 0 0 0 0
0 0 0 1 2 2 2 3 0 0 0 0 0 0 0 0 0 0 0 0 0 0 0 0 1 2 3 0 0 0 1 3
0 0 0 0 0 0 0 0 0 0 0 0 0 0 0 0 0 0 0 1 3 0 0 0 0 0 0 1 3 0 0 0
3 0 0 0 0 0 0 0 0 0 0 1 3 0 0 0 0 0 0 0 0 1 3 0 0 0 0 0 0 0 0 1
0 0 1 3 0 0 0 0 0 0 0 0 0 0 0 0 0 0 0 0 0 0 0 0 0 0 1 2 3 0 0 0
1 3 0 0 1 3 0 1 2 3 0 0 0 1 3 0 0 0 0 1 2 3 0 0 0 0 0 0 0 0 0 0
0 0 0 0 0 0 0 0 0 0 0 0 1 0 0 0 1 3 0 0 0 0 0 0 0 0 0 1 3 0 1 3
0 0 0 1 2 3 0 0 0 0 2 3 0 1 2 3 0 0 0 0 1 2 2 3 0 0 0 0 0 0 0 0
0 0 0 0 0 0 1 2 3 0 0 0 0 0 0 0 0 0 0 0 0 0 0 0 0 0 0 0 0 0 0 0
2 2 2 3 0 0 0 0 0 0 1 3 0 1 2 3 0 0 0 0 0 1 2 2 2 3 0 0 0 0 1 2
0 0 0 0 0 0 0 1 3 0 0 0 0 0 0 0 0 0 0 0 0 0 0 0 0 0 1 3 0 0 0 0
0 0 0 1 2 3 0 0 0 0 0 1 2 0 0 0 0 1 2 0 0 0 1 2 2 3 0 0 0 0 0 0
```

Figure 3.7 Configuration typical of **Example 3.6**.

If we change the temperature from $T = 1$ to $T = 10$ for the situation given by Figures 3.7 and 3.8 we obtain the configuration in Figure 3.9. It is obvious how far we are here from the local regularity ρ of **Example 3.1**. Ones occur together, the right and left endpoints appear at the wrong end, so chaos reigns. The corresponding image, Figure 3.10, is very hot; lines crowd each other (note, especially, the top) and little structure is visible. This case is used repeatedly in the following as an illustrative example. The pattern itself is of no interest, but serves

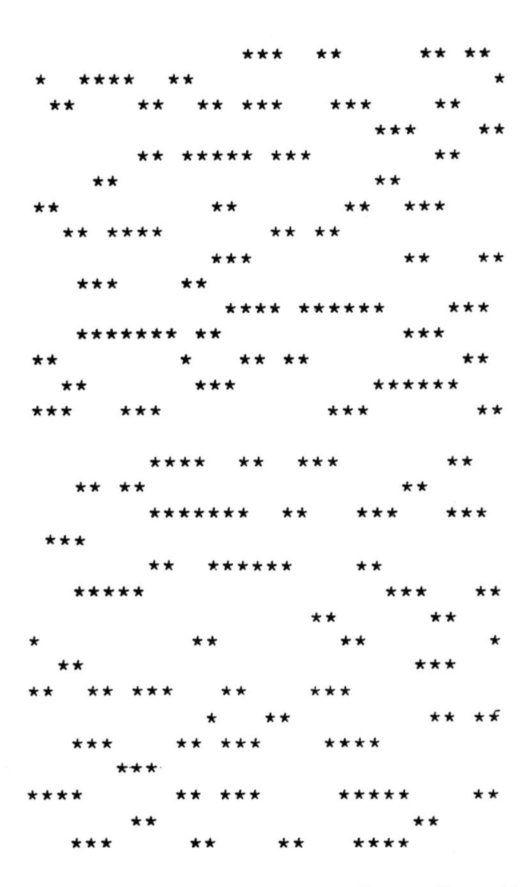

Figure 3.8 Image produced by the configuration of Figure 3.7.

a didactic purpose. We have not said anything about the method of pattern synthesis used to give these pictures. It is far from easy and is discussed later.

Deformation of images

Equation (3.7) defines a probability measure P, which serves as a ***prior*** in the Bayesian approach. In addition to the randomness described by this prior we need another source of randomness to ***deform the image***, called $\boldsymbol{D}$, a deformation mechanism:

$$D : \boldsymbol{J} \to J^{\boldsymbol{D}} = \text{a set of possible deformed images.}$$

In linear inference the most common assumption is that $\boldsymbol{D}$ acts additively: the only effect is to add random noise n to I, $\boldsymbol{I^D} = I + n$. This is called ***additive noise deformation***, and is a special case of ***contrast deformations***. Another contrast deformation is the ***symmetric binary noisy channel*** which works as

```
0 3 0 0 2 3 0 1 2 2 1 1 2 3 2 1 1 0 0 1 0 0 1 1 0 1 2 2 1 0 2 3
1 3 3 0 0 0 1 3 1 1 2 2 2 1 2 1 2 2 3 2 0 3 2 3 1 2 3 1 0 1 2 0
3 2 1 2 0 1 0 0 2 2 2 2 0 2 1 3 0 2 1 1 3 3 3 3 0 0 2 0 2 0 1 2
3 0 2 0 1 0 0 1 3 0 1 2 1 0 3 1 3 0 3 1 2 3 0 1 2 3 3 1 0 0 1 1
0 0 2 2 1 2 3 0 0 0 0 0 0 1 2 2 0 1 0 0 3 2 3 3 3 1 3 0 1 3 0 0
1 2 0 0 0 1 3 2 2 1 3 3 0 2 0 0 0 0 0 0 3 0 0 0 1 0 2 0 0 2 2 3
0 2 2 3 0 0 0 2 1 3 0 1 2 0 1 3 2 1 2 1 3 3 2 1 2 2 0 0 2 3 1 3
0 1 0 0 0 2 3 3 0 0 2 3 1 0 0 3 3 0 1 3 1 0 1 1 1 2 2 0 0 1 2 0
1 0 0 0 1 1 3 2 0 3 0 0 1 3 0 2 0 0 1 2 0 0 2 3 0 0 0 3 2 0 1 2
2 1 0 0 0 1 0 0 1 2 3 0 1 1 0 1 3 0 3 2 2 2 1 0 1 0 0 1 3 2 1 2
2 3 0 0 2 2 2 3 2 0 0 1 1 1 2 1 1 1 0 0 0 3 1 2 2 2 0 1 1 1 0 0
0 2 2 3 0 0 1 3 0 3 1 3 1 0 2 1 0 1 3 1 2 0 0 0 0 0 1 2 0 2 0 3
0 0 3 1 0 1 1 0 1 0 0 0 1 0 2 0 0 1 0 1 2 3 1 0 1 3 0 1 0 0 0 0
1 2 2 2 0 1 3 2 1 2 2 0 0 0 0 3 0 1 0 1 2 3 2 0 1 0 3 1 1 0 0 0
0 0 0 1 0 0 1 3 0 0 2 3 2 0 2 2 2 1 2 0 1 3 1 3 0 3 1 2 2 1 3 0
2 2 1 2 3 0 0 0 0 1 0 3 3 0 0 1 0 0 0 1 3 3 2 0 1 0 0 3 0 0 1 3
3 0 3 3 3 1 1 3 0 0 1 3 1 0 0 3 1 0 0 2 2 1 3 0 0 1 2 0 3 0 0 0
1 0 2 3 1 0 0 2 1 1 3 0 3 0 3 0 2 2 1 3 3 0 0 1 1 3 0 1 0 0 0 3
1 3 0 0 0 0 3 2 0 3 1 3 3 3 0 0 1 0 2 0 2 1 1 0 0 0 1 3 1 3 2 2
0 0 2 1 3 2 3 0 0 0 0 0 0 1 2 2 3 2 2 3 1 2 0 1 2 1 0 0 0 0 3 3
1 3 2 3 0 1 3 3 0 0 0 0 3 3 1 0 0 0 1 0 3 2 3 2 3 3 0 3 0 0 3 0
1 0 1 0 3 0 0 0 1 1 2 3 3 1 3 0 0 3 1 2 2 0 0 3 2 2 2 1 0 1 0 2
3 3 2 2 2 1 3 1 3 0 2 2 3 1 1 2 3 1 2 1 2 2 3 3 0 0 1 0 2 2 2 0
0 0 0 0 1 3 0 0 1 3 2 0 2 0 0 1 0 3 3 0 0 0 0 3 0 3 1 2 0 0 0 0
1 1 2 3 2 3 0 0 1 0 0 0 0 0 0 0 0 0 1 2 0 1 0 1 2 3 0 2 1 0 2 0
2 3 0 0 0 0 0 2 3 1 2 1 2 3 1 1 2 3 0 0 3 0 1 3 2 3 0 0 2 0 2 0
3 0 2 2 0 0 1 3 1 1 2 0 0 2 1 0 2 3 3 3 0 0 0 0 0 0 2 3 2 0 1 2
3 0 1 0 3 2 0 3 0 0 3 2 1 0 3 2 3 2 3 1 2 3 0 1 1 2 2 2 1 3 2 2
2 0 2 0 0 0 1 1 2 1 0 3 0 3 0 1 1 2 0 0 3 2 0 1 1 3 3 0 0 0 0 0
0 1 2 1 3 1 2 2 3 2 1 0 2 0 3 1 0 2 2 1 1 2 3 0 0 0 1 0 2 2 3 0
3 0 0 0 0 0 2 0 0 3 0 1 3 3 1 3 1 0 0 0 2 3 1 1 2 0 1 2 0 0 3 1
1 3 2 2 3 3 0 2 1 0 2 0 0 3 1 0 2 3 2 1 2 1 3 0 0 1 1 3 0 3 0 1
```

Figure 3.9 Case of Figure 3.7, but with $T = 10$.

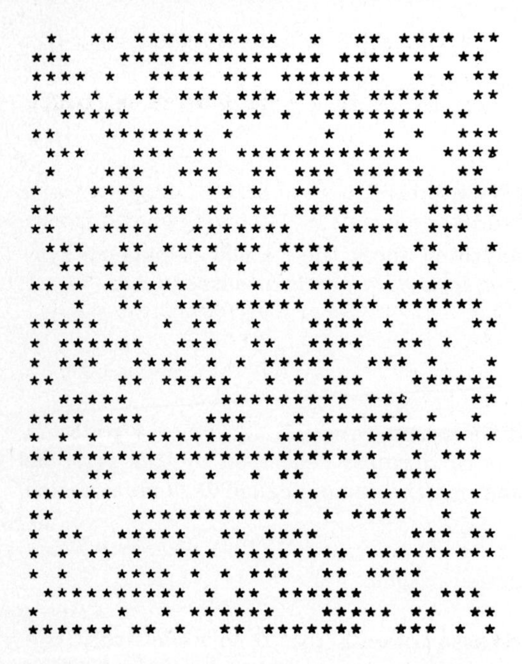

Figure 3.10 Image produced by the configuration of Figure 3.9.

follows. Let the pure image $i = I(x, y)$ be a black and white picture so that $I(x, y)$ takes only two values, say 0 and 1. Then we define, for a fixed I, that for $v = 0$ or 1,

$$P[I^D(x, y) = v \mid I] = \begin{cases} 1 - \varepsilon \text{ if } I(x, y) = v \\ \varepsilon \text{ if } I(x, y) = 1 - v \end{cases} . \tag{3.8}$$

Note that the probabilities of the error $1 \to 0$ or the error $0 \to 1$ are the same (i.e. symmetric).

Sometimes we have nonrandom deformations. An important case is ***blurring***, which acts on contrast patterns where the contrast values are in some linear space, say, the real line, $I(x, y) \in \mathbf{R}$. Then the deformed image is the convolution

$$I^D(x, y) = \sum_{\xi, \eta} w_{\xi\eta} I(x - \xi, y - \eta) . \tag{3.9}$$

Usually the range of summation is small compared to L. An example is

$$w_{\xi\eta} = \begin{cases} \dfrac{1}{(2h + 1)^2} & \text{if } 0 \le |\xi|, |\eta| \le h \\ 0 \text{ else} \end{cases} ,$$

a ***rectangular window***, but it is more common to have less sharp windows: $w_{\xi\eta}$ goes to zero smoothly as $|\xi|$ and $|\eta|$ become large. Another nonrandom D is the ***mask deformation***. If $I(x, y)$ is defined for $0 \le \xi, \eta < L$ then

$$I^D = \{I(x, y) \mid (x, y) \in M\} ,$$

where M is a sebset of $\mathbf{Z}_L^2$. Some of the observations have become hidden from the observer.

Before we discuss inference machines, let us summarize how we introduce regular structures:

Step 1 Choose a connector, σ, which is a graph with sites where the generators may reside. So far we have examined only very simple, regular connectors, but we later show how several levels within σ can be useful.

Step 2 Choose a generator space, G, with special attention given to the actual meaning of the bond values – the signals exchanged by generators. This is the crucial step, demanding inventiveness and intuition.

Step 3 Introduce the rigid regularity by a bond relation, ρ, corresponding to the type of ideal patterns that we began with.

Step 4 Describe what the ideal observer is allowed to see by an identification rule R. This is usually easy.

Step 5 Relax the regularity by replacing ρ with an acceptor function, as in equation (3.7).

Step 6 Study the deformations caused by imperfections in instrumentation and transmission.

We are now ready to study the process of pattern inference.

Introducing Pattern Inference Machines

The starting point when constructing inference machines is the question: How can we simulate probability measures of the form (3.7)? We want a systematic method for generating a random sample of i.i.d. observations from C or J obeying equation (3.5). In other words, we are attempting *pattern synthesis*.

There are two reasons why pattern synthesis plays a central role in our approach. The first is obvious – we want to synthesize images to judge whether the pattern model actually gives patterns of the type we want. The second, and this is the more important, is that all our analyses and inferences will be based on pattern synthesis. This claim, that analysis can be done by synthesis, may appear paradoxical, but we shall show that it is true.

The Markov property

The probability measure P can be said to be *regularity controlled*, since it is driven by the acceptor A, which mimics the bond relation ρ that expresses the rigid regularity. P is a very large family of measures, too large, and must be made

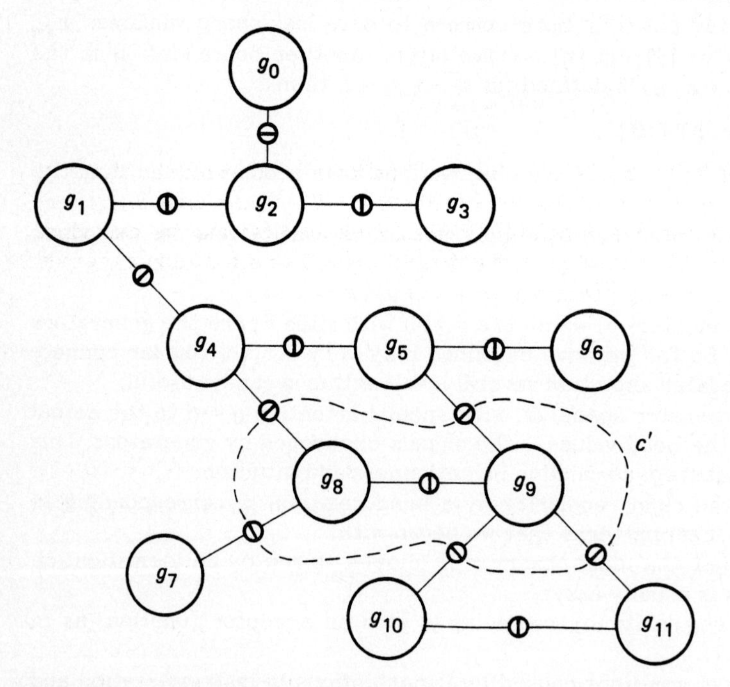

Figure 3.11 Configuration with
$c = \sigma(g_0, g_1, g_2, g_3, g_4, g_5, g_6, g_7, g_8, g_9, g_{10}, g_{11})$
$c' = \sigma'(g_8, g_9)$
$c'' = \sigma''(g_0, g_1, g_2, g_3, g_4, g_5, g_6, g_7, g_{10}, g_{11})$
$c''' = \sigma'''(g_4, g_5, g_6, g_7, g_{10}, g_{11})$

specific by choosing a particular A, or a small family of As. But before this can be done, we must first understand one fundamental property that all these Ps have in common. For simplicity we assume that:

(1) The generator space G is finite, $|G| = r < \infty$.
(2) The acceptor function (acceptor matrix) A takes only positive values.

Both of these conditions can be relaxed.

Consider the configuration diagram in Figure 3.11. In the configuration c we select a subconfiguration, c', and consider c as the connection of c' with the remaining subconfiguration c'',

$$c = \sigma_0(c', c'') .$$

Let us calculate the conditional probability using the general equation (3.5),

$$P(c'|c'') = \frac{P(c' \text{ and } c'')}{P(c'')} = \frac{P(c)}{P(c'')} .$$

The constant Z cancels since it appears in both numerator and denominator and we obtain

$$P(c'|c'') = \frac{\left\{\prod_{(k,k')\in\sigma} A[\beta_j(g_i), \beta_{j'}(g_{i'})]\right\}}{\sum\left\{\prod_{(k,k')\in\sigma''} A[\beta_j(g_i), \beta_{j'}(g_{i'})]\right\}}$$

with the notation $k = (i, j)$, $k' = (i', j)$ for the generator and bond coordinates, and where the summation extends over all generators in c'. However, a lot of factors occur in both the numerator and the denominator, and so can be canceled. Indeed, all generators that are not connected with the subconfiguration c' cancel and we, therefore, obtain (with obviously abbreviated notation),

$$P(c'|c'') = \frac{\left\{\prod_{(k,k')\in\sigma} A\right\}}{\sum_{(k,k')\in\sigma'''} A},$$

where the new subconfiguration $c = \sigma''(g_{i_1}, g_{i_2}, \ldots)$ (see Figure 3.11) consists of the outer *boundary* of c'': the set of generators that are not contained in c' but border it. In summary, we have shown that

$$P(c'|c'') = P(c'|c''') ;$$

that is, the probability depends only upon the outer boundary c''', not on all the outer generators in c''.

This is a Markov property, generalizing the idea of a Markov chain or Markov process. We are, therefore, in a general sense dealing with Markov processes (sometimes called *Markov random fields*) and our task is to invent methods of *inference in Markov processes* (partially observable).

Simulation for pattern synthesis

To provide a pattern synthesis for a given relaxed pattern P we need an algorithm that generates a random configuration from the pattern P. No direct method is known for doing this – we are forced to apply an indirect one, which is based on the Markov property discussed in the previous section. Select a site i_1 and consider the generator g_{i_1} there. Conditioned by the rest of the configuration it has some probability distribution P_{i_1}. Let us update c by replacing g_{i_1} with a random generator g'_{i_1} which has been generated from P_{i_1}. We discuss later (p 81) exactly how to do this and only remark here that P_{i_1} depends only upon the immediate neighbors of g_{i_1}.

Now we select another site, i_2, and replace g_{i_2} by some g'_{i_2} simulated under the probability law P_{i_2}. Proceeding in this way we obtain a sequence of configurations, $c_1, c_2, c_3, \ldots, c_t, \ldots$, which can be thought of as an ordinary Markov chain in time. The state space is C, probably of enormous cardinality, but finite under assumption (1) of the previous section.

We have not specified how the sequence $i_1, i_2, \ldots$ should be chosen and prefer not to do so completely yet. Suffice it to say that each site i must be visited infinitely often during the sequence for the following reasoning to hold true. It is not difficult for anyone familiar with Markov chains to see that the probability distributions of c_t converge to a unique equilibrium distribution. But the probability distribution P in equation (3.5) is certainly in equilibrium: just recall how we update c_t. Hence the limiting distribution of c_t, $t \to \infty$, is the P we want to simulate.

It should be pointed out that while the probability distributions will converge, the *configurations* c_t *themselves will not*. They will continue to vary during the updating and will not settle down to a particular one. Let us extend this: it is not necessary to update a single site at a time. The reasoning above remains valid if, at time t, we update several sites at once, so long as no two of these sites are neighbors. At time $t = 1$ we can update all sites in S_1, at $t = 2$ all sites in S_2, and so on, where the S_t, the *sweep areas*, are subsets of $\{1, 2, \ldots, n\}$.

And this is where recent advances in computer technology appear as if preordained. If we have access to a parallel computer with many processors (processing elements or PEs) working at the same time, we can apply each PE to updating each generator in a sweep area. Then we change from sweep area to sweep area. If the sweep areas are large *we can speed up the pattern synthesis drastically.*

Which sweep areas to use depends on the topology induced by the connector σ. For a square lattice with $NJ = 4$ neighbors two sweep areas are enough, one consisting of the rings and the other of the crosses in Figure 3.12(a). If $NJ = 8$ we need four sweep areas, as in Figure 3.12(b) and indicated by the four symbols used therein. For a general σ the number of sweep areas required equals the chromatic number of the graph; that is, the number of colors needed to color the graph so that neighbors have different colors.

Sometimes we violate the condition that no two sites in a sweep area be neighbors, which improves convergence but at the cost of increased computer time for the individual updating, since the above rules do not apply and have to be replaced by more complicated ones.

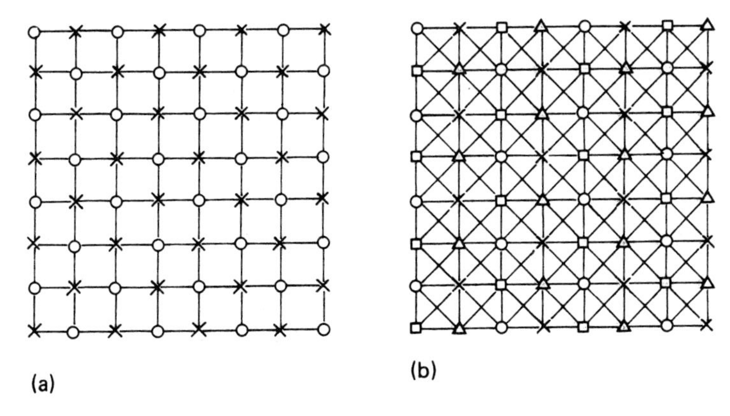

Figure 3.12 Sweep areas for connectors with (a) $NJ = 4$ and (b) $NJ = 8$.

The choice of sweep areas, the ***sweep strategy***, is not well understood at present. Moreover, only scanty knowledge is available as to the number of iterations (***the relaxation time***) needed before the probability distributions converge sufficiently to the equilibrium distribution. Many computer experiments have been performed to study this and the conclusion is that warm images are easy to synthesize; the convergence is fast. Cold images take longer to synthesize.

These empirical results have recently been supplemented by an analytical one. A theorem, which we quote without proof and without complete specification, states that the relaxation time

$$t_{\text{relax}} \geq a e^{\gamma / T} ,$$

where a and γ are positive constants. It follows that as the temperature T drops to zero, the relaxation time increases exponentially. This explains our empirical findings and should serve as a warning when highly rigid patterns are encountered. The scheme described here is called ***stochastic relaxation***.

Stochastic relaxation for pattern inference

To see how synthesis via stochastic relaxation is directly related to synthesis, examine Figure 3.13. It shows a six-level configuration, viewed from the side, so that, for example, level three actually has 16 generators but appears to have only four. The connector is fairly simple; if it were not for the horizontal connections in the level $h = 2$ it would be simply a tree. The two lowest levels have a special significance. Consider two configurations c' and c'' with the connector of Figure 3.13 and identify them, such that $c' \mathrm{R}\, c''$, if and only if they coincide at $h = 1$, $c_1' = c_1''$. This identification rule implies that the equivalence class of R-equivalent configurations is determined by the first level, so we can simply say that ***the second level*** c_2 ***is the image***.

In the computer experiment that produced Figure 3.13, the identification rule R was induced by an equivalence relation R on the generator space (see also **Example 3.6**). Hence, the acceptor function that appears in a bond couple between g_2 at $h = 1$ and g_1 at $h = 2$ can be written

$$A(\beta_1, \beta_2) = \begin{cases} 1 \text{ if } g_2 \text{ belongs to the equivalence class of } g_1 \\ 0 \text{ else} \end{cases}$$

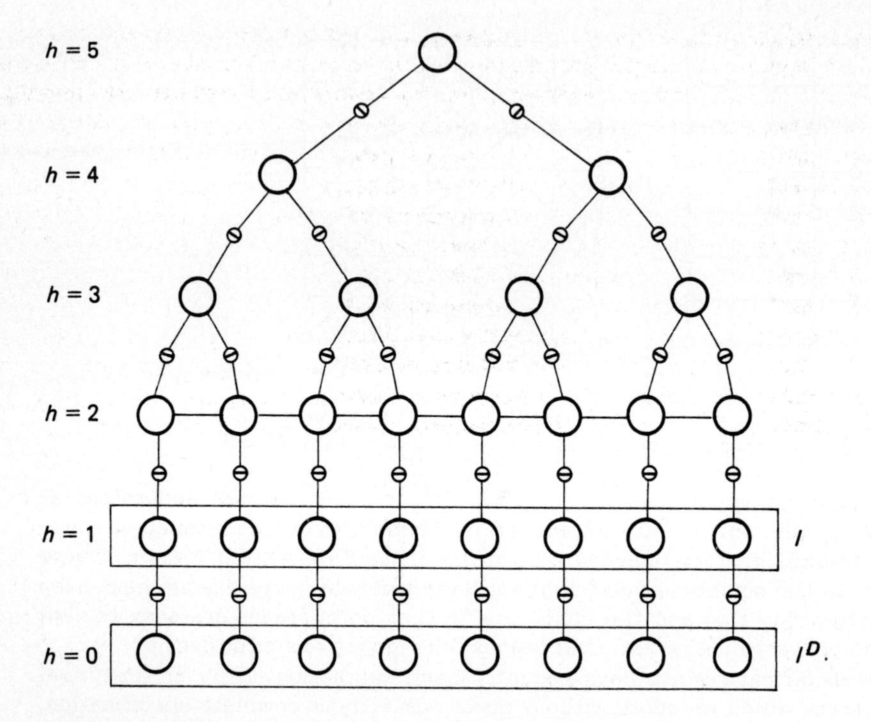

Figure 3.13 Side view of a six-level configuration.

Here the βs equal the gs; they are full information signals.

The lowest level, $h = 0$, plays the role of the deformed image, I^D, which is actually observed by the available instrumentation. In this example, a generator g_2 at $h = 1$ and a generator g_1 at $h = 0$ were related by the acceptor function

$$A(\beta_1, \beta_2) = \begin{cases} 1 - \varepsilon \text{ if } g_1 = g_2 \\ \varepsilon \text{ else} \end{cases}$$

In other words, $\boldsymbol{D}$ is a symmetric, binary noisy channel with error probability ε.

To summarize: level 0 is the deformed and observed image I^D, level 1 is the pure image I that we want to see, and the upper levels constitute the *analysis* or generation of I. Whether we want it for its own sake or not, *the analysis must be included in the inference procedure*.

The pattern inference procedure can now be run. The generators at $h = 0$ should be fixed as the values observed in I^D; the remaining levels can be given arbitrary values (but one can do better than that if speed is of importance). We then update all the sites (except for $h = 0$) in the total configuration using the updating procedure described above, with some sweep strategy. After a number of sweeps the probability distributions settle close to equilibrium. In this case, since we fix I^D, the marginal probability measure for I, $h = 1$, tends to equilibrate with P according to equation (3.9)

$$P_t(I \mid I^D) \to P(I \mid I^D), \text{ as } t \to \infty . \qquad (3.9)$$

This can be immediately used for *pattern inference of the Bayesian type*.

We started with a prior for I and simulated the posterior on the right-hand side of equation (3.9). But the Markov process in time is (asymptotically) stationary and ergodic so that *we can estimate* I *by a procedure of averaging type*. The exact form of the procedure depends upon what optimality criterion we use. If the Is are binary and we adopt Hamming distance as the optimality criterion,

$$d(I_1, I_2) = \# \{(x, y) | I_1(x, y) \neq I_2(x, y)\} \ ,$$

so that we want to find a restored image I^*, with $E[d(I, I^*)]$ a minimum, we should choose I^* using rule (3.10)

$$I^*(x, y) = \begin{cases} 1 \text{ if } f_n(x, y) \geq t_1/2 \\ 0 \text{ else} \end{cases} , \tag{3.10}$$

where we have run the inference process for the iterations $t = 1, 2, \ldots, t_1$, and $f_n(x, y)$ is the number of iterations for which we have observed a 1 at site (x, y) in $h = 1$. On the other hand, if $G = \mathbf{R}$ and we want to minimize the expected L_2 distance, we should choose the restoration procedure

$$I^*(x, y) = \frac{1}{t_1} \sum_{t=1}^{t_1} c_t'(x, y) \ .$$

Space does not allow us to discuss the refinements of such procedures. It should be mentioned, though, that substantial increases in the process speed can sometimes be obtained by *variable temperature schemes*, large sweep areas, and careful initialization.

It is important to realize that these procedures do not simply guarantee convergence. They are the best possible ones under the given conditions: *the pattern inference process exploits the available information optimality*. We have thus seen that, just as the process can synthesize patterns, it can also analyze (infer) patterns. It realizes the maxim

PATTERN ANALYSIS = PATTERN SYNTHESIS.

Let us now look at other types of pattern analysis in addition to inference. Suppose that the image has been masked from view (see the section on deformation of images, p 75): only certain sites, $(x, y) \in M^c$ can be observed (with or without pattern deformation). We then initialize the process by fixing the values at these sites to be the observed values, letting the rest be arbitrary. The process is run for a number of sweeps, level 1 observed, and the above procedure used, which results in *optimal extrapolation* from M^c to all sites.

Suppose that $\boldsymbol{D}$ blurs the image by convolution, as discussed earlier, with width 3. Then the three lowest levels of Figure 3.13 appear as in Figure 3.14 (the upper levels are not shown as they are exactly as in Figure 3.13). Remember that this is a side view; the object is really a set of planes. In Figure 3.14 the two upper levels are connected by bonds that express a deterministic relation as in equation (3.8). The two lower levels are connected by bonds that express stochastic relations, for example, a binary noisy channel.

Once it is realized that, because of the locality of blurring operations, we continue to obtain joint probabilities as before, it is clear that we can run the pattern inference process again with the lowest level, $h = 0$, fixed in order to restore the pure image at $h = 2$. In this way *the picture is simultaneously*

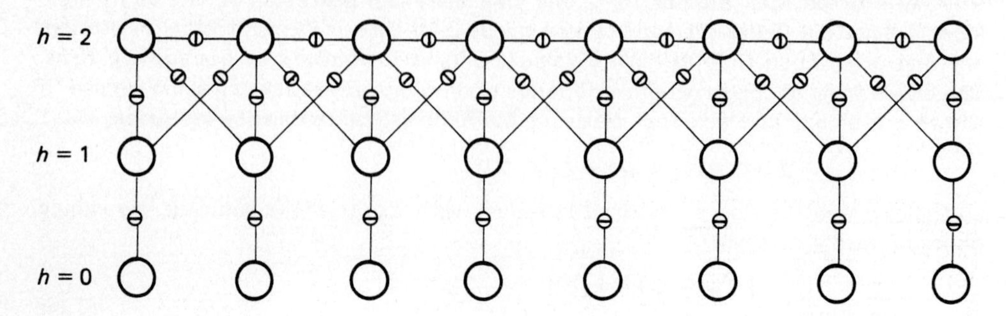

Figure 3.14 Lowest three levels of Figure 3.13 when D blurs the image by convolution with width 3.

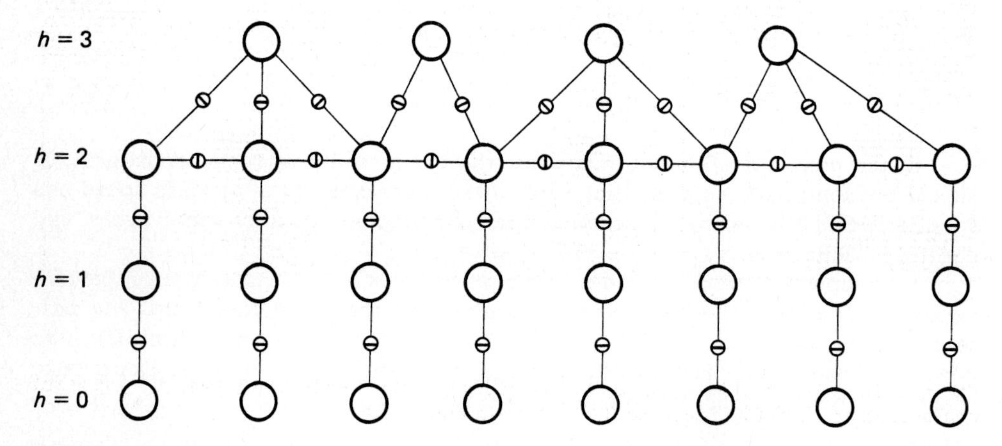

Figure 3.15 Level $h = 0$ is the observed image and the generators at $h = 3$ are binary.

unblurred and cleaned of noise, as far as is possible given the information contained in I^D. Consider Figure 3.15 in which, again, $h = 0$ is the observed image. Let the generators allowed at $h = 3$ be binary; that is, the only values are on and off.

We observe I^D, fix $h = 0$ to the values observed, run the inference process a number, t_1, of sweeps, and record how many times each of the generators at the site i were on the maximum level, $h = 3$. If such a number is greater than $t_1/2$ we say that we have recognized object i in the picture, otherwise not. This means that *the inference process achieves optimal pattern recognition* under the given conditions.

Examples of inference processes

We illustrate the general procedure described above with two simple examples. The programs are not described herein.

First, we discuss the pattern synthesis that underlies the picture shown in the section on metric pattern theory. In **Example 3.5** we used a 2-D toroidal lattice of size $L \times L$, and updated according to the scheme of rings and crosses, as shown in Figure 3.2(a); we initialized this at random. After implementing the restoration by computer programs we ran the process to obtain optimal recognition. An example is shown in Figure 3.16, where I is the pure image, I^D is the deformed image observed by the viewer, and I^* is the restored image. The results from running this pattern inference process speak for themselves.

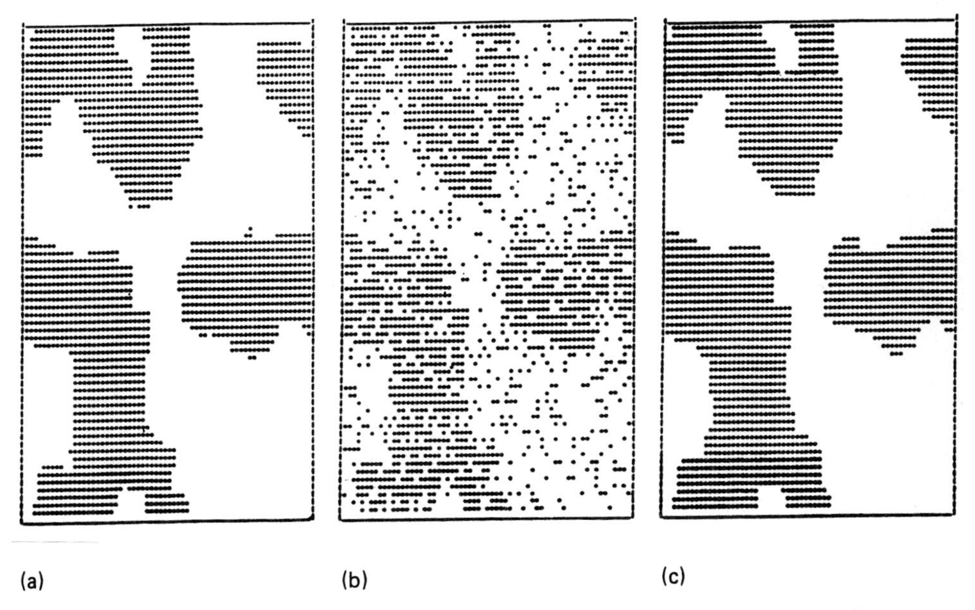

(a) (b) (c)

Figure 3.16 (a) Pure image (I), (b) deformed image observed by the viewer (I^D), and (c) restored image (I^*) for **Example 3.5** (I is as in Figure 3.6).

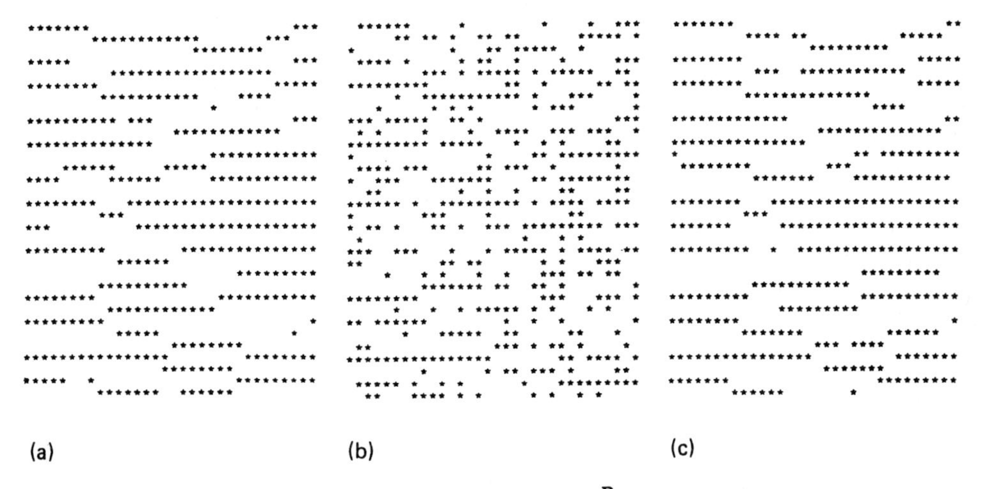

(a) (b) (c)

Figure 3.17 (a) Pure image (I), (b) deformed image (I^D), and (c) restored image (I^*) for horizontal line segments and relaxed regularity.

In our second example, in which we have horizontal line segments and relaxed regularity, we developed similar computer programs. Figure 3.17(a) shows a synthesized pure image, 3.17(b) the deformed image, and 3.17(c) the restored one. The latter has recovered much of the lost structure and the error rate has been approximately halved.

The patterns used so far as illustrative examples were synthesized from certain given generators and acceptor functions. We conclude the chapter by showing a real pattern analyzed by S. Geman, D. Geman, and D. McClure (unpublished). The image in Figure 3.18(a) was drastically deformed by additive noise, Figure 3.18(b). The restoration algorithm produced the image shown in Figure 3.18(c).

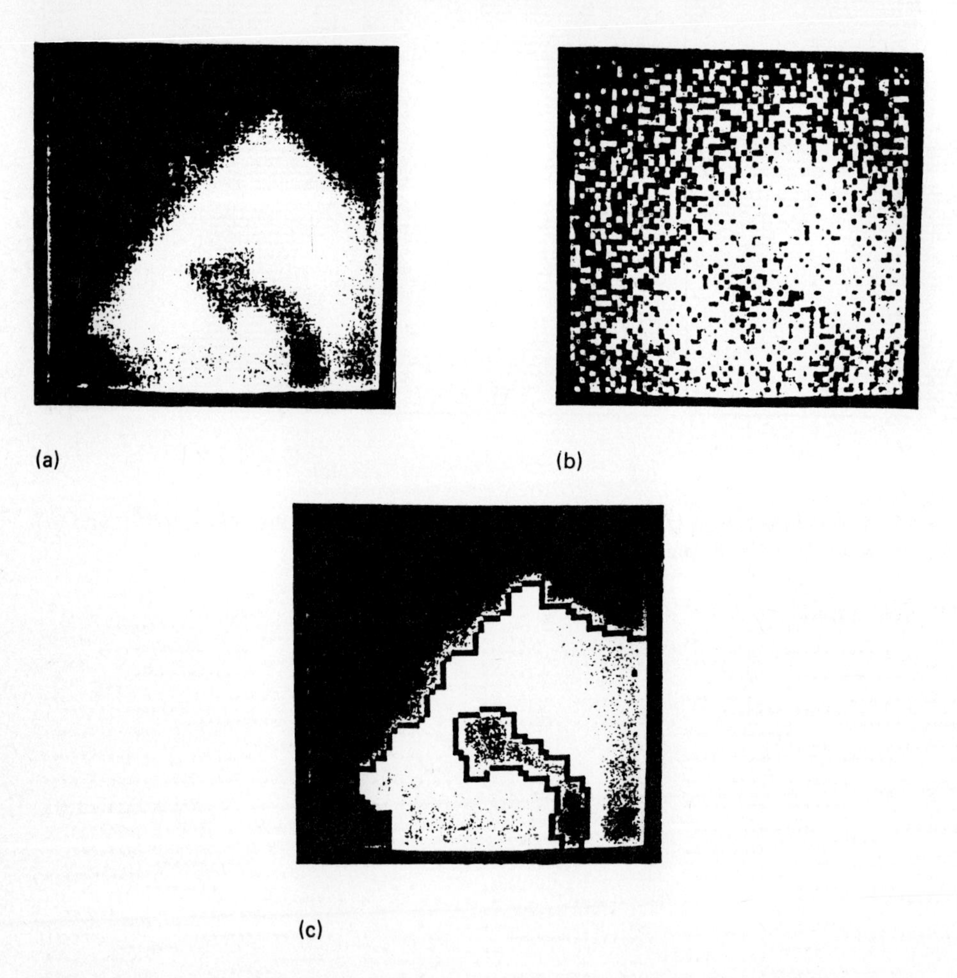

Figure 3.18 (a) Blurred image (roadside); (b) degraded image due to additive noise; and (c) restored image including the line process (1000 iterations).

The successful application of general pattern theory to picture processing, or to the analysis of other complex systems, requires a thorough examination of the basic elements: the generators, the connector graph, and the acceptor functions. In addition, the corresponding soft- and hard-ware problems must be solved, challenging tasks which have been accepted by a number of researchers. The prospects look promising, but what is absolutely essential is collaboration with other scientists who have knowlege of particular subjects: biologists, physicians, psychologists, linguists, and others.

The reader who wishes to learn more about this pattern theoretic research should consult the author's series *Lectures in Pattern Theory*, 3 volumes, Springer Verlag, 1976, 1978, 1981.

CHAPTER 4

A Survey of Replicator Equations

Karl Sigmund

Introduction

What are the units of natural selection? This question has aroused considerable debate in theoretical biology. Suggestions range from pieces of polynucleotides, genes, or gene complexes to individuals, groups, or species. It could be, however, that different answers are correct in different contexts, depending on the scale on which selection acts most decisively. This is somewhat analogous to physics, where the dominant force may be gravitational, electromagnetic, or strong or weak interparticle attractions, depending on the problem.

It is therefore convenient to consider an abstract unit of natural selection in theoretical investigations, which can be replaced by the appropriate real unit (genes, individuals, or species) in specific circumstances. This abstract unit is termed a ***replicator*** in Dawkins' book ***The Extended Phenotype*** (Dawkins, 1982). The term describes any entity which (a) can give rise to an unlimited (at least in principle) sequence of copies and (b) occurs in variants whose properties may influence the number of copies.

Biomathematical arguments support the usefulness of this concept. Indeed, the remarkable similarity of dynamical systems describing the action of selection in the most diverse fields lends weight to the notion of a common mechanism underlying these different observations. The term ***replicator dynamics*** has been applied to this mechanism (see Schuster and Sigmund, 1983). In the case of continuous time (generations blending into each other), the dynamics can be described by an ordinary differential equation $\dot{\mathbf{x}} = \mathbf{F}(\mathbf{x})$ of the type

$$\dot{x}_i = x_i [f_i(\mathbf{x}) - \Phi], \quad i = 1,\ldots,n \tag{4.1}$$

while for discrete time (separate generations) the dynamics are given by a difference equation $\mathbf{x} \to T\mathbf{x}$ with

$$(T\mathbf{x})_i = x_i \left(\frac{f_i(\mathbf{x})}{\Phi} \right), \quad i = 1,\ldots,n \quad . \tag{4.2}$$

In both cases, the term Φ is defined by

$$\Phi(\mathbf{x}) = \sum_{i=1}^{n} x_i f_i(\mathbf{x}) \tag{4.3}$$

and ensures that the state $\mathbf{x}$ of the system remains on the unit simplex

$$S_n = \{\mathbf{x} = (x_1,\ldots,x_n) \in \mathrm{R}^n : \sum_{i=1}^{n} x_i = 1, \ x_i \geq 0 \text{ for all } i\} \ . \tag{4.4}$$

The functions $f_i(\mathbf{x})$ describe the interaction of the different variants of the underlying replicator and are specified by an appropriate biological model.

In particular, first-order interaction terms, that is linear functions $f_i(\mathbf{x}) = (A\mathbf{x})_i$ defined by a matrix $A = (a_{ij})$, where

$$(A\mathbf{x})_i = \sum_{j=1}^{n} a_{ij} x_j \quad , \tag{4.5}$$

lead to dynamics which have been investigated independently in (i) population genetics, (ii) population ecology, (iii) the theory of prebiotic evolution of self-replicating polymers, and (iv) sociobiological studies of evolutionarily stable traits of animal behavior. Within these contexts, the dynamics describe the effects of selection upon (i) allele frequencies in a gene pool, (ii) relative frequencies of interacting species, (iii) concentrations of polynucleotides in a dialysis reactor, and (iv) distributions of behavioral phenotypes in a given species.

After a brief summary of the biological background in the following section, we present a survey of the mathematical aspects of continuous- and discrete-time replicator equations. There are many interesting results, in particular for the first-order case, due to the work of Akin, Hofbauer, Zeeman, and others. On pages 93–94 we are concerned with some general properties of replicator equations and on pp 94–95 we discuss the existence and stability of equilibria and present some theorems on time averages and exclusion properties. Results concerning the permanence of the biological components of the system are presented on pp 96–98. Gradient systems for replicator equations are then described, followed by an overview of the classification of low-dimensional phase portraits. Finally, we summarize the relationships between game theory and first-order replicator equations.

Biological Motivation

Population genetics

Genes are the quintessential replicators. It is therefore quite appropriate that the first systematic study of a class of replicator equations occurred in population genetics: the classic work of Fisher, Haldane, and Wright on the effects of natural selection upon the frequencies of alleles at a single locus of a diploid, randomly mating population.

Briefly, if $A_1,\ldots,A_n$ denote the possible alleles and $x_1,\ldots,x_n$ their frequencies within the adult population, then random fusion of gametes yields zygotes of genotype A_iA_j with frequency $2x_ix_j$ for $i \neq j$ and x_i^2 for $i = j$. (This is the Hardy–Weinberg law.) Let a_{ij} denote the *fitness* of genotype A_iA_j, which in this context is the probability of its survival from zygote to adulthood. The genotypes A_iA_j and A_jA_i are identical (it does not matter which parent contributes which allele) and hence $a_{ij} = a_{ji}$. Since the heterozygous genotype A_iA_j ($i \neq j$) carries one gene A_i while the homozygous genotype A_iA_i carries two such genes, the frequency $(T\mathbf{x})_i$ of allele A_i in the adult stage of the new generation is proportional to

$$\frac{1}{2}\left(\sum_{i \neq j} 2a_{ij}x_ix_j + 2a_{ii}x_i^2\right)$$

and hence to $x_i(A\mathbf{x})_i$. Thus

$$(T\mathbf{x})_i = x_i\left(\frac{(A\mathbf{x})_i}{\Phi}\right) \quad \text{with} \quad a_{ij} = a_{ji} \tag{4.6}$$

under the obvious assumption that Φ (which can be interpreted as the average fitness of the population) is not equal to zero.

The corresponding continuous-time selection equation

$$\dot{x}_i = x_i[(A\mathbf{x})_i - \Phi] \quad \text{with} \quad a_{ij} = a_{ji} \tag{4.7}$$

has been known since the 1930s. It is considerably easier to handle than its discrete counterpart (4.6), but its derivation is less clear. It is usually obtained under the assumption that the population is always in Hardy–Weinberg equilibrium, an assumption which is not strictly valid in general (see Ewens, 1979).

Thus first-order replicator equations with symmetric matrices occur in population genetics.

In the model considered here, selection acts through the different viabilities of the genotypes. Differential fecundities (where the number of offspring depends on the mating pair) lead to equations for the genotype frequencies which are not of replicator type (see Pollak, 1979). Except in some special cases (e.g., multiplicative fecundity), these equations behave rather differently from (4.6) or (4.7) (see Bomze *et al.*, 1983). The effects of mutations and (for models with several genetic loci) recombinations are also not described by replicator equations.

On the other hand, frequency-dependent fitness coefficients fall within the general framework of replicator equations. Models for haploid organisms lead to equations of the type

$$(T\mathbf{x})_i = x_i\left(\frac{a_i}{\Phi}\right) \tag{4.8}$$

or

$$\dot{x}_i = x_i(a_i - \Phi) \quad , \tag{4.9}$$

where x_i is the frequency of chromosome G_i and a_i denotes its fitness. Equations of this type are almost trivial if the coefficients a_i are constant. If they are frequency dependent (e.g., if they are linear functions of x_i), however, then interesting replicator dynamics occur.

Prebiotic evolution

Equations of type (4.8) were first studied (initially within the framework of chemical kinetics) in an important series of papers by Eigen (1971) and Eigen and Schuster (1979) on prebiotic evolution. In this context the x_i are the concentrations of self-replicating polynucleotides (RNA or DNA) in a well-stirred dialysis reactor with a dilution flow Φ regulated in such a way that the total concentration $x_1 + \cdots + x_n$ remains constant (without loss of generality we can set this concentration equal to 1). In the absence of mutations this leads to continuous-time replicator equations (generation effects do not play any part, even if the initial population of molecules reproduces in some synchronized way).

Independent replication of the polymers leads to equations (4.8) with constant reproduction rates a_i. This implies (except in the case of kinetic degeneracy) that all but one of the molecular species vanishes, with the loss of the corresponding encoded information. In their search for ways of preserving the initial amount of molecular information, Eigen and Schuster were led to study networks of catalytically interacting polynucleotides. Such interactions (and the corresponding replication rates) are usually quite complicated, but nevertheless some rather general results have been obtained. In addition, certain special cases of linear catalytic (or inhibiting) interactions, yielding the first-order replicator equations:

$$\dot{x}_i = x_i(\sum a_{ij} x_j - \Phi), \quad i = 1,\ldots,n \tag{4.10}$$

have been studied as approximations of more realistic chemical kinetics.

The hypercycle (a closed feedback loop in which each molecular species is catalyzed by its predecessor) has attracted particular attention (see Schuster *et al.*, 1979, 1980; Hofbauer *et al.*, 1980). Both the cooperation of the components within a hypercycle and the strict competition between individual hypercycles suggest that such networks may have been involved in some phases of early prebiotic evolution. The hypercycle equation is given by

$$\dot{x}_i = x_i[x_{i-1} H_i(\mathbf{x}) - \Phi], \quad i = 1,\ldots,n \quad , \tag{4.11}$$

where the indices are taken on modulo n and the functions $H_i(\mathbf{x})$ are strictly positive on S_n. If the H_i are constants, k_i, the above equation then reduces to a special case of the first-order replicator equations:

$$\dot{x}_i = x_i(k_i x_{i-1} - \Phi), \quad k_i > 0 \tag{4.12}$$

obtained if matrix $A = (a_{ij})$ in equations (4.10) is a permutation matrix:

$$A = \begin{bmatrix} 0 & 0 & \cdots & 0 & k_1 \\ k_2 & 0 & \cdots & 0 & 0 \\ 0 & k_3 & \cdots & 0 & 0 \\ \cdot & \cdot & & \cdot & \cdot \\ \cdot & \cdot & & \cdot & \cdot \\ \cdot & \cdot & & \cdot & \cdot \\ 0 & 0 & \cdots & k_{n-1} & 0 \end{bmatrix} .$$

Animal behavior

Taylor and Jonker (1978) were the first to introduce first-order replicator equations into models of the evolution of animal behavior. This approach was based on Maynard Smith's use of game theory in the study of animal conflicts within a species, equating strategies with behavioral phenotypes and payoffs with increments of individual fitness.

These investigations initially centered on the notion of evolutionary stability (see Maynard Smith, 1974), which may be interpreted as game-theoretic equilibria which are proof against the invasion of behavioral mutants. This static approach assumed certain implicit dynamics which were soon made explicit in the form of equations, once again of replicator type.

Let $E_1,...,E_n$ denote the behavioral phenotypes within a population, $x_1,...,x_n$ the frequencies with which they occur, and a_{ij} $(1 \leq i, j \leq n)$ the expected payoff for an E_i-strategist in a contest against an E_j-strategist. Then, assuming random encounters, we obtain $(A\mathbf{x})_i$ as the average payoff for an E_i-strategist within a population in state $\mathbf{x}$, and

$$\Phi = \mathbf{x}{\cdot}A\mathbf{x} = \sum x_i (A\mathbf{x})_i \tag{4.13}$$

as the mean payoff. In the case of asexual reproduction, the rate of increase $\dot{x}_i / x_i$ of phenotype E_i is given by the difference $(A\mathbf{x})_i - \mathbf{x}{\cdot}A\mathbf{x}$, which once again yields (4.10) [or, in the discrete-time case,

$$(T\mathbf{x})_i = x_i \left[\frac{(A\mathbf{x})_i + C}{\Phi + C} \right] , \tag{4.14}$$

where C is a positive constant].

The assumption of asexual reproduction at first seems rather unnatural. It can be shown, however, that in many important examples the essential features of the dynamical model are preserved in the more complicated case of sexual reproduction (see Maynard Smith, 1981; Hofbauer *et al.*, 1982; Hines, 1980; Bomze *et al.*, 1983; Eshel, 1982). Rather than introducing some sort of Mendelian machinery which, given the present state of knowledge of the genetic basis of behavior, is bound to be highly speculative, it seems reasonable to adhere to the more robust and manageable asexual model (see Schuster and Sigmund, 1984).

The corresponding replicator equations are examples of frequency-dependent sexual or asexual selection equations. Many specific types of conflict (e.g., the Hawk–Dove–Bully–Retaliator game, the War of Attrition game, and the Rock–Scissors–Paper game) have been examined within this framework (see Zeeman, 1981; Bishop and Cannings, 1978; Schuster *et al.*, 1981).

The game-dynamical aspects of the linear replicator equations (4.10) may be expected to lead to applications in fields such as psychology and economics (see Zeeman, 1981). A justification of viewing strategies as replicators is given by Dawkins (1982).

Population ecology

Equations used to model ecological systems are commonly of the form

$$\dot{y}_i = y_i f_i(y_1,\ldots,y_n), \quad i = 1,\ldots,n \quad , \tag{4.15}$$

where y_i are the densities of different populations interacting through competition, symbiosis, host–parasite, or predator–prey relationships. Such equations "live" on $\mathbf{R}^n_+$ and are usually not of replicator type. However, ***relative*** densities do yield replicator equations. In particular, Hofbauer (1981a) has shown that the classic $(n-1)$-species Lotka–Volterra equation

$$\dot{y}_i = y_i(b_{in} + \sum b_{ij} y_j), \quad i = 1,\ldots,n-1 \tag{4.16}$$

is equivalent to the first-order replicator equation (4.10) on $S_n \setminus \{\mathbf{x}: x_n = 0\}$ with $a_{ij} = b_{ij} - b_{in}$,

$$x_i = \frac{y_i}{\sum_{j=1}^{n} y_j}, \quad i = 1,\ldots,n \tag{4.17}$$

and $y_n \equiv 1$. The barycentric transformation (4.17), together with a change in velocity, maps the orbits of equations (4.16) into the orbits of equations (4.10). Which of these equations is more convenient will depend on the problem considered. Similar results hold for interactions of order higher than linear.

Sexually reproducing organisms are not replicators in the strict sense of the term, but within ecological considerations and disregarding genotypes they may be viewed as such.

General Properties

The term Φ in equation (4.3) guarantees that the continuous-time replicator equations (4.10) "live" on S_n, since $(\sum x_i)^{\cdot} = 0$ on S_n. Thus, the simplex and all its faces (which consist of subsimplices characterized by $x_i = 0$ for all i in some nontrivial subset I of $\{1,\ldots,n\}$) are invariant. In particular, the corners $\mathbf{e}_i$ are equilibria. The solutions of equations (4.1) in S_n are defined for all $t \in \mathbb{R}$.

For the discrete-time replicator equation (4.2) to have any meaning, the term Φ must be nonvanishing on S_n. It always has the same sign and we assume that the $f_i(\mathbf{x})$ are also of this sign, say positive. In this case the simplex and all of its faces are once again invariant. If a continuous- or discrete-time replicator equation is restricted to a face of S_n the resulting equation is again of replicator type.

We say that two vector fields $\mathbf{f}$ and $\mathbf{g}$ on S_n are ***equivalent*** if there exists a function $c : S_n \to \mathbf{R}$ such that $f_i(\mathbf{x}) - g_i(\mathbf{x}) = c(\mathbf{x})$ holds on S_n for all i. If $\mathbf{f}$ and $\mathbf{g}$ are equivalent then the restrictions $\dot{x}_i = x_i[f_i(\mathbf{x}) - \Phi]$ and $\dot{x}_i = x_i[g_i(\mathbf{x}) - \Phi]$ coincide on S_n. In the same way, if there exists a function $c : S_n \to \mathbb{R}^+$ such that $f_i(\mathbf{x}) = c(\mathbf{x}) g_i(\mathbf{x})$ holds on S_n for all i, then the difference equations $\mathbf{x} \to T\mathbf{x}$ with $(T\mathbf{x})_i = x_i f_i(\mathbf{x}) \Phi^{-1}$ and $(T\mathbf{x})_i = x_i g_i(\mathbf{x}) \Phi^{-1}$ coincide on S_n.

In particular, we say that the $n \times n$ matrices A and B are equivalent if the vector fields $A\mathbf{x}$ and $B\mathbf{x}$ are equivalent in the sense described above. This is the case if and only if there exist constants c_j such that $a_{ij} - b_{ij} = c_j$ for all i and j. Equivalent matrices lead to identical first-order replicator equations. Thus, without loss of generality, we may consider only matrices with zeros in the diagonal, for example, or matrices whose first row vanishes.

Another useful property is the quotient rule

$$\left(\frac{x_i}{x_j}\right)^{\cdot} = \left(\frac{x_i}{x_j}\right)[f_i(\mathbf{x}) - f_j(\mathbf{x})] \tag{4.18}$$

or, in the discrete case,

$$\frac{(T\mathbf{x})_i}{(T\mathbf{x})_j} = \left(\frac{x_i}{x_j}\right)\left(\frac{f_i}{f_j}\right) \tag{4.19}$$

for $x_j > 0$.

Losert and Akin (1983) have shown that the discrete-time first-order replicator equation induces a diffeomorphism from S_n into itself. This result is important because it excludes the chaotic behavior caused by the noninjectivity of mappings such as $x \to ax(1-x)$. However, the discrete case is still far less well-understood than the continuous one and may behave quite differently.

Equilibria and Their Stability

The fixed points of equations (4.1) or (4.2) in the interior of S_n are the strictly positive solutions of

$$f_1(\mathbf{x}) = \cdots = f_n(\mathbf{x}) \tag{4.20}$$

and

$$x_1 + \cdots + x_n = 1 \quad . \tag{4.21}$$

If equation (4.20) holds, the common value is Φ. Similarly, the equilibria in the interior of a face defined by $x_i = 0$ for some $i \in \{1,...,n\}$ are the strictly positive solutions of the analogous equations.

In particular, the inner equilibria of first-order replicator equations are the strictly positive solutions of the linear equations (4.21) and

$$\sum a_{1j}x_j = \sum a_{2j}x_j = \cdots = \sum a_{nj}x_j \quad . \tag{4.22}$$

These solutions form a linear manifold. Generically, there is either one or no interior equilibrium. In fact, there is an open, dense subset of $n \times n$ matrices such that the corresponding replicator equations admit, at most, one fixed point in the interior of S_n and in the interior of each face (Zeeman, 1980).

In many cases it is easy to perform a local analysis around a fixed point $\mathbf{p}$ by computing the eigenvalues of its Jacobian. One such eigenvalue is $\Phi(\mathbf{p})$; this corresponds to an eigenvector $\mathbf{p}$ which is not in the tangent space. Since we are studying the restriction of equations (4.1) to S_n, this eigenvalue (or more precisely, one of its multiplicities) is irrelevant. Thus, for example, the relevant eigenvalues of a corner $\mathbf{e}_i$ are the $n-1$ values of $a_{ij} - a_{ii}$ $(j \neq i)$.

For the hypercycle (4.12) there is always a unique fixed point $\mathbf{p}$ in intS_n, which is given by

$$p_i = \frac{k_i^{-1}}{\sum k_j^{-1}}$$

and the eigenvalues of the Jacobian at **p** are (up to a positive factor) the nth roots of unity, except for 1 itself (see Schuster *et al.*, 1980). It follows that **p** is asymptotically stable for $n \leq 3$ and unstable for $n \geq 5$. In fact, using Πx_i as a Ljapunov function, it can be shown that **p** is globally stable for $n \leq 4$. For $n \geq 5$, numerical computations show that a periodic attractor exists, although this has not been proved rigorously.

Linearization around the inner equilibrium of equations (4.12) allows the use of the Hopf bifurcation technique. Zeeman (1980) has shown that for $n = 3$ such bifurcations are degenerate and do not lead to periodic attractors. In fact, the equivalence of equation (4.10) for $n = 3$ with the two-dimensional Lotka–Volterra equation (4.16) implies that it admits no isolated periodic orbit. For $n \geq 4$, however, there exist nondegenerate Hopf bifurcations, the simplest of which is given by

$$\begin{bmatrix} 0 & 1 & -\mu & 0 \\ 0 & 0 & 1 & -\mu \\ -\mu & 0 & 0 & 1 \\ 1 & -\mu & 0 & 0 \end{bmatrix}$$

which, for $\mu = 0$, reduces to the hypercycle equation with globally stable interior equilibrium (see Hofbauer *et al.*, 1980). De Carvalho (1984) refined this by showing that for small $\mu > 0$, the periodic orbit is globally attracting in intS_n, except for the stable manifold of the inner equilibrium.

If there is no fixed interior point, then there exists a $\mathbf{c} \in \mathbb{R}^n$ with $\sum c_i = 0$ such that the function $\Pi x_i^{c_i}$ (which is defined on intS_n) increases along the orbits of equation (4.10) (Akin, 1980; Hofbauer, 1981b). It follows from Ljapunov's theorem that each orbit $\mathbf{x}(t)$ in the interior of S_n has its ω limit

$$\omega(\mathbf{x}) = \{\mathbf{y} \in S_n : \exists\, t_n \to +\infty \text{ with } \mathbf{x}(t_n) \to \mathbf{y}\}$$

contained in the boundary of S_n. This implies that there are no periodic, or recurrent, or even nonwandering points in intS_n if there is no fixed inner point. However, this does not mean that $\lim_{t \to +\infty} x_i(t) = 0$ for some i. Akin and Hofbauer (1982) give an example, with $n = 4$, where the ω limit of every interior orbit is a cycle consisting of the corners $\mathbf{e}_1$, $\mathbf{e}_2$, $\mathbf{e}_3$, $\mathbf{e}_4$, and the edges joining them. Conversely, if the orbit $\mathbf{x}(t)$ is periodic in intS_n or, more generally, has its ω limit in intS_n, then the time averages of this orbit

$$\lim_{T \to +\infty} \frac{1}{T} \int_0^T x_i(t)\mathrm{d}t\,, \quad i = 1,\ldots,n \tag{4.23}$$

exist and correspond to an interior equilibrium of equations (4.10) (see Schuster *et al.*, 1980). It frequently happens that an interior equilibrium is unstable and hence physically unattainable, but is nevertheless still empirically relevant as a time average.

Permanence

It is often very difficult to derive a full description of the attractors of replicator equations. [Recall that strange attractors have been observed numerically (Arneodo *et al.*, 1980), and that there is still no proof of the existence of a unique limit cycle for the hypercycles (4.12) with $n \geq 5$]. More modest results may be obtained in such situations by considering only whether the attractors are in the interior or on the boundary.

In particular, we say that the replicator equations (4.1) are *permanent* if there is a compact set in $\text{int}S_n$ which contains the ω limits of all orbits starting in $\text{int}S_n$ [or, equivalently, if there is a $\delta > 0$ such that $\lim_{t\to+\infty} \inf x_i(t) \geq \delta$ for all i, whenever $x_i(0) > 0$ for all i]. Such systems are robust in a sense which is obviously of great practical importance in ecology, genetics, or chemical kinetics. On the one hand, the state remains bounded at some distance from the boundary even if it oscillates in some regular or irregular fashion: therefore a population (or component) within this system cannot be wiped out by small fluctuations. On the other hand, if the system starts on the boundary (i.e., with one or more components missing), then mutations introducing these components (even if only in tiny quantities) spread, with the result that the system is soon safely cushioned away from the faces of the simplex.

We must make two remarks here. First, permanence is not a structurally stable property (in the same way that the asymptotic stability of a fixed point is not necessarily structurally stable). Second, a nonpermanent system does not always lead to the exclusion of some components. Zeeman (1980) has shown that there is a specific case of equation (4.10) which has an attractor on the boundary and one in the interior. It can also happen that each interior orbit remains bounded away from the faces, but by a threshold which depends on the orbit; for permanence, the threshold must be uniform.

The most useful sufficient condition for permanence is the existence of a function P defined on S_n, with $P(\mathbf{x}) > 0$ for $\mathbf{x} \in \text{int}S_n$ and $P(\mathbf{x}) = 0$ for $\mathbf{x} \in \text{bd}S_n$, such that $\dot{P} = P\Psi$, where Ψ is a continuous function with the property that, for all $\mathbf{x} \in \text{bd}S_n$, there is some $T > 0$ such that

$$\frac{1}{T}\int_0^T \Psi[\mathbf{x}(t)]\mathrm{d}t > 0 \quad . \tag{4.24}$$

We describe P as an *average Ljapunov function*. Near the boundary, P increases on average, so that the orbits move away from the boundary (Hofbauer, 1981b).

It has been shown by Schuster *et al.* (1981) and by Hofbauer (1981b) that the general hypercycle equations (4.11) have $P(\mathbf{x}) = x_1x_2\cdots x_n$ as an average Ljapunov function and are therefore permanent. This is of great importance in the realistic design of catalytic hypercycles, whose dynamics are too complex to be represented by equations (4.12).

Brouwer's fixed point theorem implies that a necessary condition for permanence is the existence of a fixed point in $\text{int}S_n$ (Hutson and Vickers, 1983). For permanent first-order replicator equations (4.10), such an equilibrium is necessarily unique. Another very useful condition for the permanence of equations (4.10) is that the trace of the Jacobian at this fixed point must be strictly negative (Amann and Hofbauer, 1984).

Amann and Hofbauer obtained a remarkable characterization of permanence for systems (4.10) with matrices A of the form

$$\begin{bmatrix} 0 & - & - & \cdots & + \\ + & 0 & - & \cdots & - \\ - & + & 0 & \cdots & - \\ \cdot & \cdot & & & \cdot \\ \cdot & \cdot & & & \cdot \\ \cdot & \cdot & & & \cdot \\ - & - & \cdots & + & 0 \end{bmatrix}$$

where + means that the corresponding element is strictly positive and $-$ means that it is negative or zero. The following conditions are equivalent for equations of this type:

(1) The system is permanent.
(2) There is a unique inner equilibrium $\mathbf{p}$ and $\Phi(\mathbf{p})$ is strictly positive.
(3) There is a vector $\mathbf{z} \in \mathbf{R}_n$, with $z_i > 0$ for all i, such that all components of $\mathbf{z}A$ are strictly positive.
(4) The matrix C obtained from A by setting $c_{ij} = a_{i+1,j}$ (taking indices of modulo n), that is, by moving the first row to the bottom, is such that its determinant and all its principal minors are strictly positive.

Note that $-\Phi(\mathbf{p})$ is just the trace of the Jacobian at $\mathbf{p}$ and that matrices such as C, which have diagonal terms strictly positive and all other terms nonpositive, play an important role in mathematical economics.

As a special case we find that the hypercycle equation (4.12) is always permanent. Another special case has been obtained by Zeeman (1980): the replicator equation (4.10) with $n = 3$ and A of the form

$$\begin{bmatrix} 0 & + & - \\ - & 0 & + \\ + & - & 0 \end{bmatrix}$$

is permanent if and only if $\det A > 0$ (in this case the inner equilibrium is a global attractor). In addition, Amann and Hofbauer (1984) have used the general theorem to characterize permanence in special types of reaction networks, such as hypercycles of autocatalysts:

$$\dot{x} = x_i(a_i x_i + b_{i-1} x_{i-1} - \Phi) \tag{4.25}$$

or superpositions of counter-rotating hypercycles:

$$\dot{x}_i = x_i(a_{i-1} x_{i-1} + b_{i+1} x_{i+1} - \Phi) \tag{4.26}$$

with $(a_i, b_i > 0)$. Hofbauer (1981b) has also proved that inhomogeneous hypercycles

$$\dot{x}_i = x_i(b_i + a_i x_{i-1} - \Phi) \tag{4.27}$$

with $a_i > 0$, are permanent if they have an interior equilibrium. This was done using $\Pi x_i^{a_i^{-1}}$ as an average Ljapunov function. More generally, Hofbauer conjectures that equation (4.10) is permanent if and only if for some $\mathbf{p}$ with $p_i > 0$, the function $\Pi x_i^{p_i}$ is an average Ljapunov function or, equivalently, if and only if for such a $\mathbf{p}$ the inequality $\mathbf{p}\cdot A\mathbf{x} > \mathbf{x}\cdot A\mathbf{x}$ holds for all fixed points $\mathbf{x}$ in bdS_n. This was proved by Amann (1984) for the case $n = 4$.

It can be shown that a necessary condition for the permanence of first-order replicator equations with $a_{ij} \geq 0$ is that an irreducible graph is obtained on drawing an arrow from j to i wherever $a_{ij} > 0$; that is, that any two vertices can be joined by an oriented graph (see Sigmund and Schuster, 1984). It would be interesting to know if such a graph is necessarily Hamiltonian; that is, contains a closed oriented path visiting each vertex exactly once. [This has been shown by Amann (1984) for the case $n \leq 4$ and $a_{ii} = 0$.]

An interesting class of examples is provided by models describing the competition between several hypercycles. If these hypercycles are disjoint then the equation is of the form

$$\dot{x}_i = x_i(k_i x_{\pi(i)} - \Phi) \quad , \tag{4.28}$$

where π is a permutation of indices containing several cycles. Such systems are not irreducible and hence not permanent. If the cycles are all of length less than 4, then one of them succeeds and the others vanish (see Schuster *et al.*, 1980). This is probably also true for larger cycles, but has not yet been proved.

Once again, the situation is much less clear in the case of discrete-time replicator equations. A sufficient condition analogous to the existence of an average Ljapunov function has been given by Hutson and Moran (1982). Hofbauer (1984) has shown that the discrete hypercycle

$$(T\mathbf{x})_i = x_i \left(\frac{k_i x_{i-1} + C}{\Phi} \right) \tag{4.29}$$

(with $k_i > 0$) is permanent if and only if $C > 0$.

Gradient Systems of Replicator Type

The evolutionary dynamics defined by the gradients of certain potential functions are of great interest because they correspond to popular notions of adaptive genotypic or phenotypic landscapes and yield biological models with extremum principles of a type familiar in theoretical physics. The action of selection in such situations drives the state uphill along the path of steepest ascent.

Gradients depend on metrics. Shahshahani (1979) provided a geometric framework for population dynamics by using a Riemann metric instead of the more usual Euclidean metric on S_n. Replicator equations which are gradients with respect to this metric are of considerable interest (see Akin, 1979).

Shahshahani defines the inner product of two vectors $\mathbf{x}$ and $\mathbf{y}$ in the tangent space $T_{\mathbf{p}} S_n$ (where $\mathbf{p} \in \text{int}S_n$) in the following way:

$$\langle \mathbf{x}, \mathbf{y} \rangle_p = \sum \frac{1}{p_i} x_i y_i \quad .$$

This introduces a notion of orthogonality which depends on $\mathbf{p}$ and a definition of distance which differs from the Euclidean distance by attaching more weight to changes occurring near the boundary of S_n. If V is a differentiable function defined in a neighborhood of $\mathbf{p}$, then the Shahshahani gradient Grad$V(\mathbf{p})$ is defined by

$$\langle \mathrm{Grad}V(\mathbf{p}), \mathbf{y} \rangle_{\mathbf{p}} = DV(\mathbf{p})(\mathbf{y}) \tag{4.30}$$

for all $\mathbf{y} \in T_pS_n$, where $DV(\mathbf{p})$ is the derivative of V at $\mathbf{p}$. The more usual Euclidean gradient grad$V(\mathbf{p})$ is defined by

$$\mathrm{grad}V(\mathbf{p}) \cdot \mathbf{y} = DV(\mathbf{p})(\mathbf{y}) \quad . \tag{4.31}$$

Using the fact that $\mathbf{y} \in T_pS_n$ if and only if $\mathbf{y} \in \mathbb{R}_n$ satisfies $\sum y_i = 0$, it can be shown that the replicator equation (4.1) is a Shahshahani gradient of V if and only if $\mathbf{f}$ is equivalent to grad$V$, in the sense outlined on pp 93–94.

The case where V is a homogeneous function of degree s is of particular interest, since this implies that $\Phi(\mathbf{x}) = sV(\mathbf{x})$, from Euler's theorem. The average fitness Φ then grows at the largest possible rate and the orbits are orthogonal (in the Shahshahani sense) to the constant level sets of Φ.

In particular, if we have

$$V(\mathbf{x}) = \sum a_i x_i \tag{4.32}$$

then the Shahshahani gradients are $x_i(a_i - \Phi)$, that is, equations (4.9). If, however, we have

$$V(\mathbf{x}) = \tfrac{1}{2}\sum_{ij} a_{ij} x_i x_j \quad , \tag{4.33}$$

where $a_{ij} = a_{ji}$, then the Shahshahani gradients are the selection equations (4.7). The corresponding extremum principles, which give conditions for the average fitness Φ to increase at the largest possible rate, have been stated by Küppers (1979) and Kimura (1958), respectively. However, they did not specify the appropriate metric. The fact that Φ increases along the orbits of equations (4.7) is Fisher's Fundamental Theorem of Natural Selection.

An immediate consequence of Fisher's theorem is that the orbits of equations (4.7) converge to the set of equilibria. In addition, each orbit converges to some equilibrium. This has been proved by Akin and Hofbauer (1982), who once again used a Ljapunov function of type $\Pi x_i^{p_i}$. Analogous results also hold for discrete-time selection equations, but are considerably harder to establish – they have been proved by an der Heiden (1975) for the case $n = 3$ and by Losert and Akin (1983) for the general case. It would be interesting to know whether this convergence holds whenever $\mathbf{f}$ is the Euclidean gradient of a homogeneous function.

First-order replicator equations (4.10) are Shahshahani gradients if and only if

$$a_{ij} + a_{jk} + a_{ki} = a_{ji} + a_{ik} + a_{kj} \tag{4.34}$$

holds for all indices i, j, and k (Sigmund, 1984). This is the case if and only if the matrix A is equivalent (in the sense described on pp 93–94) to a symmetric matrix, or equivalently, if and only if there are constants c_i such that $a_{ij} - a_{ji} = c_i - c_j$ holds for all i and j.

Equations of the type

$$\dot{x}_i = x_i[g_i(x_i) - \Phi] \tag{4.35}$$

are obviously Shahshahani gradients. If the functions g_i are monotonically decreasing, they model competition between replicators which inhibit their own growth but are otherwise independent. In this case it can be shown that there exists a unique global attractor. More precisely, we can assume without loss of generality that $g_1(0) \geq g_2(0) \geq \cdots \geq g_n(0) > 0$, in which case there exists a number K and a $\mathbf{p} \in S_n$ such that

$$g_1(p_1) = \cdots = g_m(p_m) = K \tag{4.36}$$

$$p_1 > 0, \ldots, p_m > 0, \quad p_{m+1} = 0, \ldots, p_n = 0 \quad , \tag{4.37}$$

where m is the largest integer j with $g_j(0) > K$. The point $\mathbf{p}$ is the limit, as t approaches $+\infty$, of all orbits $\mathbf{x}(t)$ for which $x_i(0) > 0$, $i = 1, \ldots, m$. A variant of this model shows that if the total concentration $\sum x_i$ is kept at a constant value c (not necessarily equal to 1) by replacing Φ by Φ/c, then the number of species that can coexist increases with increasing c (see Hofbauer *et al.*, 1981). The special cases

$$g_i(x_i) = a_i - b_i x_i \quad \text{and} \quad g_i(x_i) = \frac{1}{c_i + d_i x_i} \tag{4.38}$$

have been studied by Epstein (1979).

Classification

Except in low-dimensional cases, there is little hope of obtaining a complete classification of first-order replicator equations (4.10) up to topological equivalence. Two such equations are said to be topologically equivalent if there exists a homeomorphism from S_n onto itself which maps the orbits of one equation onto the orbits of the other equation in such a way that orientation is preserved. Two $n \times n$ matrices are described as R-equivalent if the corresponding replicator equations are topologically equivalent.

Zeeman (1980) proposed a method for the classification of stable cases. By analogy to the definition of structural stability, an $n \times n$ matrix A is said to be stable if its R-equivalence class is a neighborhood of A. Thus, small perturbations of A do not change the topological structure of the corresponding replicator equation. Zeeman conjectured that the stable matrices form an open dense set in the space of $n \times n$ matrices and are divided into a finite number of R-equivalence classes for each n. He proved this for $n = 2$ and 3, and classified all corresponding stable replicator equations. (For $n = 2$ and 3 there are 2 and 19 stable classes, respectively, up to time reversal.)

A basic requirement for the classification of equation (4.10) for $n = 3$ is that there are no limit cycles. This is a consequence of the corresponding result for two-dimensional Lotka–Volterra equations (see, e.g., Coppel, 1966) and of the equivalence between such equations and first-order replicator equations (Hofbauer, 1981a). Bomze (1983) extended Zeeman's classification to cover unstable cases, obtaining 102 types of phase portraits up to time reversal.

Little is known about stable matrices for higher dimensions, apart from the fact that stability implies that all fixed points of equations (4.10) are hyperbolic (the real parts of the eigenvalues of their Jacobians do not vanish). This was proved by de Carvalho (1984).

Recall that, without loss of generality, the diagonal of a matrix may be assumed to contain only zeros. Let Z_n denote the class of such matrices with nonzero off-diagonal terms. Two matrices A and B in Z_n are said to be ***sign equivalent*** if the corresponding off-diagonal terms have the same sign and ***combinatorially equivalent*** if A can be made sign equivalent to B by permutating the indices. Zeeman (1980) showed that A and B are combinatorially equivalent if and only if the equations obtained by restricting the corresponding replicator equations to the edges of S_n are topologically equivalent. Within Z_n, R-equivalence classes are refinements of the combinatorial classes. There are 10 such combinatorial classes for $n = 2$ and 114 for $n = 3$ up to sign reversal (Zeeman, 1980). De Carvalho (1984) has studied 19 combinatorial classes without inner equilibria as a first step towards a classification of R-stable matrices for $n = 4$. Another step in this direction was taken by Amann (1984), who characterized all 4×4 matrices which lead to permanent replicator equations.

Another interesting (although highly degenerate) class of examples is provided by circulant matrices ($a_{ij} = a_{i+1,j+1}$ for all i and j, counting indices on modulo n). A partial analysis of this class is given in Hofbauer *et al.* (1980). It is shown that the center of S_n (i.e., the point $\mathbf{m}$, where $m_i = 1/n$) is always an equilibrium; it is not hard to compute the eigenvalues of its Jacobian. If $\mathbf{m}$ is a sink, then $\mathbf{m}$ is a global attractor; if $\mathbf{m}$ is a source, then all orbits converge to the boundary. Nondegenerate Hopf bifurcations occur for $n \geq 4$.

Connections with Game Theory

It has often been remarked that game theory is essentially static. However, the replicator equations (4.10) and (4.14) offer dynamic models for normal form games which are symmetric in the sense that both players have the same strategies and the same payoff matrix A. In fact, the dynamic extension is already implicit in the notion of an evolutionarily stable state (Maynard Smith, 1974, 1982), which is a refinement of the concept of a Nash equilibrium.

A point $\mathbf{p} \in S_n$ is said to be evolutionarily stable if it satisfies the following two conditions:

(1) Equilibrium condition:

$$\mathbf{p} \cdot A\mathbf{p} \geq \mathbf{x} \cdot A\mathbf{p} \text{ for all } \mathbf{x} \in S_n \ . \tag{4.39}$$

(2) Stability condition:

$$\text{if } \mathbf{p} \cdot A\mathbf{p} = \mathbf{x} \cdot A\mathbf{p} \text{ for } \mathbf{x} \neq \mathbf{p}, \text{ then } \mathbf{p} \cdot A\mathbf{x} > \mathbf{x} \cdot A\mathbf{x} \ . \tag{4.40}$$

A game can have zero, one, or several evolutionarily stable points. As shown by Selten (1985), the notion is not structurally stable: some matrices which yield evolutionarily stable points can be perturbed into matrices which do not. In this context we also refer the reader to Bomze (1985) for a thorough analysis of the relation of evolutionary stability to the multitude of equilibrium concepts used in game theory.

It can be shown that the following four conditions are equivalent (see Hofbauer *et al.*, 1979; Zeeman, 1980):

(1) $\mathbf{p}$ is evolutionarily stable.

(2) For all $\mathbf{q} \in S_n$ with $\mathbf{q} \neq \mathbf{p}$, we have

$$\mathbf{p}\cdot A[(1-\varepsilon)\mathbf{p} + \varepsilon\mathbf{q}] > \mathbf{q}\cdot A[(1-\varepsilon)\mathbf{p} + \varepsilon\mathbf{q}] \tag{4.41}$$

provided that $\varepsilon > 0$ is sufficiently small.

(3) For all $\mathbf{x} \neq \mathbf{p}$ in some neighborhood of $\mathbf{p}$, we have

$$\mathbf{p}\cdot A\,\mathbf{x} > \mathbf{x}\cdot A\,\mathbf{x} \ . \tag{4.42}$$

(4) The function $\Pi x_i^{p_i}$ is a strict local Ljapunov function at $\mathbf{p}$ for the replicator equations (4.10); that is, strictly increasing along all orbits in a neighborhood of $\mathbf{p}$.

Condition (2) is probably the most intuitively obvious in a biological context: if the state of the population is $\mathbf{p}$, then a fluctuation introducing a small subpopulation in state $\mathbf{q}$ becomes extinct, since the $\mathbf{p}$ population fares better than the $\mathbf{q}$ population against the mixture $(1-\varepsilon)\mathbf{p} + \varepsilon\mathbf{q}$.

It follows from the equivalence of (1) and (4) that any evolutionarily stable point $\mathbf{p}$ is an asymptotically stable fixed point of equations (4.10). However, the converse is not true. In particular, (3) implies that if $\mathbf{p} \in \mathrm{int}S_n$ is evolutionarily stable, then it is an attractor for all orbits in $\mathrm{int}S_n$, and hence the unique evolutionarily stable point in S_n; however, Zeeman (1980) has shown that there exist 3×3 games with two asymptotically stable fixed points, one in the interior and the other on the boundary of S_n.

Akin (1980) has shown that equations (4.10) have no fixed point in $\mathrm{int}S_n$ if and only if there exist two strategies $\mathbf{x}$ and $\mathbf{y}$ in S_n such that $\mathbf{x}$ dominates $\mathbf{y}$ in the sense that

$$\mathbf{x}\cdot A\,\mathbf{z} > \mathbf{y}\cdot A\,\mathbf{z}$$

for all $\mathbf{z} \in \mathrm{int}S_n$. This result is supplemented by precise statements concerning the support of strategies $\mathbf{x}$ and $\mathbf{y}$ and the form of global Ljapunov functions or invariants of motion for equations (4.10).

The results obtained using the time averages (4.23) described on p 95 suggest a computational method for finding equilibria (and hence solutions) of normal form games. These results, which can easily be extended to asymmetric games (i.e., games in which the players have different payoff matrices), should be compared with the classical methods (involving differential equations) for finding the solutions of games (see, e.g., Luce and Raiffa, 1957, p. 438).

The discrete analogues of such methods involve iterative procedures. It turns out, however, that discrete-time replicator equations of the type (4.14) do not appear to lend themselves very well to game dynamics; in particular, an evolutionarily stable point need not be asymptotically stable for equations (4.14) (see, e.g., Schuster and Sigmund, 1984).

The behavior of equations (4.10) and (4.14) for zero-sum games ($a_{ij} = -a_{ji}$) is analyzed in Akin and Losert (1984). If an interior equilibrium exists, then the continuous model (4.10) has an invariant of motion. The equilibrium is stable, but not asymptotically stable, and all nonequilibrium orbits of model (4.10) in $\mathrm{int}S_n$ have ω

limits in intS_n but do not converge to an equilibrium. By contrast, if the discrete time model (4.14) has an interior equilibrium then it is unstable and all nonequilibrium orbits converge to the boundary. If there is no inner equilibrium, then all orbits converge to the boundary in both discrete and continuous cases. In the discrete case all possible attractors may be described using the notion of chain recurrence.

References

Akin, E. (1979) The geometry of population genetics. *Lecture Notes In Biomathematics* 31 (Berlin, Heidelberg, and New York: Springer-Verlag).

Akin, E. (1980) Domination or equilibrium. *Math. Biosciences* 50: 239–50.

Akin, E. and Hofbauer, J. (1982) Recurrence of the unfit. *Math. Biosciences* 61: 51–63.

Akin, E. and Losert, V. (1984) Evolutionary dynamics of zero-sum games. *J. Math. Biology* 20: 231–58.

Amann, E. (1984). *Permanence for Catalytic Networks*. Dissertation (University of Vienna).

Amann, E. and Hofbauer, J. (1985) Permanence in Lotka–Volterra and replicator equations. Forthcoming.

Arneodo, A., Coullet, P., and Tressor, C. (1980) Occurrence of strange attractors in three-dimensional Volterra equations. *Physics Letters* 79A: 259–63.

Bishop, T. and Cannings, C. (1978) A generalized war of attrition. *J. Theor. Biology* 70: 85–124.

Bomze, I. (1983) Lotka–Volterra equations and replicator dynamics: a two-dimensional classification. *Biol. Cybernetics* 48: 201–11.

Bomze, I. (1985) Non-cooperative two-person games in biology: symmetric contests, in Peschel (Ed) *Lotka–Volterra Approach in Dynamical Systems*, Proc. Conf. Wartbury (Berlin: Akademieverlag).

Bomze, I., Schuster, P., and Sigmund, K. (1983) The role of Mendelian genetics in strategic models on animal behavior. *J. Theor. Biology* 101: 19–38.

de Carvalho, M. (1984) *Dynamical Systems and Game Theory*. PhD Thesis (University of Warwick).

Coppel, W. (1966) A survey of quadratic systems. *J. Diff. Eqns.* 2: 293–304.

Dawkins, R. (1982) *The Extended Phenotype* (Oxford and San Francisco: Freeman).

Eigen, M. (1971) Self-organization of matter and the evolution of biological macromolecules. *Die Naturwissenschaften* 58: 465–523.

Eigen, M. and Schuster, P. (1979) *The Hypercycle: A Principle of Natural Self-Organization* (Berlin and Heidelberg: Springer-Verlag).

Epstein, M. (1979) Competitive coexistence of self-reproducing macromolecules. *J. Theor. Biology* 78: 271–98.

Eshel, I. (1982) Evolutionarily stable strategies and natural selection in Mendelian populations. *Theor. Pop. Biol.* 21: 204–17.

Ewens, W.J. (1979) *Mathematical Population Genetics* (Berlin, Heidelberg, and New York: Springer-Verlag).

an der Heiden, U. (1975) On manifolds of equilibria in the selection model for multiple alleles. *J. Math. Biol.* 1: 321–30.

Hines, G. (1980) An evolutionarily stable strategy model for randomly mating diploid populations. *J. Theor. Biology* 87: 379–84.

Hofbauer, J. (1981a) On the occurrence of limit cycles in the Volterra–Lotka equation. *Nonlinear Analysis* IMA 5: 1003–7.

Hofbauer, J. (1981b) A general cooperation theorem for hypercycles. *Monatsh. Math.* 91: 233–40.

Hofbauer, J. (1984) A difference equation model for the hypercycle. *SIAM J. Appl. Math.* 44: 762–72.

Hofbauer, J., Schuster, P., and Sigmund, K. (1979) A note on evolutionarily stable strategies and game dynamics. *J. Theor. Biology* 81: 609–12.

Hofbauer, J., Schuster, P., Sigmund, K., and Wolff, R. (1980) Dynamical systems under constant organization. Part 2: Homogeneous growth functions of degree 2. *SIAM J. Appl. Math.* 38: 282–304.

Hofbauer, J., Schuster, P., and Sigmund, K. (1981) Competition and cooperation in catalytic self-replication. *J. Math. Biol.* 11: 155–68.

Hofbauer, J., Schuster, P., and Sigmund, K. (1982) Game dynamics for Mendelian populations. *Biol. Cybern.* 43: 51–7.

Hutson, V. and Moran, W. (1982) Persistence of species obeying difference equations. *J. Math. Biol.* 15: 203–13.

Hutson, V. and Vickers, C.T. (1983) A criterion for permanent coexistence of species, with an application to a two-prey/one-predator system. *Math. Biosci.* 63: 253–69.

Kimura, M. (1958) On the change of population fitness by natural selection. *Heredity* 12: 145–67.

Küppers, B.O. (1979) Some remarks on the dynamics of molecular self-organization. *Bull. Math. Biol.* 41: 803–9.

Losert, V. and Akin, E. (1983) Dynamics of games and genes: discrete versus continuous time. *J. Math. Biol.* 17: 241–51.

Luce, R. and Raiffa, H. (1957) *Games and Decisions* (New York: John Wiley).

Maynard Smith, J. (1974) The theory of games and the evolution of animal conflicts. *J. Theor. Biology* 47: 209–21.

Maynard Smith, J. (1981) Will a sexual population evolve to an ESS? *Amer. Naturalist* 177: 1015–8.

Maynard Smith, J. (1982) *Evolutionary Game Theory* (Cambridge: Cambridge University Press).

Pollak, E. (1979) Some models of genetic selection. *Biometrics* 35: 119–37.

Schuster, P. and Sigmund, K. (1983) Replicator dynamics. *J. Theor. Biology* 100: 535–8.

Schuster, P. and Sigmund, K. (1985) Towards a dynamics of social behavior: strategic and genetic models for the evolution of animal conflicts. *J. Soc. Biol. Structures*. Forthcoming.

Schuster, P., Sigmund, K., and Wolff, R. (1979) Dynamical systems under constant organization. Part 3: Cooperative and competitive behavior of hypercycles. *J. Diff. Eqns.* 32: 357–68.

Schuster, P., Sigmund, K., and Wolff, R. (1980) Mass action kinetics of self-replication in flow reactors. *J. Math. Anal. Appl.* 78: 88–112.

Schuster, P., Sigmund, K., Hofbauer, J., and Wolff, R. (1981) Self-regulation of behavior in animal societies. Part 1: Symmetric contests. *Biol. Cybern.* 40: 1–8.

Shahshahani, S. (1979) A new mathematical framework for the study of linkage and selection. *Memoirs* AMS 211.

Selten, R. (1985) Evolutionary stability in extensive two-person games. *Math. Soc. Sciences*. Forthcoming.

Sigmund, K. (1985) The maximum principle for replicator equations, in Peschel (Ed) *Lotka-Volterra Approach in Dynamical Systems*, Proc. Conf. Wartburg (Berlin: Akademieverlag).

Sigmund, K. and Schuster, P. (1984) Permanence and uninvadability for deterministic population models, in P. Schuster (Ed) *Stochastic Phenomena and Chaotic Behavior in Complex Systems* (Berlin, Heidelberg, and New York: Springer-Verlag).

Taylor, P. and Jonker, L. (1978) Evolutionarily stable strategies and game dynamics. *Math. Biosciences* 40: 145–56.

Zeeman, E.C. (1980) Population dynamics from game theory, in *Global Theory of Dynamical Systems*, Lecture Notes in Mathematics 819 (Berlin, Heidelberg, and New York: Springer-Verlag).

Zeeman, E.C. (1981) Dynamics of the evolution of animal conflicts. *J. Theor. Biology* 89: 249–70.

CHAPTER 5

Darwinian Evolution in Ecosystems: A Survey of Some Ideas and Difficulties Together with Some Possible Solutions

Nils Chr. Stenseth

Introduction

> Ecology, the biological science of environment, has not produced a synthesis of environment from its broad technical knowledge of influence of external parameters on organisms. Before Darwin (1859), environment was considered an organic whole. Everything in it made some contribution and has some meaning with respect to everything else. Darwin subscribed to this view, but his emphasis, and that of his followers, on the evolving organism struggling to survive, suppressed the exploration of holistic aspects of the origin of species that might have been developed. After Darwin, the organism came into great focus, first as a comparative anatomical entity, then later with physiological, cellular, molecular, behavioural, and genetic detail. In contrast, the organism's environment blurred through relative inattention into a fuzzy generality. The result was two distinct things (dualism), organism and environment, supplanting the original unified organism–environment whole (synergism). (Patten, 1982).

In a way we may say that we have two types of

> ...ecologies today, reflecting the tension between dualism and synergism. Population ecology, descended from Darwin, focuses on organism, and ecosystem ecology, in the earlier holistic tradition, deals with environment. Melding of these two subdisciplines depends on finding the means to investigate the organism–environment complex... (Patten, 1982).

The Australian physicist and father of statistical mechanics, Ludwig E. Boltzmann, is reported (see, e.g., Maynard Smith, 1982a) to have said that the nineteenth century probably would be remembered as Darwin's Century. More than one hundred years after the death of Charles Darwin, Boltzmann seems to be right: today we remember Darwin as the person who convinced us that evolution is a fact and as the person who presented a theory – consistent with the facts – which explained the mechanism of evolution, the process of natural selection. Thus, we mean by *Darwinism* the theory that evolution has occurred as a result of natural selection. As the above quotations from Patten (1982) show, the influence of Darwin has not been *only* positive – it lead to dualism by focusing on the individual organism rather than the mutual interaction between the organism and the environment. Nevertheless, Darwin has, of course, had a tremendously positive influence on biology – he made us understand how organisms evolve in a fixed environmental setting.

What Darwin (1859) did was to transform biology into a proper science with a theoretical basis, without which the study of biology would be nothing more than an enterprise of collecting curiosa – it would be stamp collecting. The idea of natural selection has helped us to organize both our data and our thoughts as to how the living world came to be like it is today. Darwin was also one of a series of scientists who documented that our own species, *Homo sapiens*, was closely related – phylogenetically – to other living creatures of the Earth; man no longer had a unique position totally separate from all other organisms.

More than one hundred years after Darwin's death and after a long period of heavy criticism and scrutiny, the basic ideas of Darwinism still hold. Much remains, of course, to be done in the refinement of the Darwinian theory of evolution. For example, one serious difficulty is that we do not understand what really generates "selective pressure".

In this chapter I present the basic *ideas* in the Darwinian theory of evolution, concentrating on the general ideas rather than providing a detailed review of the mathematical formulation and the empirical tests of this theory; good reviews of both these aspects are already available (see, e.g., Roughgarden, 1979; Futuyma, 1979). Indeed, I want to concentrate on those aspects of Darwinism which, to my mind, are inappropriately treated in the literature. In the latter part of the chapter I discuss some of the difficulties we are faced with when trying to formulate a theory as to how selective pressure is generated. Specifically, I outline some of the mathematical difficulties, but I also discuss some possible mathematical fragments useful to a theory of evolution in ecosystems.

Evolution

Microevolution, macroevolution, and phylogenetics

The idea of a changing universe has now replaced the long unquestioned view of a static world, identical in all essentials to the creator's perfect creation. Darwin more than anyone else extended to living things, and to the human species itself, the notion that mutability, not stasis, is the natural order. He suggested that material causes are a sufficient explanation not only for physical phenomena, as Descartes and Newton already had shown, but also for biological phenomena, with all their seeming evidence of design and purpose: it was no longer necessary to refer to the will of God or to the Aristotelian final causes. As Futuyma (1979) points out,

> ...by coupling undirected, purposeless variation to the blind, uncaring process of natural selection, Darwin made theological or spiritual explanations of the life processes superfluous. Together with Marx's materialistic theory

> of history and society and Freud's attribution of human behaviour to influences over which we have little control, Darwin hewed the final planks of the platform of mechanism and materialism.

Evolution, the process of changing biological structures over time, is commonly divided into two (partly overlapping) types: *micro-* and *macro-evolution*. By microevolution we mean the patterns of change that occur within a species and by macroevolution we usually mean evolution above the species level; herein, I use *macroevolution* to refer to the processes of speciation and species extinction.

Darwin held that both micro- and macro-evolution could be understood as resulting from the same process. Hence, current Darwinists hold that macroevolution may be seen as the sum of microevolutionary changes. About this presumption there has recently been much debate and disagreement; this I return to later.

Phylogenetics is the discipline that attempts to reconstruct the genealogical relationships of living and extinct organisms; from such studies phylogenetic trees are constructed. These phylogenetic trees are, by definition, assumed to indicate the evolutionary pathways that have been followed during the history of life. Phylogeny, then, represents the evolution of a race or a genetically related group of organisms (such as a genus, a family, or an order). Examples of such phylogenetic trees are given in Figure 5.1, and these show what seems to be a trend, at least initially, of increasing numbers of species (or other taxonomic categories) with the age of the group. Similarly, a closer inspection of such phylogenetic trees often suggests a trend toward greater complexity or toward, for example, larger size (e.g., Simpson, 1953; Gould, 1981): I return to this later.

Some observations

In 1973 Van Valen reported the results of a thorough analysis of rates of extinction at the species, genus, and family level. On the basis of his observations, he concluded that for several groups of ecologically related organisms each had a constant, age-independent probability of becoming extinct (see Figure 5.2). This pattern has since become known as "The Law of Constant Extinction". There has been much debate over its empirical validity (e.g., Maynard Smith, 1975, 1976a; Hallam, 1976; Van Valen, 1976, 1977; Stenseth, 1979). The present concensus is, however, that the pattern of constant age-independent extinction appears true for many groups of organisms.

When analyzing rates of evolution (i.e., how fast phenotypic characters change) we observe that in some cases evolution proceeds at a fairly constant but slow rate, whereas in other cases evolution proceeds at an erratic, nonconstant rate (Eldredge and Gould, 1972; Gould and Eldredge, 1977; Stanley, 1979; Schopf, 1982; Schopf and Hoffman, 1983; Gould, 1983). These are facts – not speculations – and have given rise to the *idea* of distinguishing between a gradualistic and a punctualistic pattern of evolution (see Figure 5.3). This idea has, however, produced much confusion (e.g., Stebbins and Ayala, 1981); what are, for example, fast and slow evolution? The distinction between a gradualistic and a punctualistic pattern of evolution has fostered a serious controversy over mechanisms in evolution (e.g., Gould, 1980; Williamson, 1981a; Charlesworth *et al.*, 1982). This controversy is a real one since the punctualistic view explicitly implies (Williamson, 1981b) a notion of constancy and *nonevolution* over extensive periods of time.

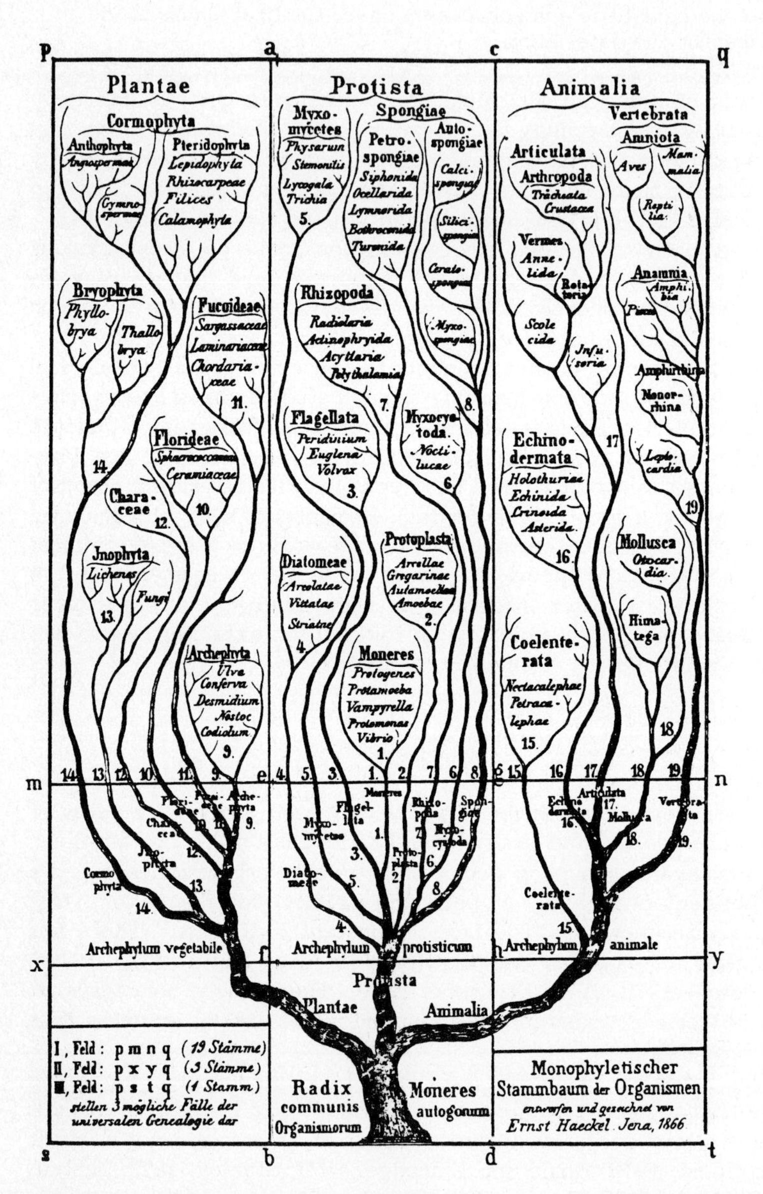

Figure 5.1 (a) The first phylogenetic tree of life, published by Ernst Haeckel (1866). The modern arrangement is quite different to the structure shown here.

We must also account for the observation that suggests some sort of convergent evolution of – or *in* – ecosystems of similar physical condition (e.g., Pianka, 1978; Orians and Paine, 1983). It is hard to imagine how this can be explained on the basis of the current Darwinian theory of evolution. I return to this later.

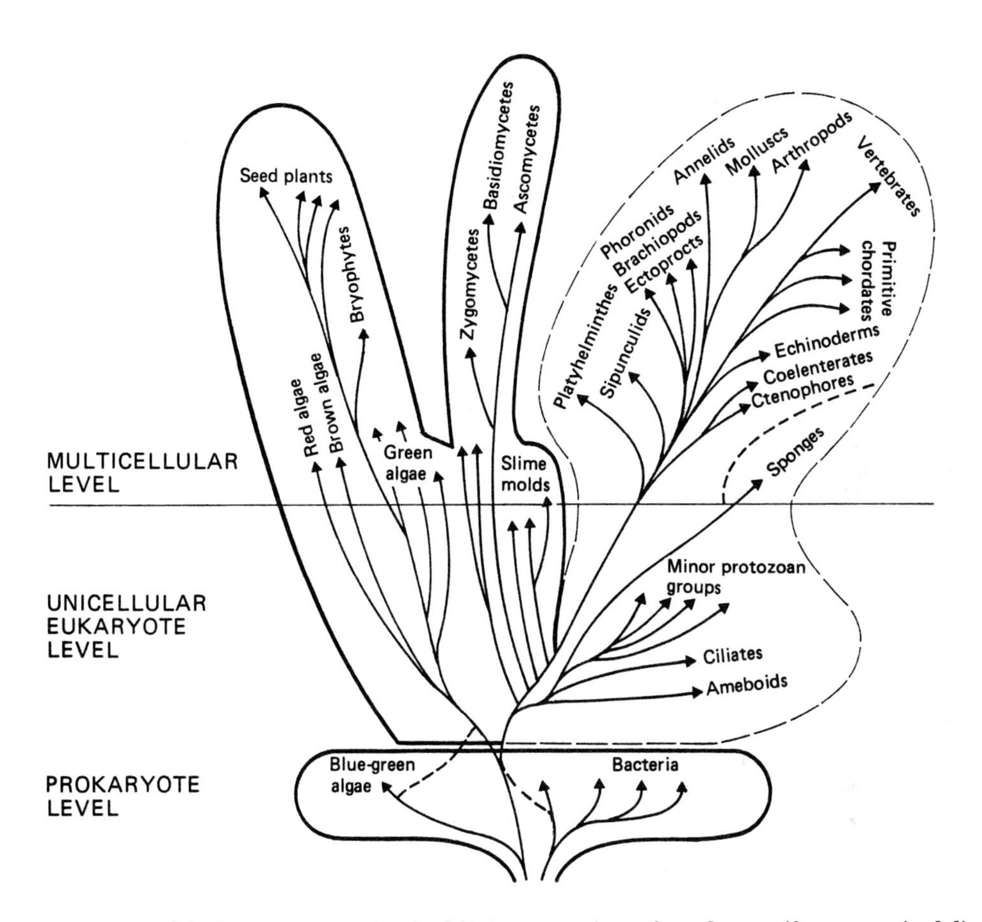

Figure 5.1 (b) Phylogenetic chart of living organisms, based upon the concept of five kingdoms. Three major grades of advancement are recognized: prokaryote, unicellular eukaryote, and multicellular. Adaptive radiation has occurred at each level. At the prokaryote level, two phyla are recognized, bacteria and blue–green algae. At the unicellular eukaryote level, extensive radiation has led to the formation of many different classes and orders, but no distinctive phyla. From some of these classes, several phylogenetically distinct groups of multicellular organisms have arisen, only five of which have advanced to the grade that includes tissue differentiation and elaboration of a distinctive form at the visible or macroorganism level. Three of these are basically photosynthetic: red algae, brown algae, and the green algal (archegoniate) seed plant line. Because they resemble each other in being autotrophic (with progressive expansion of surface area) and, in the case of most aquatic forms, have similar reproductive propagules, these three groups are placed in the plant kingdom. One line, the fungi, consists of heterotrophic organisms that absorb rather than ingest food, and thus share with plants an expansion of surface and a sessile mode of life, The third line includes the multicellular animals of Metazoa, which are heterotrophic ingestors, most of which remain compact in bodily form and develop internal rather than external membranes of absorption (based on Whittaker, 1969, and redrawn from Dobzhansky *et al.*, 1977).

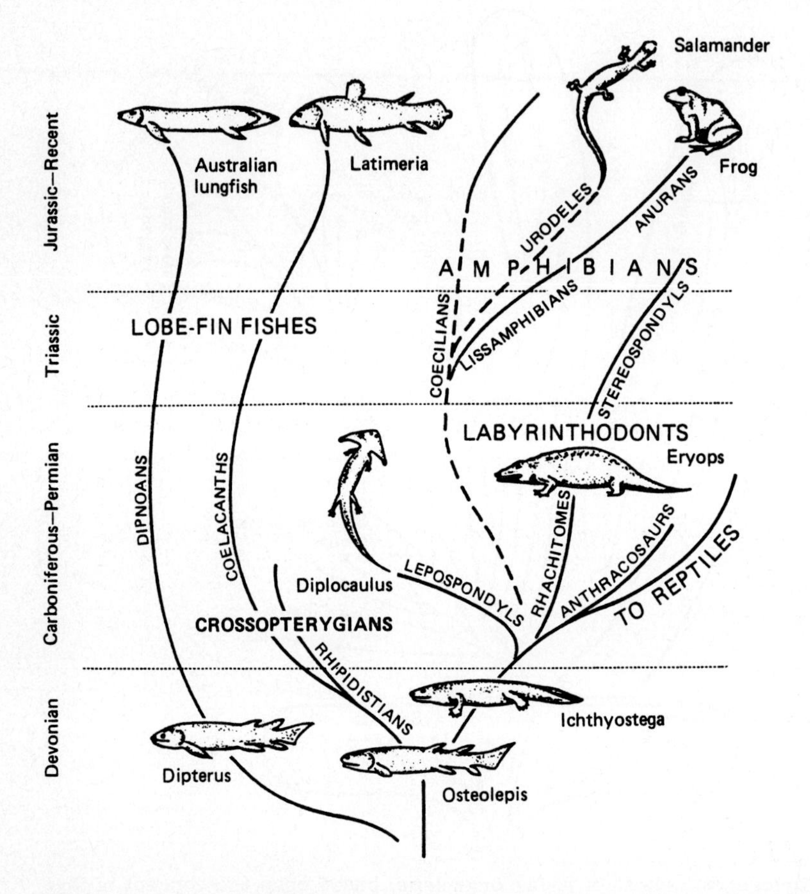

Figure 5.1 (c) Generalized phylogeny of lobe-finned fish and amphibians. Devonian lobe-fins are classed in three groups:

(1) Dipnoans, represented today by lungfishes.

(2) Coelacanths, represented today by the living fossil *Latimeria*, from deep water in the western Indian Ocean.

(3) Rhipidistians, extinct as fish although all land vertebrates are among their descendants. The tetrapods developed from primitive rhipidistians through such early amphibians as the ichthyostegods, radiating into disparate groups including large, extinct amphibians, familiar living amphibians, and primitive reptiles (after Colbert, 1969).

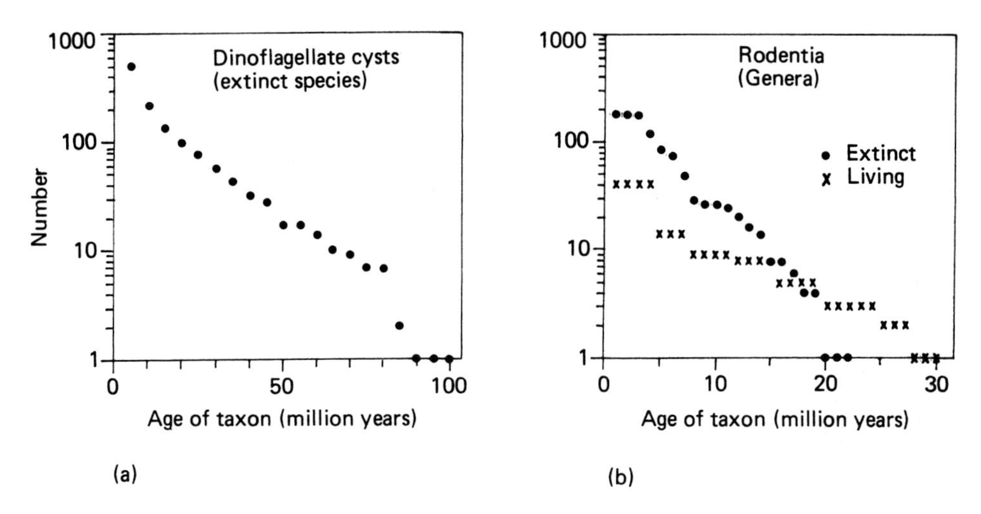

Figure 5.2 Taxonomic survival curves demonstrating a fairly constant and age-independent rate of extinction (see Van Valen, 1973; Stenseth, 1979). (a) Unicellular dinoflagellate cysts; only extinct *species* (after Van Valen, 1973). (b) Rodent *genera* – both living and extinct groups are included (after Van Valen, 1973).

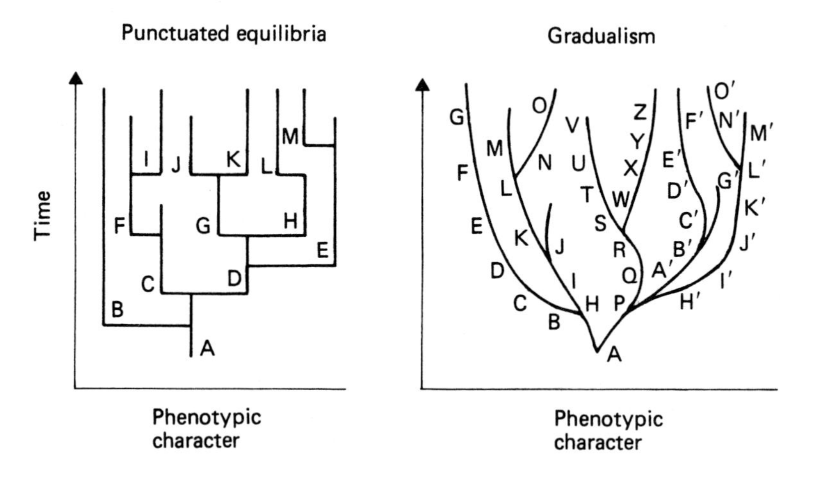

Figure 5.3 Diagrams of speciation under punctuated equilibria and gradualism. The vertical axis is geological time (a few tens of millions of years). The horizontal axis depicts morphologic and genetic divergence which, for punctuated equilibria, is believed to be focused at speciation events, with long periods of intervening stasis (species B would be a living fossil). For gradualism, morphologic and genetic divergence is believed to be a more or less continuous process through time. Hence species durations are inferred to be shorter under gradualism than under punctuated equilibria (after Schopf, 1982).

Darwinism

The core

Assume, as Darwin did, that populations consist of individuals which:

(1) On average, produce more offspring than is needed to replace them upon their death.
(2) Have offspring which resemble their parents more than they resemble randomly chosen individuals in the population.
(3) Vary in heritable traits influencing reproduction and survival (i.e., fitness).

We may refer to these three properties as ***reproduction, inheritance*** (or constancy), and ***variation***.

Intuition then suggests that, in an ecosystem with limited resources, those individuals which produce more offspring that survive until the age of reproduction are favored; other individuals will not be able to establish themselves in a population of individuals producing as many offspring as possible that survive as well as they can. It was this process that Darwin called natural selection. Mathematical models have been developed (see any modern evolution text, e.g., Roughgarden, 1979) to demonstrate that the above intuitive argument is logically correct. Hence, we do not today consider natural selection to be a hypothesis; it is now considered a logical consequence of the assumed properties of life. Hence, we need not test the idea of natural selection. What needs to be tested, by analyzing real-life data, is the assumed organization of life, as well as the patterns we deduce from the evolutionary models (for a cogent discussion of this issue, see Maynard Smith, 1978a). I give a particular example (the lemmings) later in this chapter.

Since only living things, by assumption, can evolve, we may define life by entities that have the three properties listed above; reproduction, inheritance, and variation. This is, within the Darwinian theory of evolution, *the* logical definition of life.

The form, life cycle, and so on, of all currently living species are, according to Darwinian theory, assumed molded by this process of natural selection; that is, currently existing organisms are assumed to have evolved from previously existing organisms through the process of natural selection. As pointed out above, Darwin suggested two very different ideas in his *Origin*: the phenomenon of evolution and the mechanism of natural selection – the latter was original, the former was not. By ***neo-Darwinism***, we mean the theory of evolution that occurs as a result of natural selection, but add a theory of inheritance (i.e., Gregor Mendel's contribution) and a theory of how genes spread in a population (i.e., population genetics as formulated by R.A. Fisher, J.B.S. Haldane, and S. Wright; for a review, see e.g., Futuyma, 1979; Roughgarden, 1979).

Notice that selection may occur at several levels of organization: the unit of selection may be the gene, the chromosome, the individual, the population, the community, etc. Analysis of a variety of mathematical models demonstrates, however, that selection at levels higher than the individual is far less efficient than standard Darwinian – or individual – selection (e.g., Maynard Smith, 1976a). Selection at lower levels may be of great importance; much work needs to be done, however, before we can understand how these various levels of selection are integrated (see, in this connection, Sigmund, 1985).

As already suggested, the strength and direction of the selection pressure is determined by the environment in which the evolutionary process occurs. Hence, selection of various types of individuals does take place within the framework of the ecosystem (composed of all coexisting living organisms and the physical textures). I specify this later in the chapter.

Fitness

Consider first an asexually reproducing organism. Assume that each individual gives rise to B offspring per unit time which maturate within one unit time, that each of these survives a unit time with the probability s_1, and that the adults survive a unit time with probability s_2. Then, the natural definition of ***individual fitness***, λ, is

$$\lambda = Bs_1 + s_2 \tag{5.1}$$

so that

$$N_{t+1} = \lambda N_t \tag{5.2}$$

where N_t is the density of the population at time t. Thus defined, λ is the net rate of population growth and therefore represents a formalization of Darwin's first premise [premise (1) above].

Natural selection, in the case of an asexual population, operates so as to select those strategies (defined by B, s_1, and s_2) which maximize λ. This is, in fact, a simple logical consequence of how populations grow in an otherwise stable environment [both biologically and abiotically (or physically)]. It is common to attach the label "best fit" to those individual strategies that are thus selected; these individuals are those best adapted to the environment in which they are currently living.

A necessary condition for a population to exist over some length of time is that its density either is stable over time or fluctuates within some limited bounds; so, on average, we must require some sort of ecological stability. That is, denoting the evolutionarily optimal strategy as λ_{max}, we must require that $\lambda_{max} = 1$. (As is apparent, I have in this mathematical definition tacitly disregarded the possibility of a stable limit-cycle of ecological density; I presume, however, that everything I say below can be extended to such cases – but some theoretical work is, of course, needed.)

According to the adaptationist program (see, e.g., Calow, 1983), most natural populations are close to their adaptive peak most of the time. This assumption dates back explicitly to Fisher (1930), and implicitly to Darwin (1859).

Further, it is important to realize that such a noninvadable population (see Maynard Smith, 1982b; Stenseth, 1983a; Reed and Stenseth, 1984) does not need to be homogeneous; indeed, it may be polymorphic – in which case the mathematical models become more complicated. The fitness definition given in (5.1) must then be extended to involve the average fitness, $\bar{\lambda}$, as

$$\bar{\lambda} = \sum_{i=1}^{n} p_i (B_i s_{1i} + s_{2i}) \tag{5.3}$$

where p_i is the relative frequency of each of the various coexisting strategies i, where $i = 1, 2, \ldots, n$ (see, e.g., Stenseth, 1984a). Assuming that

$$N_{t+1} = \bar{\lambda} N_t \tag{5.4}$$

where N_t is the total density at time t, natural selection maximizes $\bar{\lambda}$, but so that $\bar{\lambda}_{max} = 1$ (to comply with the constraints of ecological stability discussed above). In this case both gene-frequency changes (i.e., changes in p_i; for a general treatment, see Roughgarden, 1979; see also Sigmund, 1985) and evolutionary changes may occur (i.e., occurrence of new mutants; for a general treatment, see Reed and Stenseth, 1984).

An equivalent formulation of natural selection may easily be given for sexually reproducing populations. Thus, let N_t be the density vector of a monomorphic population with different classes and sexes. Then, an ecological model for the genetically monomorphic population with age and sex classes – analogous to model (5.2) – would, after the system has reached its stationary state (with respect to age distribution and sex ratio), be given by (see Stenseth, 1984b)

$$\mathbf{N}_{t+1} = \lambda \mathbf{N}_t \tag{5.5}$$

where λ, a scalar, is analogous to that used in equation (5.2). That the λ used in equation (5.5) is a scalar follows from Sharpe and Lotka (1911), who demonstrated that a population always returns to the same stationary age distribution – and, I presume, the same sex ratio – if it is temporarily disturbed, as long as the primary sex ratio, the age-specific reproductive rates, and the survival rates remain unchanged.

At the ecologically stable equilibrium ($\mathbf{N}^*$), $\lambda(\mathbf{N}^*) = 1$. To study the process of natural selection, we must extend model (5.5) to include ecological competition between wildtype strategy and mutant strategy (see Reed and Stenseth, 1984). Let $\mathbf{N}'_t$ be the density vector of the mutant strategy, then the extended model is

$$\begin{aligned} \mathbf{N}_{t+1} &= \Lambda_1(\mathbf{N}_t, \mathbf{N}'_t)\mathbf{N}_t \\ \mathbf{N}'_{t+1} &= \Lambda_2(\mathbf{N}_t, \mathbf{N}'_t)\mathbf{N}'_t \end{aligned} \tag{5.6}$$

where $\Lambda_1(\mathbf{N}_t, \mathbf{O})$ is equivalent to $\lambda(\mathbf{N}_t)$ and $\Lambda_2(\mathbf{N}_t, \mathbf{O})$ is equal to the zero matrix. Both Λ_1 and Λ_2 are matrices that define sexual and competitive (as well as other types of) interactions. These quantities therefore define the fitness of wildtype and mutant strategies under the prevailing conditions. For a general and far more thorough treatment of this kind, but with reference to differential equations, see Reed and Stenseth (1984); a summary is given below (pp 124–127).

When establishing the expressions for Λ_1 and Λ_2, the genetic structure of the population and, for example, kin selection arguments (Hamilton, 1964a, b) can, and should, be taken into consideration. Here it suffices to note that the ecologically and evolutionarily stable equilibrium will still be characterized by $\lambda = 1$, which means that the distribution of the various age and/or sex categories is constant over time; in particular, this implies that both sexes have the same relative fitness.

It is not necessary – nor, to my mind, always desirable – to formulate standard population genetic models for studying evolutionary processes or phenomena. Ecological models of the kind defined by equations (5.5) and (5.6) are often more desirable since they emphasize that, in order to understand natural selection, we must understand the ecological interactions (both intra- and inter-specific) between coexisting individuals (see, e.g., Stenseth and Maynard Smith, 1984); ecology is essential for understanding evolution. This formulation is not new: it dates back to the pioneers of the field (e.g., Kostitzin, 1938). I return to the importance of ecology later.

One feature of this presentation so far should be noted: I have taken an organismic point of view (favored, e.g., by Kostitzin, 1938) rather than the currently more popular genetic point of view (favored by, e.g., Fisher, 1930, and Haldane, 1932). It is my belief that the genetic point of view too easily forgets the individual organism – the only real and objective unit in natural systems – and thus is partly responsible for the difficulties facing theoretical population genetics today (see, e.g., Lewontin, 1974, 1979). To my mind, the ESS approach developed by Maynard Smith in a series of contributions (see, e.g., Maynard Smith, 1982b) does, in effect, merge the two approaches.

Natural selection operates so as to produce a population with a strategy that renders the equilibrium defined by $\mathbf{N}_t = \mathbf{N}_{t+1} = \mathbf{N}^*$, and $\mathbf{N}'_t = \mathbf{N}_{t+1} = \mathbf{0}$ asymptotically stable for any possible mutant strategy. It turns out that under a variety of biologically reasonable assumptions an asymptotically stable $(\mathbf{N}^*, \mathbf{0})$ is equivalent to maximizing λ as defined by equation (5.1) (Charlesworth, 1980; Charnov, 1982; Reed and Stenseth, 1984). Specifically, the existence of sexual reproduction does not automatically invalidate this assertion. However, several biological situations do not give equivalence between an asymptotically stable $(\mathbf{N}^*, \mathbf{0})$ and maximization of λ; one such case was discussed by Charnov (1982) and another by Reed and Stenseth (1984). However, the formulations given by equations (5.1) and (5.6) can handle a variety of biologically complicated and interesting situations.

It follows, though, that fitness is not always optimized *in the strict sense of the word*; in fact, fitness is rather difficult to define in the case of a sexually reproducing population. In both of the discussed cases, however, there is some quantity – a scalar, vector, or matrix – which is being maximized through evolution by natural selection. As has been shown above, this quantity, which may be called fitness, is well-defined mathematically and empirically.

Heredity and variation

Through observations on the breeding of cows, pigs, dogs, and so on, Darwin came to understand much about inheritance and variation [i.e., premises (2) and (3); see p 112] in living organisms. But he never really obtained a proper understanding of these features of life. One of his basic difficulties was that he had no concept of a gene; that is, he had no concept of particulate inheritance. Nor had he any thorough understanding of how new, inheritable varieties arose and spread through a population.

It was the German biologist August Weismann who provided the basis for our current understanding of these two basic features of life. Weismann is remembered for his opposition to Lamarck's premises of the inheritance of acquired characters and for having developed the theory of germ plasma (see below). On this basis, August Weismann may properly be called the greatest evolutionary biologist since Charles Darwin: in fact, owing to Weismann, natural selection is today considered not merely one of many mechanisms that adapt organisms to their environment – natural selection is considered *the* mechanism that brings about adaptations (see, e.g., Maynard Smith, 1982a).

Weismann was always a strong supporter of Darwin and wrote that the *Origin of Species* has excited "delight and enthusiasm in the minds of younger students" (Weismann, 1893). Unlike Darwin, however, Weismann firmly opposed the idea of inheritance of acquired characters. In fact, he put the matter to an empirical test in a somewhat naively conceived experiment in which he cut off the tails of mice. With painstaking thoroughness, he observed five generations of progeny from tail-less parents, 901 mice in all. Needless to say (today), they all grew normal tails.

Weismann (1886, 1893) conceived the idea, arising from his observations of the Hydrozoa, that the germ cells of animals contain "something essential for the species, something which must be carefully preserved and passed on from one generation to another": the theory of germ plasma was born. Its essence was that all living things contain a special heredity substance. The general idea is still accepted as valid (e.g., Maynard Smith, 1983), since the overwhelming majority of inherited differences between organisms are caused by differences between chromosomal genes and not cytoplasmic inheritance. Weismann, however, lacked nearly all the experimental genetic data that now exist; he filled in the details of his theory with wide-ranging – but certainly useful – speculations that at times became somewhat mystical.

Even at that time, when the writings of Gregor Mendel on genetics were lying unnoticed, Weismann saw that, since the hereditary substances from two parents mixed in the fertilized egg, there would be a progressive increase in the amount of hereditary substances unless, at some stage, there was a compensating reduction. He therefore predicted that there must be a form of nuclear division in which each daughter nucleus receives only half the ancestral germ plasma contained in the original nucleus. The cytological work of other investigators proved the correctness of this prediction and enabled Weismann, together with the others, to propose that the germ plasma was located in what were subsequently called the chromosomes of the egg nucleus. Hence, instead of germ plasma one speaks today of chromosomes, genes, and DNA.

The theory of germ plasma is the current basis for our understanding of how new, genetically determined traits (or varieties) arise and how they are transmitted to subsequent generations. Weismann assumed in his theory that any fertilized egg at an early stage gives rise to two independent populations of cells within the organism:

(1) The germ-line (or germ plasma) constituting the sex cells.
(2) The soma-line constituting the body.

Weismann hypothesized that genetic changes occurring in the germ-line are independent of genetic changes occurring in the soma-line (at least in the sense that acquired characters are not transferred to subsequent generations).

Essentially, the distinction between the germ-line and the soma-line corresponds to the distinction between *genotype* and *phenotype* (Johannsen, 1909; see also, e.g., Dobzhansky *et al.*, 1977). The genotype of an organism is its total assemblage of genes (i.e., its genome), whereas the phenotype is the organism's total assemblage of traits. The phenotype is the product of the individual's genotype as well as its environment. The fitness – or some equivalent measure – of an individual is a phenotypic property determined by both the individual's intrinsic property and the properties of the environment it is living in. Hence, natural selection operates on the phenotype. However, through this natural section acting on the phenotype, the average genotypic properties of the population are changed over generations, in response to (selective) pressures generated by the environment: that is, there is an interplay between the individual's intrinsic features and the features of the environment.

In Dawkins' (1976, 1982) terminology, the phenotype – as interpreted in this chapter – would be the survival machine or the replicator. Schuster and Sigmund (1983) and Sigmund (1985) present some rather useful and powerful mathematical results for the study of replicator dynamics in the context of evolution. They also relate the theory of replicator dynamics to the theory of games as applied to evolution (see, e.g., Maynard Smith, 1982b, 1984; see also *Open Peer Commentaries* by Barlow ***et al.***, 1984).

Weismann's idea is illustrated in Figure 5.4(a). This view, first presented in 1886, corresponds directly to what we today call the Central Dogma of molecular biology – "DNA → RNA → protein" (Crick, 1958, 1970, 1973; see also Dawkins, 1982; Stenseth, 1985a), illustrated in Figure 5.4(b). As Crick once said, "Once information has passed into proteins it cannot get out again".

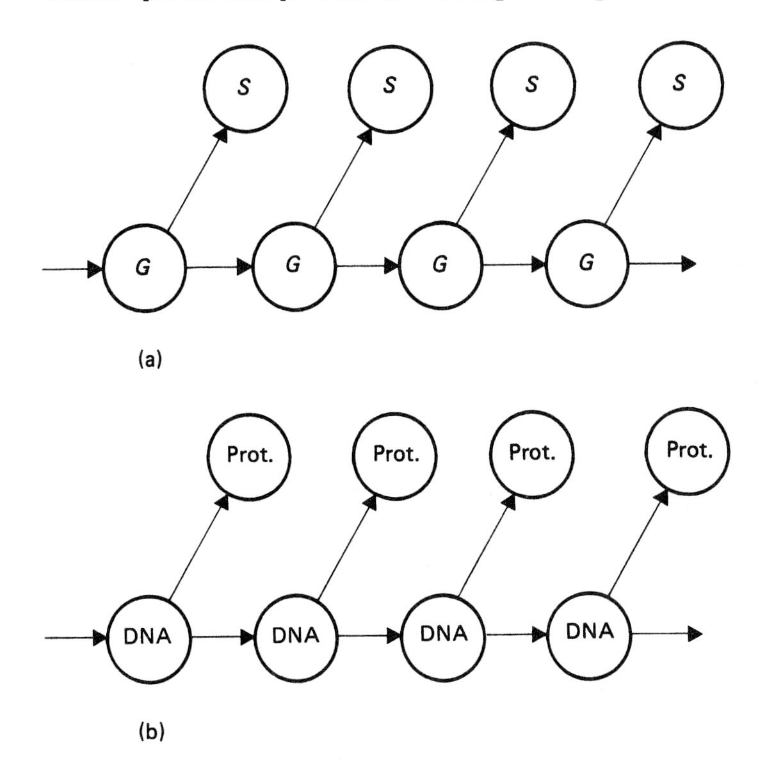

Figure 5.4 The Weismannian view of inheritance (a) The Weismannian theory of the germ plasma: the germ, G, is passed from one generation to another whereas each individual's soma, S, extinguishes when the individual dies. The process resulting in the fully developed organism from a fertilized egg is called development and is indicated by the arrow from G to S. That there is no arrow from S to G signifies that inheritance is Weismannian – and not Lamarckian (with inheritance of acquired characters). (b) The Central Dogma of molecular biology states that information is transferred from DNA to proteins (prot.) and not the other way (see text for a discussion).

The opposite of Weismannian inheritance is Lamarckian inheritance (after Lamarck, 1809) which allows acquired characters to be transmitted to subsequent generations. Most acquired characters seem to be the result of injury, disease, and old age. Therefore, a mechanism able to transmit such changes of the soma (or the phenotype) back into the germ-line would commonly lower fitness of the offspring. Maynard Smith (1983) suggested, in fact, that this is the functional explanation of why inheritance is rarely Lamarckian. He then suggested that if

> ...an organism had some way of telling which of its acquired characteristics were adaptive, a mechanism for transmitting them would be favoured. Of course, this is precisely what happens with learning and cultural inheritance.

From the Weismannian point of view, it appears correct to call Darwinism a theory of "chance and necessity", as Jacques Monod (1971) once did: it is a matter of *chance* which *new* variant – or phenotype – arises. In particular, the occurrences of these new variants are random events relative to those that are needed by the organism in order to increase its phenotypic fitness as compared with those of other, coexisting organisms. However, those genetic variants that have (on average) greater fitness or are such that (on average) the population of which they are part is noninvadable are ***necessarily*** favored and maintained by natural selection. Random gene substitutions generate new phenotypic variation – deterministic (or, at least, nonrandom) ecological interactions determine whether these newly generated variants are "chosen" by natural selection or not.

Developmental Biology

> If we are to understand evolution, we must remember that it is a process which occurs in populations, not in individuals. Individual animals may dig, swim, climb or gallop, and they do also develop, but they do not evolve. To attempt an explanation of evolution in terms of the development of individuals is to commit precisely that error of misplaced reductionism of which geneticists are sometimes accused. (Maynard Smith, 1983).

Developmental biology is the scientific study of the processes which lead to the formation of a new animal or plant from cells derived from one or more parent individuals. These are the processes that define the transition from G to S in Figure 5.4(a); these processes, by which a fertilized egg becomes a fully developed organism (i.e., the soma) we collectively call *development* or *ontogeny*. Development and ontogeny are thus the processes by which a new generation of organisms is produced from a parent generation of organisms.

In short: evolution produces phylogenies; development produces ontogenies; ontogeny is the life history of an individual.

The concepts evolution and development have always been very closely related; indeed, the two were synonymous until the 1840s (Bowler, 1975; Patterson, 1983). The change in meaning of evolution from development of the embryo to transmutation of species was initiated by von Baer (1828). Just as Darwin first promoted natural selection by analogy with artificial selection, so Haeckel (1866), who coined the word phylogeny (see p 107), first promoted it by analogy with ontogeny. Haeckel's biogenetic law, that ontogeny recapitulates phylogeny, was also a restatement of a much older tradition, the ancient idea of recapitulation or progressive ascent of a ladder-like uniserial chain of being. But Haeckel clearly differentiated his evolutionary version of recapitulation from the older, nonevolutionary version, by pointing out that the ontogeny of the individual corresponds to, or recapitulates, only a part of phylogeny. Maynard Smith (1983, pp 41–43, his Figure 2) gives a neat theoretical justification for why this is likely to be so.

Molecular and developmental biology show, without the slightest doubt, that development is exceedingly complex. Unfortunately, we are largely ignorant of the biochemical nature of events at the molecular level, the mechanisms by which they occur, and especially the control processes that govern them. Development is, in fact, one of the greatest mysteries in biology. There exists no theory of development comparable with the theory of evolution. Some (e.g., Futuyma, 1979) claim that we need to understand the complexity of development at the biochemical level before we can understand the alterations of ontogeny which are the history of evolution. Further, it is often claimed that if we obtain a better understanding of development, we automatically obtain a better understanding of evolution (see, e.g., discussions in Bonner, 1982).

After the publication of Darwin's *Origin of Species*, but before the general acceptance of Weismann's view, problems of evolution and development were inextricably bound together. However, one important consequence of Weismann's concept of the separation of germ-line and soma-line was to render possible the understanding of (population) genetics, and hence evolution, without having to understand development. This was, at least in the short term, an immensely valuable contribution, because the problems of heredity proved to be solvable, whereas those of development apparently were not so easily solvable. We do not need to understand in detail how S (in Figure 5.4) developed from G as long as S is genetically determined to some extent; the developmental process must be a mapping process. However, whether any particular S is fit or not – or how fit it is – is determined by the prevailing ecological conditions. Hence, it is my view that without a better understanding of ecology we cannot understand evolution. In fact, I dare to claim that if we obtain a better understanding of ecology, we automatically obtain a better understanding of evolution; but that I return to later.

Obviously, we will eventually require a more complete understanding of development, but we also need new concepts before we can understand it. It is comforting, meanwhile, that Weismann was right since we can progress towards understanding the evolution of adaptations without understanding *how* the relevant structures develop.

Properties such as morphology, demography, and behavior of fully developed organisms, are, in a general way, the result of interactions between gene products (i.e., proteins) in a particular environment. Thus, let g_i be the ith gene of an organism and let p_j be the jth gene product (protein); in general, one gene may give rise to several gene products; similarly, one gene product may also be the result of several genes. Some gene products may, of course, also control the rate of production of other gene products, as well as the rates of diffusion of these products between cells. In general, we can write

$$M = d(p_1, p_2, \ldots, p_n; E) \tag{5.7}$$

where M is the generalized morphology of the organism, E denotes the environment of the organism, and

$$p_i = p_i(g_1, g_2, ..., g_m; E) \quad . \tag{5.8}$$

Both d and $p_i(\cdot)$ are, in general, nonlinear functions.

Long ago, Maynard Smith (1960) and Maynard Smith and Sondhi (1960) did some interesting work on such models, based on the work of Turing (1952), which were tested on the basis of experimental work on *Drosophila* (Maynard Smith and Sondhi, 1961). [Later, several workers carried out similar analyses (e.g. Wolpert, 1969, 1983; Oster *et al.*, 1980).] They found that standing waves easily resulted for a hypothesized chemical compound produced by the gene products which could diffuse from one cell to another (Figure 5.5): hence, some sort of prepattern was assumed to be generated. The cells could then respond by developing, for example, a bristle if the value rose above some threshold. In this example then, the bristle configuration is denoted M whereas the gene products that produce the prepattern are the ps. A particular example is provided by Sondhi (1962) for *Drosophila subobscura*, and shown in Figure 5.5(d).

Because of the nonlinearity of the functions for M and p, small changes in g and p *may*, in general, cause large as well as small changes in M. Hence, we cannot conclude from the observation of a large change in M that this is caused by a large change in $(g_1, g_2, ..., g_m)$ nor in $(p_1, p_2, ..., p_n)$. René Thom (1975), of course, said this long ago. Many developmental biologists seem, however, not to have realized this simple consequence of nonlinearity in functions. Instead, they talk about developmental constraints; for a discussion, see Alberch (1980), Oster and Alberch (1982), Lander (1982), and Bonner (1982). An illustration of this is given in Figures 5.5(b) and 5.5(c); in Figure 5.5(b) a fairly large change in the threshold value produces no change in M, whereas in Figure 5.5(c) a small change in the threshold value produces a large change in M.

However, if evolution through natural selection brings about adaptation, we would expect that organisms will normally be close to their adaptive peak. This has, in fact, commonly been assumed valid since Fisher's (1930) pioneering work in population genetical theory and today forms one of the basic premises in the adaptationist research program (e.g., Calow, 1983). Thus, greatly deviating forms are, in general, characterized by lower fitness than the wildtype (or most common and, presumably, most well-adapted phenotype). Figure 5.6 illustrates this point: hopeful monsters or systemic mutations as envisaged by Goldschmidt (1940) – that is, complex new adaptations arising without selection from a restructuring of the genome – can be ruled out on probabilistic grounds; but they are not *a priori* excluded as impossible*.

This is why evolutionary biologists are not – or rather have not been – so concerned by large mutations (i.e. macromutations); recently there has, however, been some discussion, and confusion, over the importance and commonness of such macromutations. Questions of how often macromutations occur – and of how often they are of major evolutionary importance – are ultimately empirical ones. The most fruitful approach is a genetic analysis of related species which differ morphologically. Such analysis does little to suggest that mutations of large effects have been important in evolution (see, e.g., Charlesworth *et al.*, 1982). Darwinists

*In fact, we know that single mutational events can and do give rise to large phenotypic changes, as a visit to any genetic laboratory establishes. A rather exciting natural example of a successful macromutation is the gastric brooding frog (*Rheobatrachus*; Tyler, 1983): intermediate evolutionary steps seem in this particular case inconceivable.

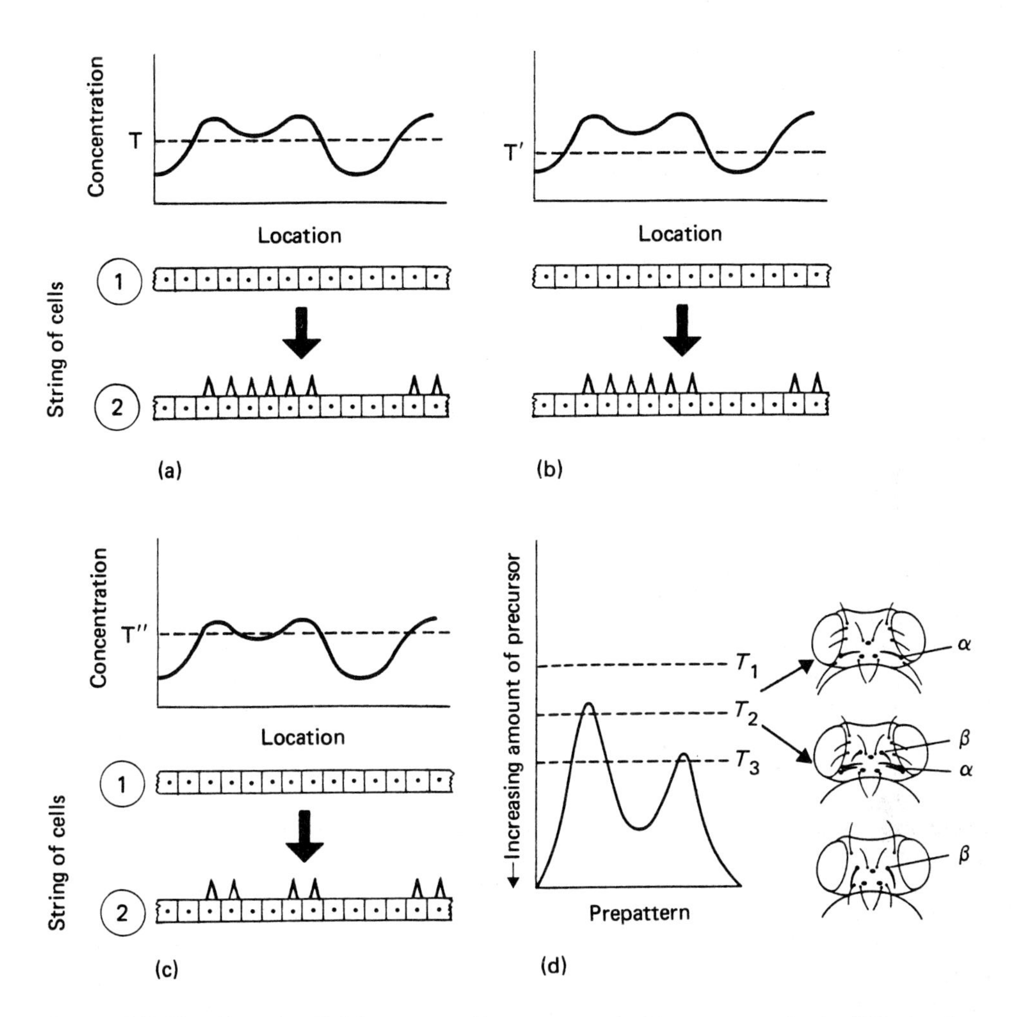

Figure 5.5 The idea of a field, or prepattern, generated by gene products diffusing from one cell to another. (a) A hypothetical case where a standing wave of some gene product is generated by the intrinsic dynamics of the system. The string of cells denoted 1 represents a hypothetical string before the developmental process has finished (which in this case is assumed to generate bristles if the concentration of the gene product is above a threshold value, T. (b) A case for which a relatively large mutational change in the threshold concentration (from T to T') has occurred. As can be seen, no phenotypic change results. (c) A case for which a relatively small mutational change in the threshold concentration (from T to T'') has occurred. As can be seen, a major phenotypic change results. The phenotypes in (a) and (c) will have distinctly different patterning of their bristles. See, for example, Maynard Smith (1983) and the text for further discussion. (d) Sondhi's model of the origin of a neomorphic (new) pattern. If the first peak of the prepattern is between threshold levels T_1 and T_2, it produces the wildtype pattern of bristles and ocelli (I) in *Drosophila subobscura*. If the second peak exceeds threshold T_3, additional bristles are formed (II): bristle α is doubled, and new bristles (β) arise. These new bristles are unknown in normal Drosophilids, but have a counterpart in *Aulacigaster leucopeza* (III), a member of a related family (after Sondhi, 1962, and Futuyma, 1979).

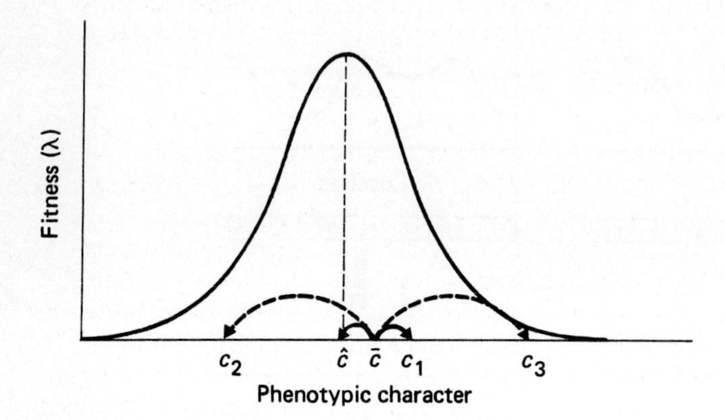

Figure 5.6 A hypothetical fitness curve relating a certain fitness (λ) to a certain phenotypic character ($c = M$). The phenotype $\hat{c}$ is, in the current environment, the optimal (or best) one. The phenotype $\bar{c}$ is the average in the current population. A small deviation (due to a mutation) from this average one (from $\bar{c}$ to either c_1 or $\hat{c}$) is as likely to produce a slightly improved phenotype as to produce a slightly deteriorated phenotype. A large mutation from $\bar{c}$ to c_3 is, because the average phenotype is assumed to be close to the optimal phenotype $\hat{c}$ in the first place, likely to produce highly deteriorated phenotypes (see text for further discussion).

also have another reason for paying most attention to mutations of small effects (i.e. micromutations): even if a large mutation should produce an integrated whole of reasonably high fitness, fine tuning by selection of mutants with *small* differences in phenotype is always required before an exact adaptation to the current environment can be achieved.

Note, however, that I have not yet said anything about how fast evolution proceeds over time. Until now I have only considered whether new variants, arising as a result of mutation, will differ much or little from their preceding parental form. The rates of evolution I return to later in the chapter.

Ecological and Evolutionary Stability

What kind of theory do we need?

Maynard Smith (1969) once pointed out that what we need to obtain a better understanding of the living world

> ... is first a theory of ecological permanence, and then a theory of evolutionary ecology. The former would tell us what must be the relationships between the species composing an ecosystem if it is to be 'permanent', that is if all species are to survive, either in a static equilibrium or in a limit cycle. In such a theory, the effect of each species on its own reproduction and on that of other species would be represented by a constant or constants...

> In evolutionary ecology these constants become variables, but with a relaxation time large compared to the ecological time scale. Each species would evolve so as to maximize the fitness of its members. If so, a permanent system might evolve to an impermanent one.

A new paradigm, essential for the study of evolutionary ecology, was introduced in 1972 with Maynard Smith's concept of ***evolutionarily stable strategies***, or ESS (see also Maynard Smith, 1982b; Sigmund, 1985). As defined by Maynard Smith and Price (1973, p 15) an ESS is "a strategy such that, if most of the members of a population adopt it, there is no 'mutant' strategy that would give higher reproductive fitness". That is, an ESS population cannot, according to this definition, be invaded by a mutant (small or large) phenotype. Only individual selection is assumed to operate (e.g., León, 1976, pp 303–304; Maynard Smith, 1982b). As in most models for studying evolutionary changes, the ESS concept requires a constant environment or a long-term, consistently changing environment. Maynard Smith (1982b) has recently reviewed many of the applications of the ESS concept.

Lawlor and Maynard Smith (1976) used this method to locate a joint ESS in a system of competing species. An ESS was, according to these authors, found at a value of the evolutionary variable (see below) such that the *actual* specific growth rate is zero, and such that if we keep the equilibrium state fixed and vary only the evolutionary variable, negative specific growth rates result for the evolving population (Figure 5.7). That is, no mutant has a higher reproductive fitness when it first arises. As pointed out by Schaffer (1977), the analysis performed by Lawlor and Maynard Smith is identical to a method provided independently by Allen (1976). Similar approaches were also described by Case and Casten (1979), Case (1982), Case and Sidell (1983), Roughgarden *et al.* (1983), and Roughgarden (1983b).

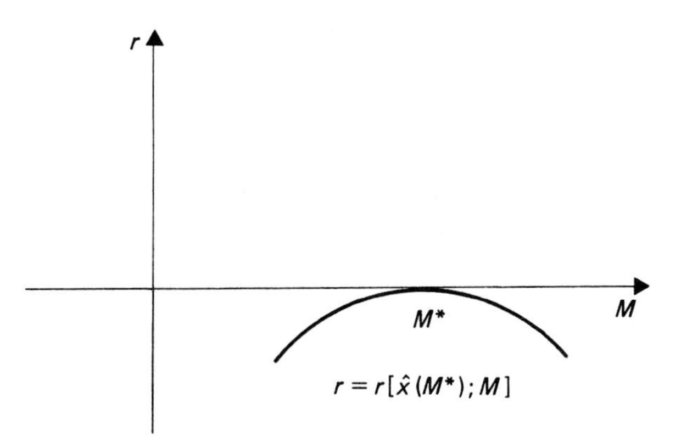

Figure 5.7 An interpretation of an ESS as found by, for example, Lawlor and Maynard Smith (1976). $r = (1/x)(dx/dt)$ as given in the main text; $\hat{x}$ is the ecological equilibrium value (found by solving $dx/dt = 0$), and M is the morphology or phenotypic character under evolution (see text and Reed and Stenseth, 1984, for further discussion).

This particular method developed by Lawlor and Maynard Smith for finding an ESS works well for the simplest cases where, for example, each evolving species is phenotypically monomorphic; see the section on fitness above (pp 113–115). However, this method is not generally applicable to more complex models, for example, in which we want to distinguish between different age classes or different genotypes of a species or when we want to consider models with two sexes (e.g., León, 1976). Obviously, we need the ESS method discussed by Lawlor and Maynard Smith to apply to such complex, but biologically very reasonable examples as well. Even though Allen (1976) applied his method to a sexually reproducing population (i.e., a polymorphic one), he treated several aspects rather superficially. Hence, together with Jon Reed, I (Reed and Stenseth, 1984) have extended the ESS method as originally presented by Lawlor and Maynard Smith (1976) in order to handle such cases as well.

A theory of noninvadability of ecologically stable ecosystems

Let $\mathbf{X} = (X_1, X_2, \ldots, X_n)$ denote the density vector of an established population on, for example, an island or continent; each element refers to suitably defined groups of organisms in terms of, for example, the density of each species. Assume, for $X_1 \geq 0$, that the dynamic behavior of these populations is described by an ecological model of the form

$$d\mathbf{X}/dt = f(\mathbf{X}) \tag{5.9}$$

characterized by a stable equilibrium at $\mathbf{X} = \hat{\mathbf{X}}$. By stable, here and in the following, I mean asymptotically stable in the sense of Liapunov stability (e.g., Arnold, 1973); thus, I only consider a local form of ecological stability. As is well known, a sufficient condition for such stability is that all eigenvalues at equilibrium have negative real parts; if at least one of these eigenvalues has a positive real part, the equilibrium is unstable. In the following, I regard this as a *necessary* condition for evaluating the stability of any particular community structure. This is reasonable because if model (5.9) does not correspond to a community with locally stable equilibria, the resulting (sometimes extensive) density variations of the species in the community are likely to cause extinction of one or more species and thereby change the entire structure of the system described (e.g., Maynard Smith, 1974). I therefore tacitly exclude the possibility of stable limit-cycles. Someone should, however, extend the following analysis to cases with limit cycles; unfortunately, I am unable to do this myself.

To require ecological stability of the kind just described is, however, not sufficient when studying evolutionary stability. This is because a new species may be able to invade the community, even though it is Liapunov stable in an ecological sense. Much empirical material demonstrates that invasion of new species into existing communities is a common ecological phenomenon (e.g., Elton, 1958; MacArthur and Wilson, 1967; Diamond, 1969; MacArthur, 1972). In addition new species may arise as a result of sympatric speciation; new species that arise as a result of allopatric speciation and then enter the community as new members are, in the present context, equivalent to a true ecological invasion. Finally, new phenotypes of the currently coexisting species may arise, either as a result of mutation *in situ* or as a result of immigration from neighboring populations.

From this it follows that we must extend the model given by equation (5.9) so that it also includes potential invading species (or phenotypic mutant forms), or groups of species. Let $\mathbf{Y} = (Y_1, Y_2, \ldots, Y_m)$ be the density vector of those forms trying to invade the established community [given by model (5.9)]. Assuming, for $X_1 \geq 0$, $Y_i \geq 0$, an extended ecological model of the form

$$\begin{aligned} d\mathbf{X}/dt &= F(\mathbf{X}, \mathbf{Y}) \\ d\mathbf{Y}/dt &= G(\mathbf{X}, \mathbf{Y}) \end{aligned} \tag{5.10}$$

where $F(\mathbf{X}, \mathbf{0}) = f(\mathbf{X})$ and $G(\mathbf{X}, \mathbf{0}) = \mathbf{0}$; that is, if there are no invaders, model (5.10) reduces to (5.9).

I assume that the invaders are rare to begin with. In arguments relating to microevolutionary changes, this seems to be a fairly reasonable assumption. Similar assumptions have previously been made by MacArthur (1972) and Roughgarden (1974) when studying species packing; the limiting similarity theory (MacArthur and Levins, 1967; MacArthur, 1972; and Roughgarden, 1979) is, in fact, a special case of the discussed model in which whether the invaders succeed or not in the established community is determined by the stability of the equilibrium given by $\mathbf{X} = \hat{\mathbf{X}}$ and $\mathbf{Y} = \mathbf{0}$ in model (5.10). If this equilibrium $(\hat{\mathbf{X}}, \mathbf{0})$ is Liapunov stable, none of the deviating forms represented by $\mathbf{Y}$ are able to invade the established community; if it is unstable, one or more of the deviating forms represented by $\mathbf{Y}$ can invade and *possibly* establish itself in the community. The resulting new community may or may not be stable in the sense of Liapunov stability; this must, of course, necessarily be analyzed by model (5.9) (see Case, 1982).

From the assumptions made for the extended model (5.10), it follows that its Jacobian matrix (see, e.g., Arnold, 1973) evaluated at the equilibrium $\mathbf{X} = \hat{\mathbf{X}}, \mathbf{Y} = \mathbf{0}$, has the triangular form

$$P = \begin{bmatrix} \dfrac{\partial \hat{F}}{\partial \mathbf{X}} & \dfrac{\partial \hat{F}}{\partial \mathbf{Y}} \\ \mathbf{0} & \dfrac{\partial \hat{G}}{\partial \mathbf{Y}} \end{bmatrix} = \begin{bmatrix} Q & S \\ 0 & R \end{bmatrix} \tag{5.11}$$

where $Q = \partial \mathbf{F}/\partial \mathbf{X} = \partial F(\hat{\mathbf{X}}, \mathbf{0})/\partial \mathbf{X} = \partial f(\hat{\mathbf{X}})/\partial \mathbf{X}$, etc. The Q-matrix is, in current ecological literature, referred to as the community or α-matrix (Levins, 1968). Reed and Stenseth (1984) suggest that R be called the ***invasion matrix***.

Let $p(\lambda)$ be the characteristic polynome for P; i.e.,

$$p(\lambda) = \det(P - \lambda I) \tag{5.12}$$

where I is the diagonal unity matrix, and λ is an eigenvalue for the system defined by model (5.10). From matrix (5.11) it follows that

$$p(\lambda) = q(\lambda) r(\lambda) \tag{5.13}$$

where q and r are the characteristic polynomes for the Q- and R-matrices; notice, however, that q corresponds to the equilibrium situation of model (5.9) defined by $\mathbf{X} = \hat{\mathbf{X}}$, whereas r corresponds to the equilibrium situation of model (5.10) defined by $\mathbf{X} = \hat{\mathbf{X}}$, $\mathbf{Y} = \mathbf{0}$. Hence, we must necessarily evaluate the eigenvalues of Q (as is usually done in ecological community studies) ***as well as*** those of R; all eigenvalues must have real parts less than zero in order to guarantee stability. [From this result it is, of course, obvious why I initially assumed $\mathbf{X} = \hat{\mathbf{X}}$ in model (5.9) to be a locally stable equilibrium point in the sense of Liapunov.] A very similar approach to the study of community stability was taken by Case (1982) for a few-species community.

An interpretation of ESS

(A) Models with evolutionary variables

Suppose that some of the populations in the considered community give rise to mutant phenotypic strategies and that these phenotypes depend on the value of an evolutionary variable, M, with domain in some subspace C of $\mathbb{R}^r$. This $C = \boldsymbol{F}$ is thought of as a constraint on the possible phenotypes; that is, a fitness set in the sense of Levins (1968) or a phenotype set in the sense of Maynard Smith (1978a).

The ***evolutionary variable*** enters the basic ecological model as an ***ecological parameter***. The ecological model therefore has the form

$$\mathrm{d}\mathbf{X}/\mathrm{d}t = f(\mathbf{X}; M) \ . \tag{5.14}$$

As above, the existence of a stable ecological equilibrium $\mathbf{X} = \hat{\mathbf{X}}(M)$, depending on M, is assumed. ***Generally, the model may have several stable equilibria for a given value of*** M; ***hence, there is a choice involved in the definition of*** $\hat{\mathbf{X}}(M)$. Thus, depending on the particular properties of the ecological equilibrium, evolution may proceed in different directions.

With the ecological parameter, M, made explicit as in model (5.14), the effect of changes in this parameter on the fitness of the evolving organisms may be studied more easily. Let $\mathbf{Y}$ represent populations with mutant phenotypes corresponding to the parameter value $M' \neq M$. Then we assume an extended ecological model, in the sense of model (5.10), of the form

$$\begin{aligned} \mathrm{d}\mathbf{X}/\mathrm{d}t &= F(\mathbf{X}, \mathbf{Y}; M, M') \\ \mathrm{d}\mathbf{Y}/\mathrm{d}t &= G(\mathbf{X}, \mathbf{Y}; M, M') \end{aligned} \tag{5.15}$$

with the equilibrium $\mathbf{X} = \hat{\mathbf{X}}(M)$, $\mathbf{Y} = \mathbf{O}$.

In any particular application of this method, the relation between F and G, if any, can be specified. Further, F and G may or may not be specified as specific growth rates multiplied by the density of the particular species. In any case, it seems impossible to define F and G without a thorough understanding of the ecological situation – just as it would be impossible to define appropriately f in model (5.14) without a thorough ecological understanding.

It should be realized that this approach actually transforms an evolutionary problem into an ecological problem; whether or not a feature evolves depends on the outcome of the (ecological) competitive situation initially facing mutant forms, when they are rare. Biologically, this is of course a rather appropriate way of viewing the problem. That is, an ecological understanding of competition seems necessary for understanding evolution – ecology is important.

(B) ESS, evolutionarily stable strategy

Given these models and equilibria we say that the phenotype corresponding to the parameter value M is ***evolutionarily stable*** if all $M' \neq M$ correspond to stable $(\hat{\mathbf{X}}, \mathbf{O})$ equilibria in model (5.15): this value of M we denote M^* (i.e. $M = M^*$). This M^* is then said to represent an ESS, if the equilibrium of model (5.15) is stable for all M' in some neighborhood region outside M in C; that is, for all M' in some subset of C. Thus, no mutant with a parameter value different from that of the wildtype should be able to invade the established community when this is at ecological equilibrium.

To my mind, this ESS method is today the most powerful technique available for studying the ***evolutionary*** maintenance of phenotypic features, M^* [see, e.g., equation (5.7)], where the asterisk denotes the ESS morphology. Usually, this method is presented as a local problem; that is, $M' = M^* \pm \varepsilon$ where ε is a small deviation (or mutation) from the wildtype, M^*. Using the approach developed by Reed and Stenseth (1984), this ESS method may easily, and properly, be extended to a ***global*** problem. Notice, however, that mathematically we do this by studying the ***local*** stability properties of the equilibrium $(\hat{\mathbf{X}}, \mathbf{0})$ of model (5.10). That is, if $(\hat{\mathbf{X}}, \mathbf{0})$ is *locally stable* for ***all*** mutant strategies, $m \in \boldsymbol{F}$, where $\boldsymbol{F}$ denotes the fitness set (see Levins, 1968; Maynard Smith, 1978a), then M^* is evolutionarily stable in a global sense.

This, then, is a precise way of analyzing – and obtaining a better understanding of – ***why*** the structure M (or f) came to dominate under the prevailing conditions; thus we may better understand ***why*** nature became ***how*** it is (see Mayr, 1961; Pianka, 1978).

Some examples

(A) Sex ratio in mammals

Why is the 1:1 sex ratio so commonly observed among, for example, mammals? A structural biologist would, probably, answer this question by referring to the X–Y chromosome system which is so common among mammals (Bull, 1983). However, this is only an answer to the ***how*** question (see, e.g., Pianka, 1978; and above); the evolutionary answer to the posed ***why*** question (see Pianka, 1978; and above), first given by Fisher (1930), is as follows. Assume the panmictic population to be composed of a majority of females. Then, it would pay, evolutionarily, to produce a surplus of sons since these would have a higher probability of finding a mate than would a daughter. In this way, the number of grandchildren – hence fitness – would be maximized. Similarly, if the panmictic population is composed of a majority of males, it would pay to produce a surplus of daughters since these would have a higher probability of finding a mate. Analysis demonstrates that the 1:1 sex ratio is, in fact, an ESS (Maynard Smith, 1978b; Charnov, 1982). As pointed out by, for example, Maynard Smith (1978b), this argument does not depend on a monogamous mating system; it is valid also for harem-forming species. These arguments depend critically on the fact that all offspring have exactly one father and one mother; that is, at any instant in time there are exactly as many mothers as there are fathers.

Furthermore, this argument depends critically on the existence of random mating (any unpaired individual may mate with any unpaired individual of the other sex): if inbreeding occurs, then a female-biased sex ratio will often be the ESS sex ratio (Hamilton, 1967). The wood lemming (***Myopus schisticolor***) and the collared lemming (***Dicrostonyx groenlandicus***) are interesting examples of this.

Wild populations of the wood lemming are characterized by exaggerated density cycles and an excess of females (70–80% females; Kalela and Oksala, 1966).

Breeding experiments show that this excess of females is due to a significant proportion of reproductively active females producing only daughters. Fredga *et al.* (1976, 1977) have hypothesized a gene on an X chromosome (denoted X^* in the following) which suppresses the male determining genes on a Y chromosome and causes a selective nondisjunction in the fetal ovary (a meiotic drive) so that only X^* carrying eggs are produced. The X^* gene is supposed to have no effect on X^*X-females. No difference in litter size or in any other phenotypic character has been observed among the different female types (in captivity); in particular, there are no detectable differences in their fertility. Although Fredga and his co-workers have not shown any evidence, other than circumstantial, for the existence of a mutant X^* gene, their genetic hypothesis is consistent with available data.

In the system hypothesis by Fredga *et al.* (1976, 1977) three types of matings are possible:

XX-females × XY-males, giving a 1:1 sex ratio.
XX^*-females × XY-males, giving 3 females to each male.
X^*Y-females × XY-males, giving only female offspring.

A similar system seems to exist for the collared lemming (see Fredga, 1983; Bull, 1983).

Maynard Smith and Stenseth (1978) found that this system may be stable against invasion by a mutant gene suppressing the X^* mutation. Hence, we should, according to this hypothesis, expect to find the female-biased system only in regions where intense inbreeding occurs. On this basis it is interesting to observe that for *Dicrostonyx groenlandicus* a female-biased sex ratio similar to that observed in *Myopus schisticolor* only seems to exist in places where the habitat is divided into small patches surrounded by transition habitats (*sensu* Stenseth, 1983b; see also Thorsrud and Stenseth, 1985) that are not inhabitable.

From these theoretical studies as well as from other observations it is reasonable to conclude that the ecological setting seems to be of essential importance in determining what the evolutionarily optimal strategy is; yet another example that ecological understanding is important for a better understanding of evolution.

(B) Sex ratios in tropical butterflies

The tropical butterfly *Acraea encedon* also demonstrates the importance of ecology in understanding evolution. In this species, two types of female exist – one ordinary sexual female producing broods with a 1:1 sex ratio on average and another abnormal female producing only daughters; the males are regular. Using standard population–genetic arguments, it has been claimed (see, e.g., Heuch, 1978) that if a population of only males and normal females is invaded by abnormal females, the latter outcompete the sexual females and, as a consequence, drive the entire population to extinction. This argument hinges, however, on the assumption of density-*in*dependent rates (Stenseth, 1985c). If the population dynamics rates are made density dependent, a stable population with both normal and abnormal females may easily result. That this is so may be seen using the model originally developed to study pseudogamy (Stenseth *et al.*, 1985; Stenseth and Kirkendall, 1985).

Let the density of normal females (as well as the density of males) be given by x; since the sex ratio is 1:1, $2x$ is the total density of the bisexual population. Let β_1 be the expected instantaneous rate of production of adult progeny (males and females) by mated females, and μ_1 the corresponding adult mortality rate. Finally, the potential number of mated females per male is given by α. Utilizing these quantities, it follows – for continuous reproduction – that the dynamics of females are given by

$$dx/dt = 0.5\alpha\beta_1 x - \mu_1 x \ . \tag{5.16}$$

Further, let y be the density of abnormal females, and let β_2 be the rate at which these females produce adult (female) progeny. The specific mortality rate for the abnormal type is μ_2. Let p and q ($q = 1 - p$) measure the relative success of normal and abnormal females in obtaining matings so that p/q is the relative pairing success of sexual females; then, $p/q = 1$ (i.e., $p = q = 0.5$) represents random mating, whereas $p > 0.5$ indicates that the two strains of females are disproportionately chosen.

The dynamics of the two types of females in a mixed population are then given by

$$\begin{aligned} dx/dt &= 0.5\beta_1 x\, \alpha[px/(xp + yg)] - \mu_1 x \\ dy/dt &= \beta_2 x\, \alpha[qy/(xp + yq)] - \mu_2 x \ . \end{aligned} \tag{5.17}$$

The quantities $px/(xp + yq)$ and $qy/(xp + yq)$ represent the *potential proportion* of mated normal and abnormal females; multiplying each quantity by α gives the *actual proportions* mated (as would be observed in the field); in equations (5.17) these proportions are multiplied by x (now intepreted as the density of males).

Both the availability of resources (i.e., the degree of density dependence) and the availability of males limit the population. This, I believe, is, in general, a proper ecological feature of sexually reproducing populations.

The stability properties of models (5.16) and (5.17) are depicted in Figure 5.8. In Figure 5.9 I further demonstrate that it is the density dependences in demographic rates and in the quantities describing behavioral features that enable stable coexistence between the two female types. However, as is obvious from Figure 5.9, the region of stable coexistence depicted in Figure 5.8 does not collapse; it is the existence of a stable ecological equilibrium that becomes impossible without density dependences in demographic rates. And it is elementary that an exponentially growing population cannot be maintained as such in a world of limited resources.

(C) Character displacement

Similarly, studies on character displacement (e.g., Lawlor and Maynard Smith, 1976; Slatkin and Maynard Smith, 1979; Roughgarden, 1979, 1983a, b; Lundberg and Stenseth, 1985) show that ecological settings are important in determining the evolutionarily optimal strategy. As this topic has been so well-covered in the ecological and evolutionary literature, I do not discuss it herein, but merely remind the reader of it.

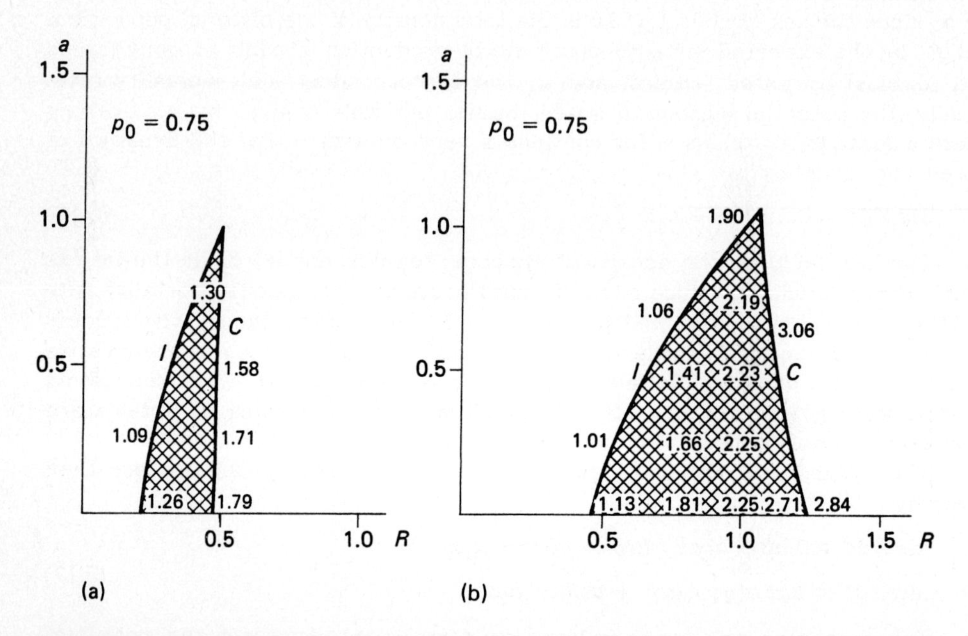

Figure 5.8 Stability properties of models (5.16) and (5.17). Let a be the degree of niche overlap between normal and abnormal females; further, let R be the relative reproductive advantage of abnormal females under extremely low densities (given that they are mated). The diagrams then show the region in the $a-R$ parameter space in which only the normal female population is both evolutionarily and ecologically stable, and in which the mixed population is ecologically stable. To the right of the I curves, abnormal females can invade an equilibrium population with normal females only. To the left of the C curves, a population with both types of female exhibits an ecologically stable equilibrium. Hence, the shaded region depicts an attainable stable equilibrium with coexistence. To the right of both the I curves and the C curves, a normal population at equilibrium is invadable by abnormal females; in this region factors other than those of (local) population dynamics have to be considered in order to explain the maintenance of all-female broods. Numbers in the diagrams represent the sex ratios $(1 + y^*/x^*)$. (a) No density dependence in p (i.e., no selective advantage for normal females in acquiring matings) and the same degree of weak density dependence in reproductive and survival rates (i.e., demographic rates – β and μ). (b) Density dependence in both p and the demographic parameters; at low density normal females have an advantage in acquiring matings. For the parameters not shown in the diagram, the following values have been assumed: $\alpha = 1$, $\mu_1(0) = \mu_2(0) = 0.1$, and $\beta_1(0) = 1$ (after Stenseth, 1985c).

Ecology – the Template of Evolution

The abiotic and the biotic environment

I think it fair to say that at present we have a reasonably well-developed theory – the Darwinian theory of evolution – to help us understand what the optimal (or most fit and hence expected) phenotypic structures are, *given a certain environmental condition*. However, the environmental situation has not been of concern to evolutionarily oriented laboratory biologists in general, nor to

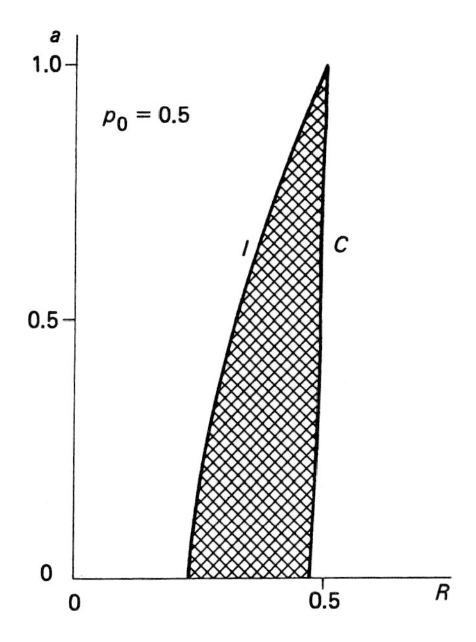

Figure 5.9 The stability properties of models (5.16) and (5.17) in the absence of density dependence in demographic rates and in p. For a fixed p_0, this stability region (see Figure 5.8) remains unchanged as do the sex ratios in various parts of this region, regardless of the degree of density dependence (as long as this degree is equal in p as well as in β and in μ). However, actual densities increase in the shaded region to infinity as the degree of density dependence diminishes to zero.

population geneticists in particular. Specifically, this was not a problem for the pioneers in the field, such as Ronald Fisher and J.B.S. Haldane: they considered, in most cases, only one species in an otherwise constant environment – and they almost never treated species–species interactions.

As a result, we have at present no convincing theory of ecology. This is unfortunate since, as I have pointed out above, the total environment of a given species consists partly of its abiotic environment and partly of its biotic environment, where the biotic component comprises other coexisting species, each of which plays the same game; they are all evolving as a result of natural selection. The challenges of evolving species are faced with because of a changing abiotic environment are likely to be of a qualitatively different kind to those of a changing biotic environment (due to evolution of coexisting species). Furthermore, these evolutionary issues will probably be studied using quite different methods.

Following Maynard Smith (1982c), what we need

> ... is a theory which says something about natural selection, and hence about the environment. Since the major component of the environment of most species – in most cases – consists of other species in the ecosystem, it follows that we need a theory of of ecosystems in which the component species are evolving by natural selection.

What we need in ecology is something like Weismann's general theory of development; but such a theory is hard to formulate. The building blocks must, I am convinced, be the Darwinian theory of evolution and a theory of ecology; the former we have, the latter we do not have. The data for testing such a theory would, I am sure, have to be provided by paleontology and ecology.

A question, two answers, and a possible solution

To demonstrate how difficult is the situation that evolutionary biology is in, it is worthwhile to note that it is now far easier to ask the following question then to answer it: Will evolution cease if we stop all physical disturbances – regular and irregular – of the environment due to factors like seasonal changes and randomly changing weather conditions? It turns out, indeed, that very competent biologists might give quite opposite answers to this question – and that each of them will defend their position rather strongly. Unfortunately, it is at present rather difficult to say who is right. And that suggests to me that something is wrong, and that something important is missing in current Darwinism. I have no doubt that it is a more solid understanding of ecology and a more synthetic theory of ecology that are missing.

How, then, can ecology more properly be incorporated into evolutionary arguments (and into the theory of evolution)? Van Valen (1973) suggested (in the same paper in which he presented the "Law of Constant Extinction") the Red Queen hypothesis: the Red Queen, you will remember, explained to Alice in Wonderland (Carroll, 1871) that

> ...it takes all the running *you* can do, to keep in the same place. If you want to get somewhere else, you must run at least twice as fast as that.

Van Valen's Red Queen view of evolution then asserts that any evolutionary change in any species is experienced by coexisting species as a change of their environmental conditions. Hence, a species must evolve as fast as it can in order to continue its existence. If the species does not evolve as fast as it can, it will become extinct. An "arms race" (Dawkins and Krebs, 1979) results: this I call the Red Queen type of evolution.

Darwin (1859) had very similar ideas:

> the most important of all causes of evolution is one which is almost independent of...altered physical conditions, namely, the mutual relation of organism to organism...

Furthermore, Darwin said:

> ...if some of these species become modified and improved, others will have to be improved in a corresponding degree or they will be exterminated.

Van Valen (1973) claimed that his Red Queen hypothesis would explain the "Law of Constant Extinction", but his argument was, however, not convincing (e.g., Maynard Smith, 1976a; Stenseth, 1979).

A Darwinian Theory of Evolution in Ecosystems

Previously, I have reviewed several models for the Red Queen hypothesis (Stenseth, 1979, 1985a) – some developed specifically for it and others appropriate for analyzing it – and I do not repeat this discussion herein. However, I do emphasize the potential prospects of the models due to Kerner (1957, 1959) on "the statistical mechanics of population dynamics" even though there are several difficulties with this approach (see, e.g., Maynard Smith, 1974; Stenseth, 1979, 1985a; Case and Casten, 1979). In addition to the difficulties mentioned in my earlier writings, there is yet another: the species number in the community of interacting organisms should be allowed to vary as a result of speciation, immigration, and extinction. I do not know whether this is at all possible, but some competent mathematician should think about it – if advancements are made, they could benefit evolutionary ecology greatly.

The lag load

Maynard Smith (1976a) specified the Red Queen hypothesis by introducing the evolutionary lag load concept. Let L_i be the lag load of the ith species in an area. Further, let $\bar{W}_i = (1/x_i)(dx/dt) = f_i(x)$ be the ith species' current mean fitness [see equation (5.9); also Stenseth, 1983a] and let $\hat{W}_i$ be the maximal possible fitness of the species in the current environment which it would have if all possible favorable alleles, whether or not they are yet segregating, are incorporated. As pointed out by Stenseth and Maynard Smith (1984), $\bar{W}_i / \hat{W}_i$ is, in principle, a measurable quantity; hence, L_i is measurable. Using these concepts, L_i is defined as

$$L_i = (\hat{W}_i - \bar{W}_i)/\hat{W}_i \ . \tag{5.18}$$

A community with a fixed number of species

Assume first that a fixed number of species coexist in the community. Let δL_i denote the change in unit time of the ith species' lag load. As explained by Maynard Smith (1976a), L_i may be expressed as

$$\delta L_i = \delta_e L_i - \delta_g L_i \tag{5.19}$$

where $\delta_e L_i$ is the change in the ith species' lag load due to (micro-) evolution of the $S-1$ coexisting species (i.e., change in its biotic environment) and $\delta_g L_i$ is the change in the ith species' lag load due to its own (micro-) evolutionary approach to its adaptive peak as defined by its current environment.

Let β_{ij} measure the increase in the ith species' lag load caused by a unit change in the jth species' lag load. Then we may rewrite equation (5.19) as

$$\delta L_i = \sum_j \beta_{ij} \delta_g L_i - \delta_g L_i \tag{5.20}$$

(where, of course, $\beta_{ii} = 0$). Assuming that $\delta_g L_i$ may be expressed as $k_i L_i$ (which, in fact, is a generalized form of Fisher's fundamental theorem of natural selection;

see Stenseth and Maynard Smith, 1984) and putting $\bar{L} = (1/S)\sum_i L_i$, we obtain

$$d\bar{L}/dt = (k/S)[\sum_j (L_j \sum_i \beta_{ij}) - \sum_j L_j] . \quad (5.21)$$

As pointed out by Maynard Smith (1976a), model (5.21) only has a stationary equilibrium point if $\sum_i \beta_{ij} \equiv 1$ for all j; if $\sum_i \beta_{ij} < 1$ for most j, $\bar{L}$ would decrease and evolution in the community would be in what Maynard Smith called the ***convergent mode***; if, however, $\sum_i \beta_{ij} > 1$ for most j, $\bar{L}$ would increase and evolution in the community would be in what Maynard Smith called the ***divergent mode***. As pointed out by Stenseth (1979), these conclusions do not refer to the global behavior of the model since both β and S are treated as constants – which they, of course, are not. Hence, the conclusion reached by Maynard Smith (1976a), that the Red Queen hypothesis seems to be implausible, is not necessarily correct (see Stenseth, 1979; and the next section).

A community with a changing number of species

Stenseth (1979) pointed out that changing the number of species in the system (as a result of speciation, immigration, and/or extinction) would, in general, change the values of β_{ij} and thus the value of $\sum_i \beta_{ij}$ for the various j. In order to study the effects of changing S, Stenseth and Maynard Smith (1984) studied the dynamic behavior of the following system:

$$\begin{aligned} d\bar{L}/dt &= (a + b\bar{L} + cS)\bar{L} \\ dS/dt &= h + (d - e)\bar{L} + (f - g)S \end{aligned} \quad (5.22)$$

in which they argued that $b < 0$, $h > 0$ and $(f - g) < 0$, but with a, c, and $d - e$ of uncertain sign. The complete dynamic behavior of model (5.22) is summarized in Figure 5.10 (for details, see Stenseth and Maynard Smith, 1984; and also Stenseth, 1985a). As can be seen, both a Red Queen type of continued evolution and a stationary state without any evolutionary changes may occur in a physically stable environment: a similar distinction also emerges in the case of some background, low-level physical noise (see Stenseth, 1985a). Stenseth and Maynard Smith (1984) suggest that the Red Queen type of coevolution corresponds to the gradualistic pattern of evolution (see pp 107–108), whereas the stationary state corresponds to the punctualistic pattern of evolution (see above).

Obviously, model (5.22) is too general to be of much help in interpreting the fossil record in any detail. However, it does demonstrate that a reasonable Darwinian model may be formulated which predicts ***both*** the gradualistic and the punctualistic pattern, depending on the values of the parameters. This conclusion is, to my mind, important since it shows that it is not necessarily correct to say, as, for example, Gould (1980, 1982), Stanley (1975, 1979, 1982), and Stanley *et al.* (1983) do, that the punctualistic pattern of the fossil record cannot be predicted on the basis of Darwinism. Specifically, the analysis carried out by Maynard Smith and myself demonstrates clearly that one cannot draw conclusions regarding

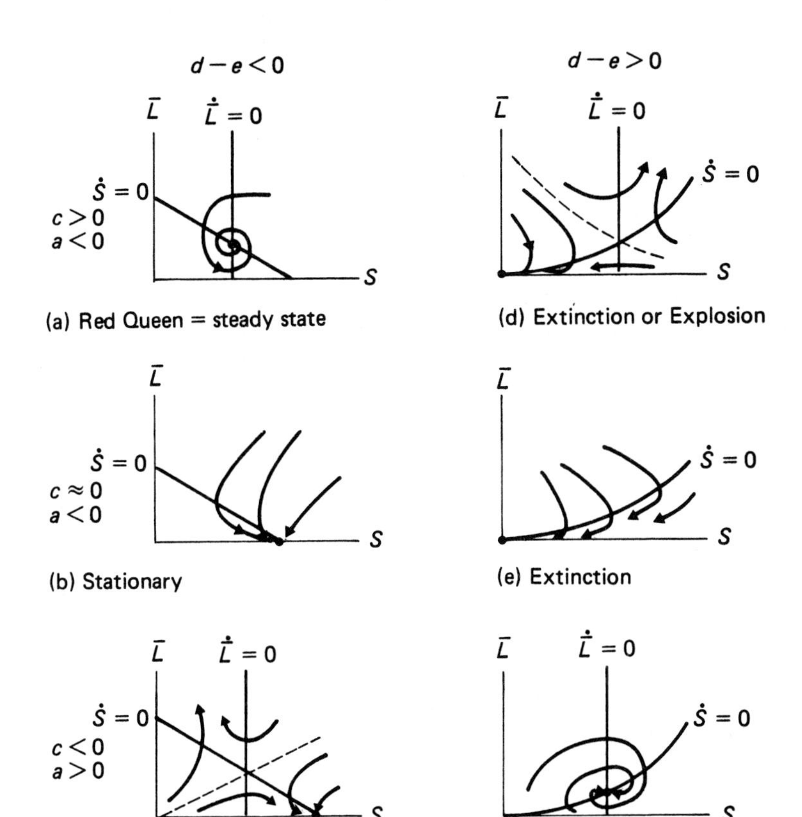

Figure 5.10 The dynamic behavior of model (5.22) for various combinations of parameters. As can be seen, both the gradual Red Queen type of evolution and the stasis-plus-punctuation type may result. The dynamics of the evolutionary system for various values of a, c, and $d - e$ are shown. Notice that the $d\bar{L}/dt$ isocline in (e) does not intersect the dS/dt isocline since the latter approaches an asymptote far to the left of the $d\bar{L}/dt$ isocline.

mechanisms on the basis of observed patterns (see Maynard Smith, 1982c, for a relevant discussion).

Several other conclusions can be drawn from this analysis: first, the Red Queen hypothesis is plausible; as explained above, depending on the parameters of the system the Red Queen mechanism may lead to either a Red Queen type of evolution or to stagnation. Second, the "Law of Constant Extinction" follows from the Red Queen hypothesis. Several other applications are discussed by Stenseth (1985a). Most important in this context is the suggestion that if evolution is of the Red Queen type and continues for ever even in a physically stable environment, increasingly more complex forms (defined in some proper way) tend to evolve.

Two more general conclusions about evolution may be drawn from model (5.22):

(1) I have elsewhere (Stenseth, 1985b) argued on the basis of this model that the tropics may properly be considered the cradle of most of the world's biota; the tropics are, however, a bad museum since the rate of extinction is high because of the higher species diversity.

(2) I have further suggested (Stenseth, 1985d) on the basis of this model that evolutionary novelties are more likely to arise in environmentally unstable regions than in environmentally stable regions. Essentially, this is so because the pattern depicted in Figure 5.10 (b) is more likely to occur in physically unstable regions, whereas the pattern depicted in Figure 5.10 (a) is more likely to occur in physically more stable regions. Some observations suggest the validity of this prediction (Jablonski *et al.*, 1983; Hickey *et al.*, 1983; Zinmeister and Feldman, 1984; see also Stenseth, 1985d).

Finally, it seems plausible that this kind of Red Queen model for coevolution in ecosystems should be able to explain observations that suggest convergent evolution at the community level (e.g., Orians and Paine, 1983). Much work remains to be done in this area, though.

Environs – an alternative approach

Based on the system ecology approach (e.g., Odum, 1969, 1971), Patten and his co-workers have taken another route (e.g., Patten, 1975, 1978, 1981, 1982, 1983, 1985; Patten and Auble, 1980, 1981; Patten and Odum, 1981; Patten *et al.*, 1976); Wilson's (1980) approach is along similar lines.

As in the case of the Red Queen approach (e.g., Maynard Smith, 1976a; Stenseth and Maynard Smith, 1984), evolution is, as far as I understand, assumed to proceed by natural selection of individual organisms due to both direct and indirect effects (generating the selective pressure) of the environment. Also, as in the model of Stenseth and Maynard Smith (1984), the ecological niche (e.g., Grinnell, 1917, 1928; Elton, 1927; Hutchinson, 1957; Pianka, 1978; Roughgarden, 1979) is assumed to represent the point of direct contact between organisms and their environment (both its abiotic and biotic parts). Patten and his co-workers (e.g., Patten, 1981, 1983; Patten and Auble, 1980, 1981) suggest, however, a new theoretical concept, the *environ*, which is the unit of organism–environment coevolution. At least to me, this approach seems exceedingly complex, described in a highly developed system-theory jargon. However, I have a very strong feeling that it might be important in helping us out of some of the difficulties we are faced with in evolutionary theory today. This approach, linked with Van Valen's (1973) Red Queen view of evolution might provide an important advancement and should be attempted by someone better qualified than I.

But to hypothesize that evolution favors linearity at the ecosystem level, or that undesirable effects of nonlinearity at the ecosystem level are selected against, as Patten (1975, 1983) suggests, is – I believe – a wrong track: such a selection seems only to be possible if ecosystems, or the like, are the unit of evolution. And as demonstrated by Maynard Smith (1976b), group selection of this type is likely to be far less efficient than evolution at the individual level – or at some lower level of organization. See Stenseth (1984c) for a general discussion of ecosystem evolution.

What kind of models do we need?

Previously (Stenseth, 1983a, 1985a; see also 1984c) I have discussed the kinds of mathematical models we would need in order to understand the operation of natural selection in a dynamic community which changes with respect to both number and phenotypic properties of the coexisting species. This is, to my mind, the most important technical problem to solve before we can develop improved models for evolution in ecosystems.

Consider a model of the kind

$$(1/x_i)(dx_i/dt) = x_i f_i(x) \tag{5.23}$$

where x_i is the density of the ith species ($i = 1, 2, \ldots, S$, where S is the total number of coexisting species), and f_i is the specific growth rate of the ith species [see equation (5.9)]. For a fixed number of species, S, and nonchanging f_i, model (5.23) is a standard population dynamics model referring to ecological time (*sensu* Stenseth and Maynard Smith, 1984); here x_i are the only variables. As discussed above, the f_i-functions, however, change as a result of microevoluion occurring on what might be called a population genetic time scale (*sensu* Stenseth and Maynard Smith, 1984; see also Lawlor and Maynard Smith, 1976). For a community with a fixed number of species, f_i may reach an ESS or a CSS state (Schaffer and Rosenzweig, 1978; see also Roughgarden, 1979). Several theorists contend that it is sufficient to study few-species assemblages (e.g., Beddington and Lawton, 1978). Models of many-species communities – maybe with more tightly packed subsystems – are, however, important to analyze since we know that diffuse competition does occur fairly frequently (see, e.g., Pianka, 1978, 1980).

We have mathematical techniques available for analyzing both the population dynamic and microevolutionary changes. However, as pointed out in the previous subsection, S is also a dynamic variable changing on a fairly slow time scale as a result of speciation and extinction (see Stenseth and Maynard Smith, 1984). As far as I know, there is no standard analytical technique available for analyzing model (5.23) with S as a dynamic variable (but see the section on environs above)*. The development of such techniques would, I am sure, greatly benefit evolutionary biology.

I am at present unable to recommend what sorts of mathematics should be used for analyzing the problems discussed in this chapter. However, one thing is important to remember whan choosing which mathematics to use in various fields of the natural sciences: living material is, by definition, characterized by units having the properties of ***reproduction***, ***inheritance***, and ***variation*** (see p 112). Nonliving material is everything else. As is apparent, I have in this chapter on living material been using mathematics primarily developed for the nonliving world. That might be a serious shortcoming of all the studies I have reviewed.

I am afraid I have been very vague in my recommendations; I wish I could be more precise and specific.

---

*Several good discussions of the modeling of such evolving ecosystems exist, however (e.g., Beddington and Lawton, 1978; Roughgarden, 1979); see also the interesting and relevant discussion presented by Johnson (1981).

A Synopsis of Darwinism

The basic premises in the Darwinian theory of evolution (i.e., Darwinism) are *reproduction*, *inheritance*, and *variation*. If a population of individuals is characterized by these properties, evolution *may* occur as a result of natural selection; if the environmental situation changes, evolution occurs, in general, as a result of natural selection adapting the organism to the new environmental conditions. In this way the population evolves closer to its adaptive peak. Where this adaptive peak is located (i.e., which phenotype corresponds to the adaptive peak) is determined by the coexisting species (i.e., the biotic properties of the environment) as well as by the abiotic properties of the environment.

These microevolutionary changes brought about by natural selection due to (phenotypic) fitness differences are reflected in genotypic changes (i.e., gene frequency changes) in the population. Notice, however, that many genotypic changes may occur without any corresponding phenotypic change: this we call non-Darwinian evolution (Kimura, 1983).

Referring to Figure 5.11, we may then say that the shape of the fitness curve (and in particular the location of the adaptive peak along the phenotype axis) is determined by the environment (both biotic and abiotic); since we presently have only vague ideas about ecology, we have only vague ideas about how this fitness curve is determined. The location of any individual along the phenotype axis is determined by the *developmental* processes; even though we have very vague ideas at present about developmental biology, we know that for any particular genotype there is, in a particular environment, a particular phenotype that develops. This process is a many-to-one mapping. Which genotype is selected is

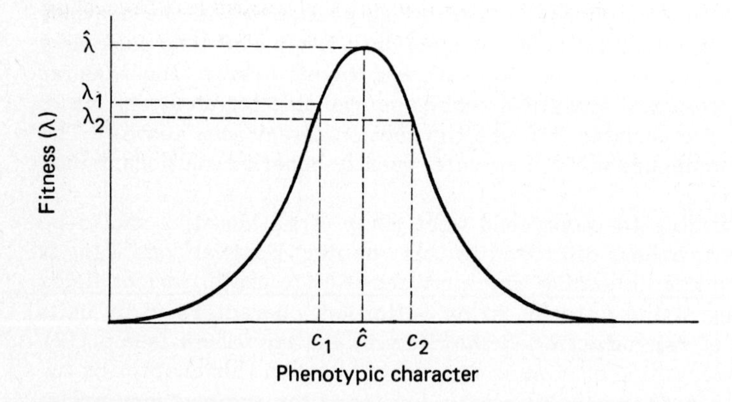

Figure 5.11 A synopsis of Darwinian evolution in ecosystems. The phenotype $\hat{c}$ is the optimal one under current environmental conditions (see text for further discussion). The positions of the fitness curve and $\hat{c}$ are determined by the environment. The positions of the two depicted organisms' phenotypes (c_1 and c_2) are determined by their genotypes and environments. Which one is favored in the process of evolution is determined by the Darwinian process of natural selection (comparing λ_1 and λ_2).

then determined by the fitness (or some analogous property) of this phenotype compared with other phenotypes' fitnesses. According to the theory of natural selection, the one with the highest fitness is selected.

Since we do not understand development (and, in particular, since we do not have a theory for development) we cannot predict the course of evolution. But even without such a theory for development, we can predict that evolution occurs if the fitness curve changes. But without a theory for ecology we are unable to say what is needed to change the fitness curve.

After more than one hundred years of research into the Darwinian theory of natural selection, we have a fairly good understanding of how natural selection operates for a *given* selective pressure. However, this is of little use when trying to predict the long-term patterns of the fossil record, since we do not understand how the fitness curve changes as a result of evolution in coexisting species. That is, we do not understand what generates the selective pressure. For that reason, study ecology! And even more, try to develop a general theory for ecology.

Then, having such a theory for ecology, we might be able to solve several of the current controversies in biology today and assist the solution of practical problems like pest control and resource management (see, e.g., Stenseth, 1983a, 1984c).

Today there is certainly no indication that Darwinism ought to be rejected. The only solid experimental threats to the Darwinian view are the observations of nonchromosomal inheritance (such as, e.g., the phenomena of cortical inheritance in ciliates; e.g., Steel, 1979; see also Brooks, 1983). Darwinists should, as Maynard Smith (1983) emphasized, not be allowed to forget these observations – doing that would be to reject critical evidence against one's view. However, the very reason for not rejecting the Darwinian view in spite of such observations is that the overwhelming majority of inherited differences between organisms are caused by chromosomal genes – hence the Weismannian assumption holds true in the majority of cases.

Contrary to some writers' views (Gould, 1980, 1982; Stanley, 1979; Stanley *et al.*, 1983), the pattern of stasis-plus-punctuation seen in the fossil record [Figure 5.3(a)] certainly constitutes no reason to reject the Darwinian approach (see above): at best, such observations only suggest that we have to refine existing theory (see Maynard Smith, 1982c, for a relevant discussion).

Above all, we have to bring ecology (back) into evolutionary theory – or rather, we have to study the organism–environment as a synergistic whole. Ecology and the Darwinian evolutionary theory should no longer

> ...remain...separate disciplines, travelling separate paths while politely nodding at each other as they pass. (Lewontin, 1979).

I hope that some of the studies reviewed in this chapter may help to bring ecology and evolution into a closer, truly mutual relationship. Much remains to be done – but the seeds for a unified theory of evolutionary biology are there. Most of all, we must weed our flower-bed. Unfortunately, this is rather hard when we do not know for sure which seedlings are weeds and which seedlings are the flowers we want to grow. I only wish I knew how to tell them apart.

Acknowledgements

My work on evolutionary biology has been generously funded by the Norwegian (NAVF) and Swedish (NFR) science foundations, the Nansen Foundation, and the Nordic Council for Ecology. I am grateful to the organizers of the Abisko meeting (Anders Karlqvist, Pehr Sällström, and Uno Svedin) for inviting me to present my thoughts on the subject – hence forcing me to think through the entire field once more. Discussions with Stig Omholt were essential for summarizing my ideas into one figure, Figure 5.11. Tove Valmot is thanked for perfect secretarial work in finishing the manuscript under rather exceptional conditions: without her help, there would have been no chapter.

References

Alberch, P. (1980) Ontogenesis and morphological diversification. *Am. Zool.* 20: 653–67.

Allen, P.M. (1976) Evolution, population dynamics and stability. *Proc. Natl. Acad. Sci. (USA)* 73: 665–8.

Arnold, V.I. (1973) *Ordinary Differential Equations* (Cambridge, MA: MIT Press).

von Baer, K.E. (1828) *Uber Entwicklungsgeschichte der Tiere. Beobeachtung und Reflexion* (Königsberg: Bornträger).

Barlow, G.W. *et al.* (1984) Open peer commentary – 18 separate replies to Maynard Smith's (1984) article. *Behav. Brain Sci.* 7: 101–17.

Beddington, J.R., and Lawton, J.H. (1978) On the structure and behaviour of ecosystems. *Colloque C5*, suppl. no 8; 39: 39–43.

Bonner, J.T. (Ed) (1982) *Evolution and Development* (New York: Springer Verlag).

Bowler, P.J. (1975) The changing meaning of 'evolution'. *J. Hist. Ideas* 36: 95–114.

Brooks, D.R. (1983) What's going on in evolution? A brief guide to some new ideas in evolutionary theory. *Can. J. Zool.* 61: 2637–45.

Bull, J.J. (1983) *Evolution of Sex Determining Mechanisms* (Menlo Park, CA: Benjamin/Cummings Publ. Co.).

Calow, P. (1983) *Evolutionary Principles* (Glasgow: Blackie).

Carroll, L. (1871) *Through the Looking-Glass and what Alice found there* (Reprinted in 'Alice in Wonderland', W.W. Norton Co. Inc. NY, 1971).

Case, T.J. (1982) Coevolution in resource-limited competition communities. *Theor. pop. Biol.* 21: 69–91.

Case, T.J. and Casten, R.G. (1979) Global stability and multiple domains of attraction in ecological systems. *Am. Nat.* 113: 705–14.

Case, T.J. and Sidel, R. (1983) Pattern and change in the structure of model and natural communities. *Evolution* 37: 832–49.

Charlesworth, B. (1980) *Evolution in Age Structured Populations* (Cambridge, UK: Cambridge University Press).

Charlesworth, B., Lande, R., and Slatkin, M. (1982) A neo-Darwinian commentary on macroevolution. *Evolution* 36: 474–98.

Charnov, E.L. (1982) Parent–offspring conflict over reproductive effort. *Am. Nat.* 119: 736–7.

Colbert, E.H. (1969) *Evolution of the Vertebrates*, 2nd edn (New York: Wiley).

Crick, F.H.C. (1958) The biological replication of macromolecules. *Symp. Soc. Exp. Biol.* 12: 138–63.

Crick, F.H.C. (1970) Central Dogma of molecular biology. *Nature* 227: 561–3.

Crick, F.H.C. (1973) Project K: The complete solution of *E. coli*. *Perspectives in Biol. and Medicine* 17: 67–70.

Darwin, C. (1859) *On the Origin of Species by Means of Natural Selection, or the Preservation of Favoured Races in the Struggle for Life* (London: John Murray).

Dawkins, R. (1976) *The Selfish Gene* (Oxford: Oxford University Press).
Dawkins, R. (1982) *The Extended Phenotype* (San Francisco: Freeman).
Dawkins, R. and Krebs, J.R. (1979) Arms races between and within species. *Proc. Roy. Soc. Lond. B* 205: 489–511.
Diamond, J.M. (1969) Avifaunal equilibria and species turnover rates on the Channel Island of California. *Proc. Natl. Acad. Sci. (USA)* 64: 57–63.
Dobzhansky, T., Ayala, F.J., Stebbins, G.L., and Valentine, J.W. (1977) *Evolution* (San Francisco: Freeman).
Eldredge, N. and Gould, S.J. (1972) Punctuated equilibria: an alternative to phyletic gradualism, in T.J.M. Schopf (Ed) *Models in Paleobiology* (San Francisco: Freeman) pp 82–115.
Elton, C.S. (1927) *Animal Ecology* (London: Sidgewick & Jackson).
Elton, C.S. (1958) *The Ecology of Invasions by Animals and Plants* (London: Methuen).
Fisher, R.A. (1930) *The Genetical Theory of Natural Selection* (Oxford: Clarendon Press).
Fredga, K. (1983) Aberrant sex chromosome mechanisms in mammals. Evolutionary aspects. *Differentiation* 23 (Suppl.): S20–S30.
Fredga, K., Gropp, A., Winking, H., and Frank, F. (1976) Fertile XX- and XY-type females in the wood lemming (*Myopus schisticolor*). *Nature* 261: 225–7.
Fredga, K., Gropp, A., Winking, H., and Frank, F. (1977) A hypothesis explaining the exceptional sex ratio in the wood lemming (*Myopus schisticolor*). *Hereditas* 85: 101–4.
Futuyma, D.J. (1979) *Evolutionary Biology* (Sunderland, MA: Sinauer Assoc.).
Goldschmidt, R.B. (1940) *The Material Basis of Evolution* (New Haven, CT: Yale University Press).
Gould, S.J. (1980) Is a new and general theory of evolution emerging? *Paleobiology* 6: 119–30.
Gould, S.J. (1981) Paleontology plus ecology as paleobiology, in R.M. May (Ed) *Theoretical Ecology: Principles and Applications*, 2nd edn (Sunderland, MA: Sinauer Assoc.) pp 295–317.
Gould, S.J. (1982) Darwinism and the expansion of evolutionary theory. *Science* 216: 380–7.
Gould, S.J. (1983) Punctuated equilibria in the fossil record (a reply). *Science* 219: 439–40.
Gould, S.J. and Eldredge, N. (1977) Punctuated equilibria: the tempo and mode of evolution reconsidered. *Paleobiology* 3: 115–51.
Grinnell, J. (1917) The niche relationships of California thrasher. *Ank* 21: 364–82.
Grinnell, J. (1928) The presence and absence of animals. *Univ. Calif. Chronicle* 30: 429–50.
Haeckel, E. (1866) *Generelle Morphologie der Organismen*, 2 vols (Berlin: G. Reimer).
Haldane, J.B.S. (1932) *The Causes of Evolution* (London: Longmans, Green & Co.).
Hallam, A.A. (1976) The Red Queen dethroned. *Nature* 259: 12–3.
Hamilton, W.D. (1964a) The genetical evolution of social behaviour I. *J. theor. Biol.* 7: 1–16.
Hamilton, W.D. (1964b) The genetical evolution of social behaviour II. *J. theor. Biol.* 7: 17–52.
Hamilton, W.D. (1967) Extraordinary sex ratios. *Science* 156: 477–88.
Heuch, I. (1978) Maintenance of butterfly populations with all-female broods under recurrent extinction and recolonization. *J. theor. Biol.* 75: 115–22.
Hickey, L.J., West, R.M., Dawson, M.R., and Choi, D.K. (1983) Arctic terrestrial biota: paleomagnetic evidence of age disparity with mid-northern latitudes during the late coctaceous and early tertiary. *Science* 221: 1153–6.
Hutchinson, G.E. (1957) *A Treatise on Limnology; Vol. 1, Geography, Physics and Chemistry* (New York: Wiley).
Jablonski, D., Sepkoski, J.J., Bottjer, D.J., and Sheehan, P.M. (1983) Onshore–offshore patterns in the evolution of Phanerozoic shelf communities. *Science* 222: 1123–5.

Johannsen, W. (1909) *Elemente der exakten Erblichkeitslehre* (Jena: Gustav Fisher).
Johnson, L. (1981) The thermodynamic origin of ecosystems. *Can. J. Fish. Aquat. Sci.* 38: 571–80.
Kalela, O. and Oksala, T. (1966) Sex ratio in the wood lemming, *Myopus schisticolor* (Lilljeb.) in nature and captivity. *Ann. Univ. Turkuensis, Ser. AII*, 37: 1–24.
Kerner, E.H. (1957) A statistical mechanistics of interacting biological species. *Bull. Math. Biophysics* 19: 121–46.
Kerner, E.H. (1959) Further considerations of the statistical mechanistics of biological associations. *Bull. Math. Biophysics* 21: 217–35.
Kimura, M. (1983) *The Neutral Theory of Molecular Evolution* (Cambridge, UK: Cambridge University Press).
Kostitzin, V.A. (1938) Equations différentielles générales du problème de sélection naturelle. *Comptes rendus* 203: 156–7.
Lamarck, J.B. (1809) *Philosophie Zoologique* (Paris: Librairie Schleicher Frères).
Lander, G.V. (1982) Historical biology and the problem of design. *J. theor. Biol.* 97: 57–67.
Lawlor, L.R. and Maynard Smith, J. (1976) The coevolution and stability of competing species. *Am. Nat.* 110: 79–99.
León, J.A. (1976) Life histories as adaptive strategies. *J. theor. Biol.* 60: 301–35.
Levins, R. (1968) *Evolution in Changing Environments* (Princeton, NJ: Princeton University Press).
Lewontin, R. (1974) *The Genetical Basis of Evolutionary Change* (New York: Columbia University Press).
Lewontin, R. (1979) Fitness, survival and optimality, in D.J. Horn, G.R. Stairs, and R.D. Mitchell (Eds) *Analysis of Ecological Systems*. (Columbia: Ohio State University Press) pp 3–21.
Lundberg, S. and Stenseth, N.C. (1985) Coevolution of competing species: convergent and divergent evolution and crossover in character displacement. *Theor. pop. Biol.* (in press).
MacArthur, R.M. (1972) *Geographical Ecology: Patterns in the Distribution of Species* (New York: Harper & Row).
MacArthur, R.M. and Levins, R. (1967) The limiting similarity, and character displacement in a patchy environment. *Proc. Natl. Acad. Sci.* 51: 1207–10.
MacArthur, R.M. and Wilson, E.O. (1967) *The Theory of Island Biogeography* (Princeton, NJ: Princeton University Press).
Maynard Smith, J. (1960) Continuous, quantized and modal variation. *Proc. Roy. Soc., Lond. B* 152: 397–409.
Maynard Smith, J. (1969) The status of neo-Darwinism, in C.H. Waddington (Ed) *Towards a Theoretical Biology, Vol. 2, Sketches* (Edinburgh: Edinburgh University Press) pp 82–9.
Maynard Smith, J. (1972) *On Evolution* (Edinburgh: Edinburgh University Press).
Maynard Smith, J. (1974) *Models in Ecology* (Cambridge, UK: Cambridge University Press).
Maynard Smith, J. (1975) *The Theory of Evolution*, 3rd edn (Harmondsworth: Penguin Books).
Maynard Smith, J. (1976a) A comment on the Red Queen. *Am. Nat.* 110: 331–8.
Maynard Smith, J. (1976b) Group selection. *Quart. Rev. Biol.* 51: 227–83.
Maynard Smith, J. (1978a) Optimization theory in evolution. *Ann. Rev. Ecol. Syst*, 9: 31–56.
Maynard Smith, J. (1978b) *The Evolution of Sex* (Cambridge, UK: Cambridge University Press).
Maynard Smith, J. (1982a) Introduction, in J. Maynard Smith (Ed) *Evolution Now* (London: MacMillan) pp 1–6.
Maynard Smith, J. (1982b) *Evolution and the Theory of Games* (Cambridge, UK: Cambridge University Press.).
Maynard Smith, J. (1982c) Evolution – sudden or gradual? (Introduction to ch. 5) in J. Maynard Smith (Ed) *Evolution Now* (London: Macmillan) pp 125–8.

Maynard Smith, J. (1983) Evolution and development, in B.C. Goodwin, N. Holder, and C.C. Wylie (Eds) *Development and Evolution* (Cambridge, UK: Cambridge University Press) pp 33–46.

Maynard Smith, J. (1984) Game theory and the evolution of behaviour. *Behav. Brain Sci.* 7: 94–101 (see reply to "Open Peer Commentary" pp. 117–125).

Maynard Smith, J. and Price, G.R. (1973) The logic of animal conflict. *Nature* 246: 15–8.

Maynard Smith, J. and Sondhi, K.C. (1960) The genetics of pattern. *Genetics* 45: 1039–50.

Maynard Smith, J. and Sondhi, K.C. (1961) The arrangement of bristles in *Drosophila*. *J. Embryol. exp. Morph.* 9: 661–72.

Maynard Smith, J. and Stenseth, N.C. (1978) On the evolutionary stability of the female-biased sex ratio in the wood lemming (*Myopus schisticolor*): the effect of inbreeding. *Heredity* 41: 205–14.

Mayr, E. (1961) Cause and effect in biology. *Science* 134: 1501–6.

Monod, J. (1971) *Chance and Necessity: an Essay on the Natural Philosophy of Modern Biology* (New York: Knopfer).

Odum, E.P. (1969) The strategy of ecosystem development. *Science* 164: 262–70.

Odum, E.P. (1971) *Fundamentals of Ecology*, 3rd edn (Philadelphia: Saunders).

Orians, G.H. and Paine, R.T. (1983) Convergent evolution at the community level, in D.J. Futuyma and M. Slatkin (Eds) *Coevolution* (Sunderland, MA: Sinauer Assoc.) pp 431–59.

Oster, G. and Alberch, P. (1982) Evolution and bifurcation of developmental programs. *Evolution* 36: 444–59.

Oster, G., Odell, G., and Alberch, P. (1980) Mechanics, morphogenesis and evolution. *Lect. Math. Life Sci.* 13: 165–255.

Patten, B.C. (1975) Ecosystem linearization: an evolutionary design problem. *Am. Nat.* 109: 529–39.

Patten, B.C. (1978) Systems approach to the concept of environment. *Ohio J. Sci.* 78: 206–22.

Patten, B.C. (1981) Environs: the superniches of ecosystems. *Am. Zool.* 21: 845–52.

Patten, B.C. (1982) Environs: relativistic elementary particles for ecology. *Am. Nat.* 119: 179–219.

Patten, B.C. (1983) Linearity enigmas in ecology. *Ecol. Modelling* 18: 155–70.

Patten, B.C. (1985) Further developments toward a theory of the quantitative importance of indirect effects in ecosystems, in B.C. Patten (Ed) *System Analysis and Simulation in Ecology*, vol. 00 (in press).

Patten, B.C. and Auble, G.T.(1980) Systems approach to the concept of niche. *Synthese* 43: 155–81.

Patten, B.C. and Auble, G.T. (1981) Systems theory of the ecological niche. *Am. Nat.* 118: 345–69.

Patten, B.C. and Odum, E.P. (1981) The cybernetic nature of ecosystems. *Am. Nat.* 118: 886–95.

Patten, B.C., Bosserman, R.W., Finn, J.T., and Cale, W.G. (1976) Propagation of cause in ecosystems, in B.C. Patten (Ed) *System Analysis and Simulation in Ecology, Vol. 4* (New York: Academic Press) pp 457–579.

Patterson, C. (1983) How does phylogeny differ from ontogeny? in B.C. Goodwin, N. Holder, and C.C. Wylie (Eds) *Development and Evolution* (Cambridge, UK: Cambridge University Press) pp 1–32.

Pianka, E.R. (1978) *Evolutionary Ecology* (New York: Harper & Row).

Pianka, E.R. (1980) Guild structure in desert lizards. *Oikos* 35: 194–201.

Reed, J. and Stenseth, N.C. (1984) On evolutionary stable strategies. *J. theor. Biol.* 108: 491–508.

Roughgarden, J. (1974) Species packing and the competition function with illustrations from coral reef fish. *Theor. pop. Biol.* 5: 163–86.

Roughgarden, J. (1979) *Theory of Population Genetics and Evolutionary Ecology: An Introduction* (New York: Macmillan).

Roughgarden, J. (1983a) The theory of coevolution, in D.J. Futuyma and M. Slatkin (Eds) *Coevolution* (Sunderland, MA: Sinauer Assoc.) pp 33–64.

Roughgarden, J. (1983b) Coevolution between competitors, in D.J. Futuyma and M. Slatkin (Eds) *Coevolution* (Sunderland, MA: Sinauer Assoc.) pp 383–403.

Roughgarden, J., Herckel, D., and Fuentes, E.R. (1983) Coevolutionary theory and the biogeography and community structure of *Anolis*, in R. Huey, E.R. Pianka, and T. Schoener (Eds) *Lizard Ecology: Studies of a Model Organism* (Cambridge, MA: Harvard University Press) pp 371–410.

Schaffer, W.M. (1977) Evolution, population dynamics, and stability. *Theor. pop. Biol.* 11: 326–9.

Schaffer, W.M. and Rosenzweig, M.L. (1978) Homage to the Red Queen. I. Coevolution of predators and their victims. *Theor. pop. Biol.* 14: 135–57.

Schuster, P. and Sigmund, K. (1983) Replicator dynamics. *J. theor. Biol.* 100: 533–8.

Schopf, T.J.M. (1982) A critical assessment of punctuated equilibria. I. Duration of taxa. *Evolution* 36: 1144–57.

Schopf, T.J.M. and Hoffman, A. (1983) Punctuated equilibria in the fossil record. *Science* 219: 438–9.

Sharpe, F.R. and Lotka, A.J. (1911) A problem in age distribution. *Phil. Mag.* 21: 435–8.

Sigmund, K. (1985) A survey of replicator equations (this volume, Chapter 4).

Simpson, G.G. (1953) *The Major Features of Evolution* (Columbia: Columbia University Press).

Slatkin, M. and Maynard Smith, J. (1979) Models of coevolution. *Quart. Rev. Biol.* 54: 233–63.

Sondhi, K.C. (1962) The evolution of pattern. *Evolution.* 16: 186–91.

Stanley, S.M. (1975) A theory of evolution above the species level. *Proc. Natl. Acad. Sci (USA)* 72: 646–50.

Stanley, S.M. (1979) *Macroevolution: Pattern and Process* (San Francisco: Freeman).

Stanley, S.M. (1982) Macroevolution and the fossil record. *Evolution* 36: 460–73.

Stanley, S.M., van Valkenburgh, B., and Steneck, R.S. (1983) Coevolution and the fossil record, in D.J. Futuyma and M. Slatkin (Eds) *Coevolution* (Sunderland, MA: Sinauer Assoc.) pp 328–49.

Stebbins, G.L. and Ayala, F.J. (1981) Is a new evolutionary synthesis necessary? *Science* 213: 967–71.

Steel, E.J. (1979) *Somatic Selection and Adaptive Evolution* (Toronto: Williams-Wallace Prod. Intern).

Stenseth, N.C. (1979) Where have all the species gone? On the nature of extinction and the Red Queen Hypothesis. *Oikos* 33: 196–227.

Stenseth, N.C. (1983a) A coevolutionary theory for communities and food web configurations. *Oikos* 41: 487–95.

Stenseth, N.C. (1983b) Causes and consequences of dispersal in small mammals, in I.R. Swingland and P.J. Greenwood (Eds) *The Ecology of Animal Movement* (Oxford: Oxford University Press) pp 63–101.

Stenseth, N.C. (1984a) Testing evolutionary predictions: a reply to Bleken and Ugland. *Oikos* 43: 126–8.

Stenseth, N.C. (1984b) Fitness, population growth rate and evolution in plant-grazer systems: a reply to Nur. *Oikos* 42: 414–5.

Stenseth, N.C. (1984c) Why mathematical models in evolutionary ecology?, in J.H. Cooley and F.B. Golley (Eds) *Trends in Ecological Research for the 1980s* (New York: Plenum Press) pp 239–87.

Stenseth, N.C. (1985a) Darwinian evolution in ecosystems – the Red Queen view, in P.J. Greenwood, M. Slatkin, and P. Harvey (Eds) *Evolution* (Cambridge, UK: Cambridge University Press) (in press).

Stenseth, N.C. (1985b) The tropics: cradle or museum? *Oikos* (in press).

Stenseth, N.C. (1985c) A new hypothesis for explaining the maintenance of the all-female broods in the African butterfly *Acraea encedon* (in review).

Stenseth, N.C. (1985d) Origin of species in physically stressed environments (in review).

Stenseth, N.C. and Kirkendall, L.R. (1985) On the ecological and evolutionary stability of pseudogamy; the effect of niche-overlap between sexual and asexual females (in review).

Stenseth, N.C. and Maynard Smith, J. (1984) Coevolution in ecosystems: Red Queen evolution or stasis? *Evolution* 38: 870–80.

Stenseth, N.C., Kirkendall, L.R., and Moran, N. (1985) On the evolution of pseudogamy. *Evolution* (in press).

Thom, R. (1975) *Structural Stability and Morphogenesis: An Outline of a General Theory of Models* (Reading, MA: W.A. Benjamin).

Thorsrud, Ø. and Stenseth, N.C. (1985) On the evolutionary stability of the female biased sex ratio in the wood lemming (*Myopus schisticolor*) and the collared lemming (*Dicrostonyx groenlandicus*): the effect of dispersal in a patchy environment (manuscript for publication).

Turing, A.M. (1952) The chemical basis for morphogenesis. *Phil. Trans. Roy. Soc. Ser. B* 237: 37–72.

Tyler, M.J. (1983) Evolution of gastric brooding, Ch 10 in M.J. Tyler (Ed) *The Gastric Brooding Frog* (London: Croom Helm).

Van Valen, L. (1973) A new evolutionary law. *Evol. Theory* 1: 1–30.

Van Valen, L. (1976) The Red Queen lives. *Nature* 260: 575.

Van Valen, L. (1977) The Red Queen. *Am. Nat.* 111: 809–10.

Weismann, A. (1886) *Essays upon Heredity and Kindred Biological Problems* (Oxford: Clarendon Press).

Weismann, A. (1893) *The Germ-Plasma: A Theory of Heredity* (Oxford: Clarendon Press).

Whittaker, R.H. (1969) Evolution of diversity in plant communities. *Brookhaven Symp. Biol.* 22: 178–96.

Williamson, P.G. (1981a) Paleontological documentation of speciation in Cenozoic molluscs from Turkana Basin. *Nature* 293: 437–43.

Williamson, P.G. (1981b) Morphological stasis and developmental constraints: real problems for neo-Darwinism. *Nature* 294: 214–5.

Wilson, D.S. (1980) *The Natural Selection of Populations and Communities* (Menlo Park, CA.: Benjamin/Cummings).

Wolpert, L. (1969) Positional information and the spatial pattern of cellular differentiation. *J. theor. Biol.* 25: 1–47.

Wolpert, L. (1983) Constancy and change in the development and evolution of pattern, in B.C. Goodwin, N. Holder, and C.C. Wylie (Eds) *Development and Evolution* (Cambridge, UK: Cambridge University Press) pp 47–58.

Zinmeister, W.J. and Feldman, R.M. (1984) Cenozoic high latitude heterogeneity of southern hemisphere marine fauna. *Science* 224: 281–3.

CHAPTER 6

On System Complexity: Identification, Measurement, and Management

John L. Casti

Complexity and Simplicity[†]

> I have yet to see any problem, however complicated, which, when you looked at it the right way, did not become still more complicated.
>
> *Poul Anderson*

The notion of system complexity is much like St. Augustine's description of time: "What then is time [complexity]? If no one asks me, I know; if I wish to explain it to one that asks, I know not." There seem to be fairly well-developed, intuitive ideas about what constitutes a complex system, but attempts to axiomatize and formalize this sense of the complex all leave a vague, uneasy feeling of basic incompleteness, and a sense of failure to grasp important aspects of the essential nature of the problem. In this chapter we examine some of the root causes of these failures and outline a framework for the consideration of complexity that provides a starting point for the development of operational procedures in the identification, characterization, and management of complex processes. In the process of developing this framework for speculation, it is necessary to consider a variety of system–theoretic concepts closely allied to the notion of complexity: hierarchies, adaptation, bifurcation, self-organization, and reductionism, to name but a few. The picture that emerges is that of complexity as a *latent* or *implicate* property of a system, a property made explicit only through the interaction of the given system with another. Just as in baseball where some pitches are balls and some are strikes, but "they ain't nothin'" until the umpire calls them, complexity cannot be thought of as an intrinsic property of an isolated (closed) system; it is only made manifest by the *interaction* of the system with another, usually in the process of measurement and/or control. In this sense, it is

[†]Notes and references relevant to each section are given at the end of the chapter.

probably more meaningful to consider complexity more as a property of the interaction than of the system, although it is clearly associated with both. The exploration and exploitation of this observation provides the starting point for an emergent *theory* of complex processes.

Before embarking upon a detailed consideration of complexity in natural and human phenomena, it is useful to consider for a moment why a deeper understanding of complexity, *per se*, is of either theoretical or practical importance. The basic reason is the seemingly inherent human need to simplify in order to understand and direct (control). Since most understanding and virtually all control is based upon a *model* (mental, mathematical, physical, or otherwise) of the system under study, the simplification imperative translates into a desire to obtain an equivalent, but reduced, representation of the original model of the system. This may involve omitting some of the original variables, aggregating others, ignoring weak couplings, regarding slowly changing variables as constants, and a variety of other subterfuges. All of these simplification techniques are aimed at reducing the degrees of freedom that the system has at its disposal to interact with its environment. A theory of system complexity would give us knowledge as to the limitations of the reduction process. For example, it is well known that the three-body problem of celestial mechanics cannot be resolved in analytic terms; however, the two-body problem is completely solvable, but a sequence of two-body problems cannot be combined to solve the three-body problem. Thus, the complexity of the three-body problem is intrinsically greater than any sequence of two-body problems and there is an irretrievable loss of information in passing to such a reduced representation. A useful theory of system complexity would provide conditions under which such a decomposition would work and perhaps even suggest novel, nonphysical, simpler representations that would be valid when the "natural" simplifications fail.

What are the distinguishing structural and behavioral characteristics of those systems we intuitively think of as being complex? Perhaps the easiest way to approach this question is to consider its converse: what features do we associate with *simple* systems? Some of the most evident properties of simple systems are:

- *Predictable behavior*. There are no surprises: simple systems exhibit a behavior pattern that is easy to deduce from knowledge of the external inputs (decisions) acting upon the system. If we drop a stone, it falls; if we stretch a spring and let it go, it oscillates in a fixed pattern; if we put money into a fixed-interest bank account it grows to a sum according to an easily understood and computable rule. Such predictable and intuitively well-understood behavior is characteristic of simple systems.

  Complex processes, on the other hand, display counter-intuitive, seemingly acausal behavior full of unpredictable surprises. Taxes are lowered and unemployment and stagflation persist; low-cost housing projects generate slums worse than those the housing replaced; construction of freeways results in unprecedented traffic jams and increased commuting times. For many people, such unpredictable and seemingly capricious behavior *defines* a complex system.

- *Few interactions and feedback/feedforward loops*. Simple systems generally involve a small number of components, with self-interaction dominating the mutual interaction of the variables. For instance, primitive barter economies involving only a small number of goods (food, tools, weapons, clothing) are generally much simpler and easier to understand than the developed economies of industrialized nations, in which the pathway between raw material inputs and finished consumer goods follows a byzantine route involving large numbers of interactions between various intermediate products, labor, and capital inputs.

  Besides involving only a few variables, simple systems generally have very few feedback/feedforward loops. Such loops enable the system to restructure, or at least modify, the interaction pattern of its variables, thereby opening-up the possibility of a wider range of potential behavior patterns. As an illustration, imagine a large organization characterized by the variables: employment stability, substitution of work by capital, and level of individuality (personal level). Increased substitution of work by capital decreases the human level in the organization, which in turn may decrease employment stability. Such a feedback loop exacerbates any initial internal stresses, potentially leading to a collapse of the process. This type of collapsing loop is especially dangerous for social resilience and is a common feature of complex social phenomena.
- *Centralized decision-making*. Power in simple systems is generally concentrated in one or, at most, a few decision-makers. Political dictatorships, privately owned corporations, and the Roman Catholic Church are good examples of such systems. These systems are simple because there is very little interaction, if any at all, between the lines of command. In addition, the effect of the central authority's decision upon the system is usually rather easy to trace.

  By contrast, complex systems display a diffusion of *real* authority. There is generally a nominal, supreme decision-maker, where the buck stops, but in actuality the power is spread over a decentralized structure, with the actions of a number of units combining to generate the system behavior. Typical examples include democratic governments, labor unions, and universities. Systems exhibiting distributed decision-making tend to be somewhat more resilient and more stable than centralized structures, as they are more forgiving of mistakes by any one decision-maker and are more able to absorb unexpected environmental fluctuations.
- *Decomposable*. Typically, a simple system involves weak interactions among its constituent components. Consequently, if we sever some of these interactions the system behaves more-or-less as before. Relocating American Indians to reservations produced no major effects on the dominant social structure in Arizona, for example, since, for cultural reasons, the Indians were only weakly coupled to the local social fabric. Thus, the simple social interaction pattern could be further decomposed and studied as two independent processes, the Indians and the settlers. A similar situation occurs for the restricted three-body problem, involving the Sun, Earth, and Moon. For some purposes, this system can be decomposed by neglecting the Moon and so studied as a simpler two-body problem.

On the other hand, a complex process is irreducible. Neglecting any part of it or severing any connection usually irretrievably destroys essential aspects of the system's behavior or structure. We have already mentioned the unrestricted three-body problem in this regard. Other examples include the tripartite division of the US government into executive, judicial, and legislative subsystems, an RLC electrical circuit, and a Renoir painting.

The picture that emerges from the foregoing considerations of simple systems is a notion of complex phenomena characterized by counter-intuitive behavioral modes that are unpredictable from knowledge of environmental inputs; by relatively large numbers of variables interacting through a rich network of feedback/feedforward connections; by decentralized decision-making structures and a high level of functional indecomposability. Since such features are characteristic of many of the human systems of modern life, it is necessary to develop effective procedures for managing and planning the future course of such processes. Let us briefly consider some of the issues involved in obtaining a handle on complex systems.

Management of the Complex

> Some problems are just too complicated for rational, logical solutions. They admit of insights, not answers.
>
> *J. Wiesner*

We have already noted that system complexity is a contingent property arising out of the interaction I between a system S and an observer/decision-maker O. Thus, any perception and measure of complexity is necessarily a function of S, O, and I. Conditioned by the physical sciences, we typically regard S as the active system, with O being a passive observer or disengaged controller. Such a picture misses the crucial point that generally the system S can also be regarded as an observer of O and that the interaction I is a two-way path. In other words, for a given mode of interaction I, the system S displays a certain level of complexity relative to O, while at the same time O has a level of complexity relative to S. For the sake of definitiveness, let us denote the former as ***design complexity*** and the latter as *control complexity*. It is our contention that the behavior of S becomes uncontrollable when these two complexity levels are too far apart; hence the "golden rule" for management of complex systems is to arrange matters so that

design complexity = control complexity.

The distinction between design and control complexity has been blurred in the natural sciences because of the almost universal adoption of the tacit assumption that the interaction I is one-way, from O to S. When S is a system of macroparticles as in, say, the observation of an oscillating pendulum in mechanics, it is defensible to argue that the pendulum cannot "see" O or, at least, the pendulum has no awareness of O as a system with which it is in interaction. Hence, there is no notion of control complexity and the regulation and management of S by O proceeds according to classical principles. But when we pass to the microscopic and quantum levels or to the global and cosmic levels, the assumption of no control complexity becomes increasingly difficult to defend. And by the time we move to systems possessing even primitive levels of self-awareness in biology and the social sciences, we can no longer neglect the inherent symmetry in the interaction I. The first step in addressing management issues for complex systems is the explicit incorporation of control complexity into the modeling and decision-making framework.

To illustrate the above points, consider the structure associated with representative government at the regional or national level. Here we have a system S composed of the political leaders (mayor, governor, etc.) interacting with a system O consisting of the general public. If the complexity of S as perceived by O is high, then the public sees its leaders as taking incomprehensible actions; they see a byzantine and unwieldy governmental bureaucracy and a large number of independent decision-makers (government agencies) affecting their day-to-day life. In short, what would be observed is exactly what is seen in most countries today. On the other hand, if the political leadership were to perceive the public as being very complex, what would their observations be? They would see a seemingly fickle, capricious public, composed of a large number of independent self-interest groups clamoring for more and more public goods and services. Furthermore, there would be a perception that the public interest groups were connected together in a rather elaborate network that could not be decomposed into simpler subgroups. Consequently, actions or decisions taken to address the interests of one group could not be isolated in their effect, which may possibly be contrary to the interests of another. Or, even worse, because of the dense web of interconnections and feedback loops comprising the public structure, unpredictable and unpleasant side effects may emerge from actions taken to satisfy some subgroups. It goes without saying that these observations form part of the everyday life of most public officials in the western world (and, most likely, the eastern, too).

From the above considerations, we can conclude that the crux of the problem of modern government *versus* its citizenry is that both the public and the governing officials regard each other as complex systems. If either recognized the other as simple, much of the tension and dissatisfaction with contemporary political structures would disappear. The ideal situation would be for each to perceive the other as simple, in which case both parties would be happy. Failing this, simple government with a complex public or complex government with a simple public would at least reduce the difficulties and tensions in one direction, but with possibly increased tensions in the other. Local administration in a small, rural community would be representative of the former, while a political dictatorship of some sort would be typical of the latter situation. Unfortunately, at the regional and national level throughout most of the western world, we have the complex/complex case, which requires a deeper consideration of how each side comes to attach the label "complex" to the other, before the question of complexity management can be meaningfully addressed.

As emphasized earlier, complexity as a system property emerges from the interaction of a given system with another. If a system S can interact with O in a large number of *nonequivalent* ways, then S regards O as complex; conversely, if S has only a small number of modes of interaction with O, then O appears simple. In the governmental context, a dictatorship appears more complex to the public, because the public has many different modes of interaction with the government since, in such situations, most of the agencies of day-to-day life (police, military, communications, transport, agriculture, etc.) are directly in governmental hands. Such centrally planned structures require a high level of control complexity to maintain and are perceived as complex by other systems which have to interact with them.

A system is counted as simple if there are only a small number of non-equivalent ways to interact with it. The pen I used to write this manuscript is a simple system to me. The only mode of interaction with it that I have available is to use it as a writing instrument; however, if I were, say, a chemical engineer, then many more modes become available. I could analyze the plastic compound of which it is made, the composition of chemicals forming the ink, the design of the writing ball at its tip, and so forth. So, for a chemical engineer my ballpoint pen becomes a far more complex object than it is for me.

If we adopt the position of this chapter that effective management of complexity consists of arranging systems so that design and control complexity are approximately equal, preferably at a relatively high or low absolute level, then we operationally face the question of how to formally characterize the idea of a system, an interaction between two systems, and the notion of equivalent interactions.

Systems, Observables, and Models

For the things of this world cannot be made known
without a knowledge of mathematics.
Roger Bacon

To progress beyond the obvious and trivial, it is necessary to formalize the common language and linguistic terms used earlier to describe system complexity and its management. Only through such a formalization can we transfer these intuitive, but fuzzy, terms into a mathematical setting that provides the possibility of gaining operational insight into the way complexity is generated and suggests how procedures can be developed to cope with the complex.

For us, a ***system*** S is composed of an abstract set of states Ω, together with a collection of real-valued ***observables*** $f_i : \Omega \to R$. For example, let the system S consist of the rotational symmetries of an equilateral triangle. There are then several candidates for the abstract state space Ω, as shown in Figure 6.1. Thus, there is nothing sacred about the state space Ω; it is just a collection of elements that ***name***, or ***label***, the possible positions of the triangle. A typical observable for this system would be the map f, which assigns to the state $\omega \in \Omega$ the minimal number of rotations through $2\pi/3$ needed to reach ω from the state [a, b, c]. Thus, $f : \Omega \to \{0,1,2\} \subset R$. In this case, if we take $\Omega = \Omega_3$, then $f(\omega) = \omega$, but if we use $\Omega = \Omega_1$ or Ω_2, then $f(\omega) \in \Omega_3$. Consequently, for the observable f it is possible to code any of the states in Ω_2 or Ω_3 by an element of Ω_3; in a certain sense, Ω_3 is a ***universal*** state space for this system, relative to the observable f.

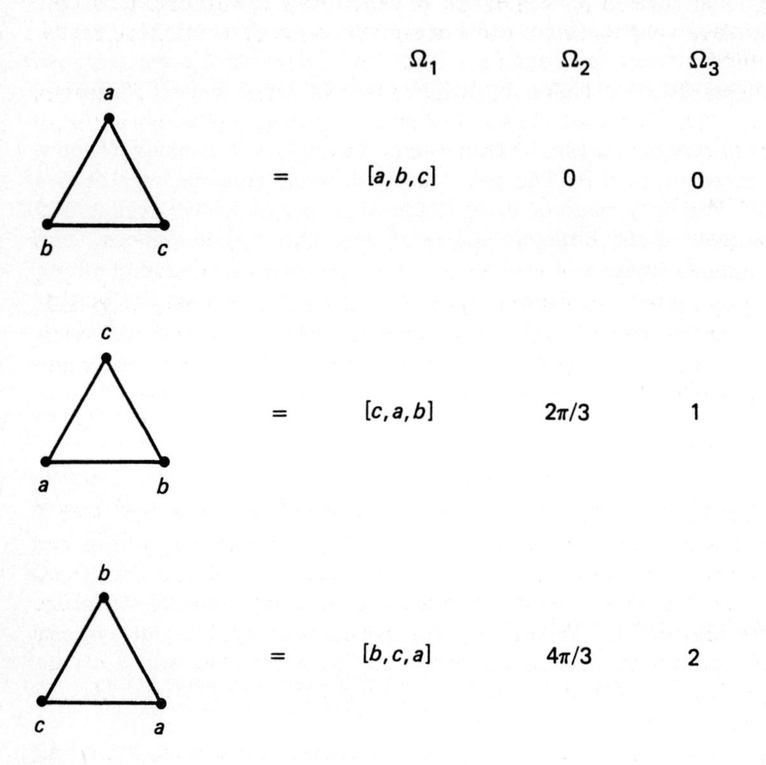

Figure 6.1

In physics and engineering, it has become common practice to use $\Omega = R^n$ as a universal state space for a system involving n observables, $\{f_i\}_{i=1}^n$. In fact, a good deal of the art behind mathematical modeling in the physical sciences lies in a judicious *choice* of observables $\{f_i\}$, so that the points of R^n serve as a universal coding scheme for the actual abstract states of S. It is both remarkable and unfortunate that this procedure works as well as it does: remarkable since there is no *a priori* reason to expect that the natural world is constructed so as to uniformly lend itself to such an encoding scheme; unfortunate, since the successes in physics and engineering have generated a certain sense of unjustified confidence that a similar procedure will work equally well in the social and behavioral sciences. It does not, which accounts for a great deal of the difficulties found in many attempts to mimic the methods of physics when modeling human affairs. All that having been said, let us return to the formalization of system descriptions and complexity.

From the (possibly infinite) set of all observables characterizing S, we select a subset (usually finite), $F = \{f_1, f_2, \ldots, f_N\}$, and call F an ***abstraction*** of S. Associated with the abstraction F is a relation, or a set of relations, Φ, between the observables f_i of F,

$$0 = \Phi(f_1, f_2, \ldots, f_N) \; .$$

Such a relationship Φ is termed an ***equation of state*** or a ***description*** for the system S. Since the observables are all real-valued functions of Ω, if there are m relations, $\Phi : R^n \to R^m$.

As a simple illustration of the preceding ideas, let the system S be the citizenry of a country. The abstract states Ω of such a system might characterize the political mood of the populace. For this, we could take

$$\Omega = \{\omega_1, \omega_2, \omega_3, \omega_4, \omega_5\} ,$$

where ω_1 = very content, ω_2 = weakly content, ω_3 = divided, ω_4 = some dissatisfaction, ω_5 = great unrest. Two (of many) observables for this system could be f_1, the fraction of the population favorably disposed to the political party in power, and f_2, the fraction neutral or opposed to the current regime. The actual numerical values of f_1 and f_2 when the system is in any state, $\omega \in \Omega$, need to be determined on empirical grounds. However, we always have the equation of state

$$\Phi(f_1, f_2) = f_1 + f_2 - 1 = 0 ,$$

for any $\omega \in \Omega$.

In the above situation, there is no notion of causality. The observables of F and the equation of state Φ are simply quantities that represent our view of the system S; they compactly summarize our experimental and observational knowledge of S; that is, the data. The common manner in which a causal structure is imposed upon the observables is through the recognition that in all systems there are noticeably different time-scales according to which the values of the observables change. We can employ (tacitly or directly) these time-scales to induce a notion of order, or a causal structure, upon F.

To see how a causal structure can be introduced, imagine a system S characterized by an abstraction $F = \{f_1, \ldots, f_N\}$ involving N observables. Further, assume that observation has shown that the observables change on three time-scales, slow, medium, and fast, for example. For the sake of exposition, let the observables be labeled so that

$$a = \{f_1, \ldots, f_k\} = \text{slow} ,$$

$$u = \{f_{k+1}, \ldots, f_q\} = \text{medium} ,$$

$$y = \{f_{q+1}, \ldots, f_N\} = \text{fast} .$$

Let A, U, and Y represent the range of values of the observables a, u, and y, respectively. By the preceding argument, we have $A \subset R^k$, $U \subset R^n$, and $Y \subset R^m$, where $n = q - k$ and $m = n - q$. The causal relationship is induced by invoking the principle that slow dynamics force, or cause, fast dynamics. Thus, we regard a and u as causing y. In common parlance, the slow variables a are generally termed ***parameters***, while the medium-speed, causal variables u are termed ***inputs*** (***controls, decisions***). The response variables y are the system ***outputs***.

Usually, there is a feedback effect in that u, and sometimes a, is modified by the output y. But the important point here is that when we think of some observables causing others, it is the rate-of-change of the observables that produces the temporal ordering which we assign to the system. Thus, causality is not necessarily a natural or ***intrinsic*** aspect of S, but rather is introduced by the way the observer perceives the various time-scales at work in the system. In the classical physical sciences, this point is not usually particularly important and becomes significant only at cosmic and quantum levels; however, in the social and behavioral sciences it is an issue at the very outset, and partially accounts for the difficulties in economic and social modeling of deciding what causes what, a question which lies at the heart of any sort of predictive modeling.

A better intuitive understanding of the partitioning of the system observables is obtained if we employ an evolutionary metaphor. The slow variables a can be thought of as specifying the system genotype; that is, the aspects of S that enable us to recognize the system as S and not some other system S'. For instance, in an urban environment, a might code information about the local geographic, cultural, political, and economic structure that allows us to know we are in Omsk rather than Tomsk. The medium-speed observables u correspond to the system's *environment*. Thus, u represents either natural environmental factors or those created by decision-makers. Finally, the outputs y characterize the morphostructure, or form, of S, the so-called system *phenotype*. For many social systems, y represents the behavioral responses of S to genetic mutation (change of a) and/or environmental fluctuation (change of u). In the urban context, u may reflect various actions by policymakers, such as imposition of zoning restrictions, urban renewal legislation, and the like, while y would then display the effects of those environmental decisions, together with the given genotype (city), as new housing developments, modifications of transport channels, redistribution of industry, and so forth. The important point is the relative time-scales of the processes.

Now let us turn to the central question of this section: how to decide whether two descriptions, or models, of the same system are equivalent. In the above terminology, we have the description

$$\Phi_a : U \to Y \ ,$$

and the description

$$\hat{\Phi}_{\hat{a}} : U \to Y \ ,$$

both purporting to describe the same system S, and our question is whether the two descriptions convey the same information about S or, what amounts to the same thing, do Φ and $\hat{\Phi}$ provide independent descriptions of S?

Mathematically, the descriptions Φ_a and $\hat{\Phi}_{\hat{a}}$ are ***equivalent*** if there exist maps g and h, depending on a and $\hat{a}$, such that the following diagram commutes:

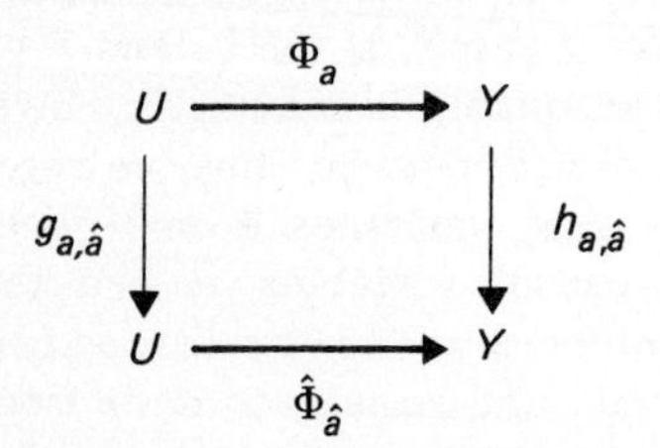

The existence, properties, and construction of the maps $g_{a,\hat{a}}$ and $h_{a,\hat{a}}$ depend strongly upon the mathematical structure assumed for the sets U and Y and the descriptions Φ_a and $\hat{\Phi}_{\hat{a}}$. We do not discuss these matters here. A purely mathematical treatment of the above question forms the core of singularity theory, which is covered in detail by Golubitsky and Guillemin (1973), Lu (1976), and Gibson (1979). The systems view of singularity theory as outlined above is treated in Casti (1984).

It is worthwhile to pursue, for a moment, the implications of system equivalence. If Φ_a and $\hat{\Phi}_{\hat{a}}$ are equivalent, it means that a change of the parameter a to $\hat{a}$ can be neutralized, or cancelled out, by a corresponding relabeling of the elements of the sets U and Y. Speaking metaphorically, if we regard S as an organism described by Φ_a, then the genetic mutation $a \to \hat{a}$ can be made invisible by an appropriate modification of the environment U and the phenotype Y. When put in such terms, the notion of system equivalence is strongly reminiscent of the theory of biological transformations originally developed by d'Arcy Thompson in the early 1900s. In that theory, an attempt was made to show that a common genetic structure in the past could be inferred from phenotypic equivalence in the present. In other words, two species $(y, \hat{y})$ with different genotypes $(a \neq \hat{a})$ in the *present*, would be considered to have arisen from a common ancestor $(a = \hat{a})$ in the *past*, if there is a phenotypic transformation h which transforms one species into the other. This is clearly a special case of our diagram when the environment U is held fixed (g = identity).

For given genotypes a and $\hat{a}$, it may be that there exist no transformations g and h which enable us to pass from Φ_a to $\hat{\Phi}_{\hat{a}}$. In this case, there exist mutations $\hat{a}$ near a that result in qualitatively different phenotypic structures. Such a situation forms the underlying basis for a theory of ***bifurcation*** and ***catastrophes***, which we consider in more detail below.

The Emergence of Complexity

The electron is not as simple as it looks.
Sir William Bragg

The complexity of a system S is a contingent property, depending upon the nature of the observables describing S, the observables characterizing the system O measuring S, and their mutual interactions. Imagine that O sees S in an operational mode which O describes by the equation of state Φ_a. Further, suppose that at another time O sees S in the mode $\hat{\Phi}_{\hat{a}}$. If Φ_a and $\hat{\Phi}_{\hat{a}}$ are equivalent, in the sense described above, O concludes that S is manifesting essentially the same behavior in the two modes, and O is able to use equally well either description to characterize ***both*** modes of S. On the other hand, if $\Phi_a \not\sim \hat{\Phi}_{\hat{a}}$ (i.e., they are not equivalent), O is unable to reduce one description to the other and regards the operation of S as being more *complex*, since O sees more variety in the possible modes of S's behavior. This simple idea forms the nucleus of our main thesis that

complexity of S = the number of nonequivalent descriptions
(relative to O) $\quad \Phi_a$ that O can generate for S.

Interchanging the roles of S and O, the complexity of O relative to S is defined in a similar manner. Let us denote these two complexities as $C_O(S)$ and $C_S(O)$, respectively. Thus, $C_O(S)$ is what we earlier termed design complexity, while $C_S(O)$ is the control complexity of the joint system S and O.

A crucial aspect of our notion of system complexity is that it is a ***comparative*** concept: there is a tacit assumption that in order to compute $C_O(S)$, O must have available a ***family of descriptions*** of S and a method for deciding whether or not two descriptions from the family are equivalent. If Q denotes the family of descriptions, the above procedure defines an equivalence relation on Q, thereby partitioning it into appropriate equivalence classes. Since, by definition, all descriptions belonging to a given class are equivalent, the number $C_O(S)$ is just equal to the number of classes that Q is separated into by our concept of system equivalence. To operationally implement this procedure, the following steps are needed:

(1) Beginning with a fixed description S construct a ***family*** Q of descriptions containing S as a member. One fairly standard way of doing this has already been described above, when we begin with the description $\Phi(f_1,...,f_N)$ and isolate some observables as parameters α. The values of α then provide a parameterized family of descriptions of S.
(2) Partition Q into equivalence classes in accordance with the equivalence relation "$\sim$" described earlier. To accomplish this task, it is necessary to employ the machinery of singularity theory, once the mathematical character of Q and the equivalence relation are fixed.
(3) Calculate $C_O(S)$ = card $Q/\sim$ = the number of classes into which Q is split by the relation $\sim$.

In terms of management and decision-making, it is O who must select the family Q and the relation $\sim$; different selections lead to different levels of complexity as perceived by O. Similar remarks apply to the view of O as seen by S.

A simple example in which the above concepts are explicitly displayed is when $\Phi: U \to Y$ is linear with $U = R^n, Y = R^m$. In this case, Φ can be represented by an $m \times n$ matrix, once bases are chosen in U and Y. In order to parameterize the description Φ, let us suppose that we regard the first diagonal element of Φ as a parameter; that is $\alpha = [\Phi]_{11}$. Then the family $Q = \{\Phi_\alpha : R^n \to R^m, \alpha \in R\}$. Now let P and Q be linear coordinate transformations in U and Y, respectively, and suppose we consider an alternative description, $\Phi_{\hat{\alpha}}$; that is, we change the value of the element $[\Phi]_{11}$ from α to $\hat{\alpha}$. We ask if $\Phi_\alpha \sim \Phi_{\hat{\alpha}}$ or, what is the same thing, does the diagram

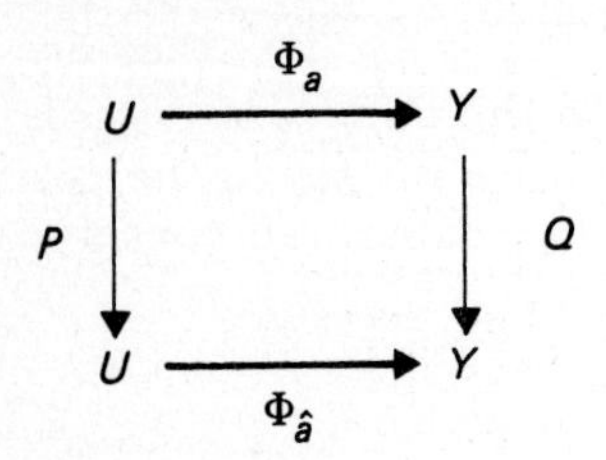

commute? Well-known results from matrix theory tell us that in this case $\Phi_\alpha \sim \Phi_{\hat{\alpha}}$ if and only if

$$\text{rank } \Phi_\alpha = \text{rank } \Phi_{\hat{\alpha}} .$$

Thus, if we let $\alpha = \min \{m, n\}$, we can assert that

$$\text{complexity } \Phi_a \leq \alpha + 1 \ .$$

The exact complexity, of course, depends upon the structure of the fixed elements of Φ_a. If, for example, rank Φ_a is constant for all $a \in R$, then complexity $\Phi_a = 1$. Thus,

$$\text{complexity } \Phi_a = \text{number of different values that rank } \Phi_a \text{ assumes as } a \text{ ranges over } R.$$

In passing, we note that the points $a^* \in R$ at which Φ_a changes rank are what we earlier termed bifurcation points. They represent places where the inherent information in the description Φ_a (here represented by the number of linearly independent rows of Φ_a, for example) is different from that in Φ_{a^*} for a near a^*. We return to this point in a more general context later.

In summary, complexity emerges from simplicity when alternative descriptions of a system are not reducible to each other. For a given observer, the more such inequivalent descriptions he or she generates, the more complex the system appears. Conversely, a complex system can be simplified in one of two ways: reduce the number of potential descriptions (by restricting the observer's means of interaction with the system) and/or use a coarser notion of system equivalence, thus reducing the number of equivalence classes. The first strategy is exemplified by a decision-maker who listens to only a few advisors before making a decision rather than gathering a full spectrum of views on a particular issue; a failure to dig deep enough to get all the facts surrounding a situation before taking action would be representative of the second approach to simplification. Both approaches are considered in more detail below, but first let us examine some of the ways in which the complexity of a system can change in a *natural* manner.

The Evolution of Complexity

> In short, the notion of structure is comprised of three key ideas: the idea of wholeness, the idea of transformation, and the idea of self-regulation.
>
> *J. Piaget*

It has been recognized, at least since the work of Turing and von Neumann on self-reproducing machines, that in order for a system to evolve to a higher level of complexity, it is necessary for the system to contain its own self-description. We might well ask why it would not be possible to design a self-reproducing system with given functional characteristics using hardware alone, without also requiring an internal linguistic description of what it is doing. The answer lies in the conditions for reliability, adaptation, growth, and evolution that we use to characterize complex systems; we are not interested in a system whose natural tendency is to degenerate or lose its function. Systems that contain their own genetic description are one known type of organization that allows survival and evolution despite errors within the system, or even errors in the description. In general, we have only a feeble understanding of the explicit conditions for the linguistic descriptions needed to achieve the threshold of reliability and adaptability necessary for survival and evolution.

In the above view, a complex system is a composite consisting of a physical structure (the hardware) carrying out functions under the instructions of an internal description of itself (the software). This situation would be well understood, as it is in computer science, if it were not for the fact that in most systems of interest the hardware and software are contained in the same physical structure. A key problem in the understanding of complex processes is the way in which the dynamic modes of the system interact with the linguistic modes, and the manner in which these complementary modes are combined to provide an external observer with some level of complexity, as outlined earlier. If we regard a *measurement* process as a physical structure that executes a rule which relates a system to an element of its description, then the encoding of dynamical processes to linguistic structures is very closely related to measurement. On the other hand, the decoding and physical execution of a genetic description is a problem of *interpretation*.

The measurement/interpretation complementarity can be very easily demonstrated by examining ordinary human speech. We can either say what we mean or we can examine how we have said it, but we can't do both simultaneously. We can represent physical structures as descriptions only when we recognize that the structures are obeying a coherent set of rules, which we call a language. And it is in this language that we formulate our concepts of complexity or simplicity. The irony in this picture is that the natural language we use to identify complexity may cause us to interpret inherently simple events, as seen by the internal language of our self-describing system, as complex messages in our interpretative natural language. An important component in the management of complexity is the institution of procedures to bring the internal and natural languages much closer, and so to prevent the external observer from receiving a message that is not really in the system itself.

Considerations of structure and description also bear heavily upon the emergent complexity arising out of lower level, simpler processes. If we think of the evolutionary process, in general, as a mapping of environmental variety and constraints into the structure of the evolving system in the form of organizing principles and coded information, then it is possible to distinguish three quite distinct evolutionary strategies: the *phylogenetic*, *ontogenetic*, and *sociogenic*. Let us consider these strategies in light of our earlier remarks.

- *Phylogenetic*. This strategy involves random genetic mutations and gene mixing which are tested in their phenotypic forms by interaction with environmental stresses. The successful structures (if any) result in the blind, natural selection of the corresponding genotypes. In terms of our earlier formalism, the map $\Phi_a : U \to Y$ is modified by purely random changes in a with future mutations of a entirely unaffected by the resulting phenotypes $y_a(u)$. Such a strategy is enormously profligate and slow, permitting rapid environmental fluctuations to reduce the viability of species before the phylogenetic mapping can catch up as, for example, with the extinction of the dinosaurs.

- *Ontogenetic*. If the system has some means of storing the results of mutations in α, for example, with some neurophysiological structure like a brain, then instead of random genetic changes, we have selective trial-and-error probings of the environment. In short, the genetic changes are directed by what has gone before in a process called ***learning***. Such an ontogenetic strategy permits a more rapid and refined process of information generation about the environment; there is an adaptive mechanism by which successful phenotypic characteristics are fed back to the gene pool to promote further genotypic changes. We might think of this feedback or learning mechanism as embodied in the neural code of the system, as opposed to its genetic code. However, this strategy also has its drawbacks, principally the fact that the information is stored in the system and goes out of existence with its death.
- *Sociogenic*. This strategy is associated with systems that are not only social, as in various insect societies, but also sociocultural, which involves not only a permanent social organization, but also an arbitrary symbolic coding of the role relationships in the society. At this level, the sociogenic strategy of evolution involves an additional code, the ***normative*** code, which is stored outside the physical system itself. Thus, the information about the environment does not die with the system and, in fact, can be passed on to new systems without their having to first directly experience the actual environmental interactions. In this strategy, besides the advantage of extra-somatic storage of information, there is the possibility of the system restructuring itself very rapidly when environmental pressures become great enough.

In the sociogenic strategy, we pass from a variation of the genetic code to mutations of the normative code, which guides the social and psychological development of new generations. Instead of a gene pool comprising the system's stock of coded information, there is an idea pool which is a reservoir of the culture's templates for the coordination and integration of individual actions and interactions. New ideas or ideologies are continually generated as mutations, subject to various selection pressures, with reproductive success measured by the perpetuation of one normative system and social structure as opposed to others.

As a simple illustration of sociocultural evolution, consider the development of societal regulatory mechanisms; that is, the dominant political structures. The appearance of democratic forms of social regulation represents, from the purely objective point of view of cybernetics, the evolution of a more adaptive political structure. For example, a more extensive idea pool, fuller information and feedback channels in the system, and a more extensive mapping of the internal as well as external states of the system and environment.

Of special importance is the balance between those institutional structures and processes designed to maintain a given structure and those designed to enable better adaptation to environmental conditions. The former structures are much more strongly incorporated into the micro- and macrostructure of the political system than the latter; hence, pressures tend to mount until the old structure can be changed only through potentially destructive revolution – a singularly poor strategy for evoluton.

Our previous consideration of system complexity as a property of the interaction between a system and its observer/regulator applies at each level of the above evolutionary scheme. However, we can also think of the emergence of a new *type* of system complexity as we pass from the phylogenetic to sociogenic strategies. This is an evolution not of the complexity displayed by a fixed system, but rather a qualitative change of the type of system from individual, nonlearning units to social collections of adaptive units, each system type requiring its own complexity concept. We touch on some of these distinctions in the next section which deals with the interrelationships between system complexity and the concepts of adaptation, hierarchy, and bifurcation.

Complex Systems: Adaptation, Hierarchy, and Bifurcation

> There is nothing in the whole world that is permanent. Everything flows onward; all things are brought into being with a changing nature; the ages themselves glide by in constant movement.
> *Ovid* (Metamorphoses)

Treatments of complexity often place great emphasis upon various behavioral or structural characteristics of a system, which, if present, offer supposed *prima facie* evidence that the system is complex, by whatever interpretation the author is advocating. Three of the most commonly cited characteristics are:

- *Adaptability.* The capacity for the system to monitor its environment and to reconfigure itself on the basis of its observations in order to more effectively perform its function.
- *Hierarchy.* The tendency for the system to be structurally organized in a stratified manner so that information and activities at lower levels are combined as inputs to higher levels, while overall direction and control passes from higher to lower levels.
- *Bifurcation and novelty.* The tendency for complex processes to spontaneously display a shift from one behavioral or structural mode to another, as levels of organization increase. These surprises or emergent novelties represent points of bifurcation where a previous description of the system breaks down and a new description, not reducible to the old, is required.

While it should be clear by now that we do not hold to the view that any of the above features is an infallible indicator of complexity, it certainly is true that many complex phenomena *are* hierarchically structured, *do* display emergent behavioral modes, and *can* adapt to new situations. Consequently, it is of interest to examine how well these system properties can be accommodated to the complexity concept introduced earlier in this chapter.

Adaptation

Consider the capability of a system to *adapt* to changing conditions in the environment. This is a functional concept involving at least some subsystems changing their functional behavior to accommodate the new environment. A political system granting voting rights to women in response to egalitarian social currents, as in Switzerland in recent times, is the type of adaptive change a complex system can often make. So is the way in which banks have been introduced into modern economic structures as an adaptation to provide for intertemporal exchanges in disequilibrium. Here, a subsystem whose previous function was only to act as a storehouse of wealth, has changed its function to provide credit and other services which allow an economy to sustain a continual state of disequilibrium. One might say, even, that all adaptation arises as a result of a principle of function change, whereby subsystems created for one function begin to perform a quite different function when the system perceives the new function to be evolutionarily more advantageous than the old. The classical biological example of this kind of shift is the evolution of the human eye, which cannot confer any survival advantage until it sees and cannot see until it is highly evolved and complex. Thus, it is difficult to imagine how such an organ could arise as the result of minute differential changes in a fixed organ, even over millions of years. It is much more reasonable to suppose that originally the eye performed a function quite different from sight and an accidental feature of this proto-eye was that it was photosensitive. As time wore on, the photosensitivity feature became more and more evolutionarily advantageous and the original function of the eye was lost.

The picture of adaptation as being a system response to changed circumstances leads to the basic evolutionary equation

$$\text{variation} + \text{selection} = \text{adaptation},$$

expressing the fact that, in order to adapt, the system must have many potential modes of behavior and a procedure for evaluating the relative fitness of the various alternatives in a given environment. One of the difficulties with complex human social systems is that redundancy at the genetic level, which gives the capacity for independent variations, is too limited. As a result, there is too little room for trying new approaches and for exploring alternative pathways to a given functional goal when operating circumstances change. Systems such as large nuclear power plants, national economies, major ecosystems, and the like have little, if any, degrees of freedom in their structure or design with which to experiment. The consequences of a failure are too great to allow the evolutionary equation to operate effectively, at least in its natural mode. In our view, until more resilient design policies are employed for such large-scale systems, the only possible way to escape this prison of hypotheticality is by way of mathematical models and computer exploration of alternative systems, rather than by relying upon nature's trial-and-error. On balance this is probably a better strategy anyway, since we don't have millions or even hundreds of years to find solutions to our energy, economic, and environmental problems. But the potential Achilles heel in the computer simulation strategy is that it is totally dependent upon the existence of faithful models of reality, expressible in mathematical terms. Thus, the weight of the entire edifice is concentrated upon the need to develop a science of modeling and effective procedures for the identification of "good" models of human and natural phenomena.

To incorporate the above ideas into our earlier formalism, we must introduce a feedback mechanism through which environmental fluctuations are sensed by the system and used to generate exploratory variations in the system's "genomes". Recalling that the basic description (or model) of the system is given by a family of relations

$$\Phi_a : U \to Y ,$$

inclusion of adaptive capabilities requires two steps:

(1) *Feedback/feedforward loops*. The system genome a is now thought of as being at least partially determined by either current and past states of the environment (feedback), in which case $a = a[u(t - \tau)]$ and/or upon predicted future states of the environment (feedforward). In the latter event, $a = a[\hat{u}(t + \tau)]$. Here τ is some time-lag, while $\hat{u}$ denotes the predicted future environmental state. There are good arguments for both feedback and feedforward mechanisms in adaptive structures and, most likely, any truly self-organizing complex structure develops both modes for coping with environmental change.

(2) *Selection procedure*. Implicit in the above feedback/feedforward mechanism is a selection procedure; the environment is sensed and predicted and a rule is applied which tells the system how to modify its genome to best fit the changed circumstances. Thus, the feedback/feedforward loops represent both random and directed search in the space of the genomes, together with a procedure to weed out the "good" genetic patterns from the "bad".

At this point it is useful to note the distinction between the adaptive capability of an individual system and the effect that the association of individuals in a society has on this capacity. Basically, the adaptive capacity of an individual is reduced, but group adaptive capacity is increased as individuals join together in cellular societies. The key point here is that the group capacity is increased, but on a much longer time-scale than that for individuals. Thus, individual companies join together to form a multinational conglomerate, thereby gaining a group ability to respond to global economic fluctuations that no individual member could easily accommodate, but on a much longer time-scale than the reaction time of a typical firm. It is probably fair to say that higher-level associations only arise through defects in the adaptive capability of individuals. More than any other factor, it is this limited adaptive capacity of individuals that gives rise to the hierarchical organizations so typically present in complex systems.

Hierarchy

The failure of individual subsystems to be sufficiently adaptive to changing environments results in the subsystems forming a collective association that, *as a unit*, is better able to function in new circumstances. Formation of such an association is a *structural* change; the behavioral role of the new conglomerate is a *functional* change; both types of change are characteristic of the formation of hierarchies. It has been argued by Simon (1969, 1981), as well as others, that evolution favors those systems that display stable, intermediate *levels of structure*. Furthermore, a complex system is incomprehensible unless we can simplify it by using alternative *levels of description*. A digital computer illustrates both types of hierarchies, where we have structural or hardware levels from microchips to functional units like disc drives, terminals, processors, and so on. On the descriptive side, we have the system software which describes what the structural levels are to do, using a series of descriptive levels from machine languages to high-level, natural-language programming languages.

In a hierarchical structure, the various levels of organization refer primarily to different ways in which it is possible for us to interact with the system, i.e. nonequivalent types of state descriptions generate different hierarchical levels. It is not possible, for instance, to understand the machine language operations represented by a particular BASIC statement without moving away from the level of BASIC to the more microscopic level of machine instructions. The two descriptions are incompatible in much the same way that it is impossible to understand a biological organism by studying its individual atoms and molecules. Of course, the same situations occur repeatedly in economics under the rubric micro–macro problems, as well as in urban studies, psychology, sociology, and many other areas.

It is interesting to note that in hierarchical organizations, the organizational characteristics *look the same* at each level, in that the dynamics and structural interactions at each level appear to be models of each other. This feature was noted long ago by Haeckel in his bioenergetic law – "ontogeny recapitulates phylogeny", expressing the observation that each organism carries the entire history of the phylum within itself. Other examples of this principle abound: computer programs and their subroutines, a symphony and its various movements, a neural network and the associated network of genetic control, a book and its component chapters, and so on. Some of these hierarchies are structural, while others are functional, and it appears safe to say that the central problem of hierarchy theory is the understanding of the relation between the structural and the descriptive (or functional) levels. Most of the classical physical sciences have concentrated upon structural decompositons, culminating in today's multimillion-dollar searches for the ultimate particles of matter. This is suitable for the study of physics, but for an understanding of living systems (biological, human, social) it is necessary to look for *functional* decompositions: the new reductionism will be based upon units of function and description, not units of structure.

How can the preceding concepts of hierarchical levels be incorporated into our mathematical formulation? At the structural level, the atoms of our modeling formalism are the real-valued observables $f_i : \Omega \to R$, where Ω is the system's set of abstract states. In a loose sense, $\{f_i\}$ are the state variables of the model. Structural hierarchies are formed by combining these state variables, either by aggregation or disaggregation, into new quantities. Imagine that we have n observables that can be collectively written $f = (f_1,\ldots,f_n)$. A hierarchy is formed by prescribing a rule for combining these quantities into m new observables, $\hat{f} = \{\hat{f}_1,\hat{f}_2,\ldots,\hat{f}_m\}$; that is each $\hat{f}_i = \hat{f}_i(f_1,\ldots,f_n)$. Diagrammatically, we have

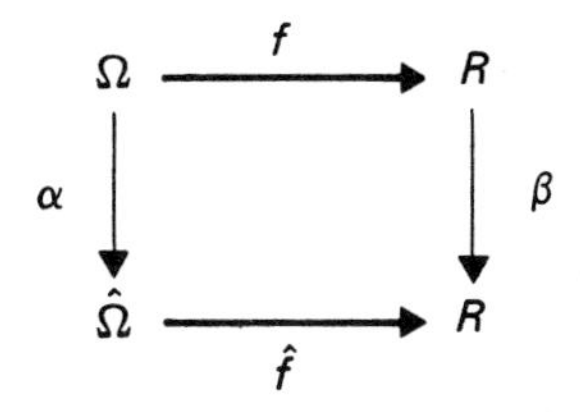

in which, the map β is either an imbedding or a projection of $R^n \to R^m$, depending upon whether $n < m$ or $n > m$. The interesting part of the diagram involves the map α and the new state space $\hat{\Omega}$. Since $(\hat{\Omega}, \hat{f})$ represent a different hierarchical level than (Ω, f), it is generally the case that $\Omega \neq \hat{\Omega}$; that is, the set of states appropriate for characterizing the system at a given level is not generally the state set appropriate for another level. But the diagram makes it clear that there is some flexibility in passing from Ω to $\hat{\Omega}$. We can either choose α, thereby fixing the new state set $\hat{\Omega}$, or we can choose $\hat{\Omega}$ and then determine α from the relation $\beta \circ f = \hat{f} \circ \alpha$. The picture sketched above provides a prototypical framework for all structural stratifications that involve the introduction of hierarchies through aggregation and disaggregation.

The descriptive stratification proceeds on the basis that the system activity is determined by the equation of state that links its observables. Thus, the function that the system performs is described by the rule

$$\Phi(f_1, \ldots, f_n) = 0 \ .$$

Earlier, we subdivided the observables using cause-and-effect arguments and wrote this relationship as

$$\Phi_a : U \to Y \ .$$

Now let us consider what is implied when the system passes to a new descriptive level at which a new function is performed. In our context it can mean only one thing: the equation of state Φ has been modified to a new equation $\hat{\Phi}$, possibly (but not necessarily) with a change of observables from $f \to \hat{f}$; that is, in diagrammatic form

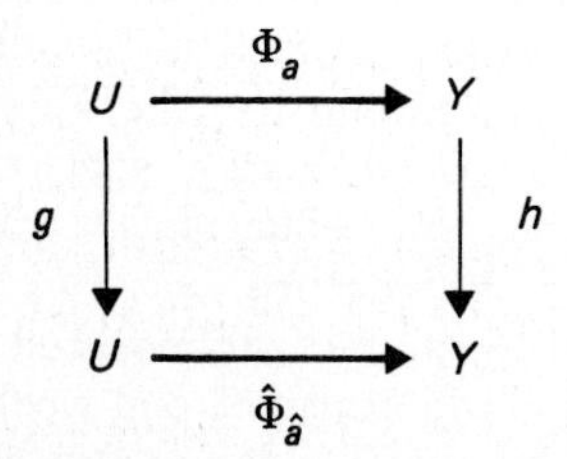

We have already discussed the ramifications of this diagram and note here only that the appearance of a new functional hierarchical level is abstractly the same as the occurrence of a bifurcation in the system description. Consequently, the emergence of new functional hierarchies is completely intertwined with the concept of system bifurcation, and an understanding of the system's functional levels of organization can only occur through a deeper investigation of the number and type of its bifurcation points.

Bifurcation, Error, and Surprise

Earlier, we considered the situation in which there were two descriptions of a given system, say Φ_a and $\hat{\Phi}_{\hat{a}}$, and addressed the question of when we could meaningfully say that Φ_a was *equivalent* to $\hat{\Phi}_{\hat{a}}$. It was argued that $\Phi_a \sim \hat{\Phi}_{\hat{a}}$ if maps g and h could be found such that the diagram above commutes. In other words,

$\Phi_a \sim \hat{\Phi}_{\hat{a}}$ if a change of genotype $a \to \hat{a}$ can be neutralized by appropriate changes, g and h, of the environment and phenotype, respectively. If no such g and h exist (within some appropriate class of maps), then a is called a ***bifurcation point*** for the description Φ (or, equivalently, $\hat{a}$ is a bifurcation point for the description $\hat{\Phi}$). We then ***define*** the complexity of the system in terms of the number of bifurcation points. So in this sense, a system S is more complex than a system S' if our description of S contains more bifurcation points than our description for S'. Thus, the concept of system complexity and the idea of a bifurcation are intimately linked at the very outset of our theory: increased complexity can only emerge at a bifurcation point and, conversely, every bifurcation point gives rise to a new mode of system behavior that is not reducible (i.e. understandable) in terms of the old. Now let us consider a quite different way in which bifurcations can generate emergent behavior when two systems are made to interact with each other.

Consider the simple situation in which we have Ω = the real numbers R, and the observables $f = (f_1, \ldots, f_n)$, are defined as

$$f_i : R \to R \qquad i = 1,2,\ldots,n.$$
$$r \to i\text{th coefficient in the decimal expansion of } r$$

Then, clearly, $r_1, r_2 \in R$ are equivalent with respect to the observables f when r_1 and r_2 agree in the first n terms of their decimal expansions. Now choose numbers r_1', r_2' such that

$$r_1 \sim_f r_1' \quad , \quad r_2 \sim_f r_2' \quad .$$

Now we let the 1-system interact with the 2-system through multiplication; that is, we form the products $r_3 = (r_1 r_2)$ and $r_3' = (r_1' r_2')$ and find that, in general, $r_3 \nsim_f r_3'$; that is, the equivalence classes under f are split by the interaction (i.e., by the dynamics). In other words, the interaction generates a bifurcation of the f-classes, a bifurcation that we usually call round-off *error*, in the above context. It is instructive to examine the source of this so-called error.

To see the way the error is introduced in the above situation, let us consider a numerical example. Let $r_1 = 123$, $r_1' = 124$, $r_2 = 234$, and $r_2' = 235$, and use $f = (f_1, f_2)$; that is, the equivalence relation generated by f is such that two numbers are equivalent if they agree in the first two places. Here we have $r_1 r_2$ (= 28782) $\nsim_f r_1' r_2'$ (= 29140), a discrepancy with our expectation based on the f-equivalence. Our surprise at finding $r_1 r_2 \nsim_f r_1' r_2'$ occurs because the set of observables $f = (f_1, f_2)$ is too limited, thereby causing an unrealistic expectation concerning the interaction between the 1- and 2-systems. If we had expanded the set of observables to the set $\hat{f} = (f_1, f_2, f_3)$, then no such discrepancy would have occurred, since there would be no equivalence, at all, of r_1, r_1' under $\hat{f}$. So, the entire source of our observed error is purely from the incompleteness in the description of the system.

The preceding arguments are entirely general: error (or surprise) always involves a discrepancy between the objects (systems) ***open*** to interaction and the abstractions (models, descriptions) ***closed*** to those same interactions. The remedy is equally clear, in principle: just supplement the description by adding more observables to account for the unmodeled interactions. In this sense, error and surprise are indistinguishable from bifurcations. A particular description is inadequate to account for uncontrollable variability in equivalent states and we need a new description to remove the error.

It is interesting to note that since bifurcation and error/surprise are identical concepts, and that complexity arises as a result of potential for bifurcation, we must conclude that complexity implies surprise and error; that is, to say a system displays counter-intuitive behavior is the same as saying that the system has the capacity for making errors, although the error is not *intrinsic* to an isolated system, but occurs when a system interacts with another.

Models, Complexity, and Management

> The man who draws up a program for the future is a reactionary.
> *Karl Marx*

It has been said that the reason we construct models is to be able to say "because". Coping with complexity involves the creation of faithful models of not only the system to be managed, but also of the management system itself. As we have continually emphasized, complexity, its identification and control, is an interactive concept between the system and its manager and it is impossible for the management system to effectively regulate the controlled system without having a concept (read: model) of itself, as well as of the system to be managed. This self-description is essential if the management system is to survive in the face of inevitable error and environmental disturbances of the type discussed above. In our earlier terms, effective complexity management reduces to the simple prescription

design complexity = control complexity.

But, what is involved in reaching this state of system–theoretic nirvana?

One aspect we can be certain of is that the search for effective management of complexity does not necessarily involve simplifying the process to be regulated. As Einstein pointed out, things should be as simple as possible, but no simpler, which we could translate as reducing the design complexity to the level of the control complexity, but no lower. Turning this argument around, we can also think of *increasing* the complexity of the management system to bring it into line with the design complexity of the system. Thus, effective complexity management may involve *either* simplifying or complexifying, depending upon the circumstances. But, in either case, it is first necessary to have means for assessing the levels of complexity of the two interacting systems. Thus, we must begin to develop the framework for a *theory of models*, one that includes effective methods for identifying the complexity of interacting systems and the means by which the conflicting complexity levels can be brought into harmonious balance.

Imagine, for a moment, that such a theory of models already exists and consider the types of mangement strategies that would serve to balance design and control complexities at some acceptably high level. First, we note that it is not sufficient simply to equalize the complexity levels of the system and its observer/controller. They must be balanced at a sufficiently high level: if I simplify a Chopin piano sonata by requiring that it be played only on the white keys, I have certainly reduced its complexity level to the level of my observational ability (complexity) to understand the piece. However, I obtain very little pleasure from this kind of complexity balance; the variety that makes the piece interesting has been destroyed and I would probably benefit more with no system at all to observe. In this situation, it is far more reasonable to raise the complexity level of my observing system to match the level of the piece, which presumably already exists at a high enough level to perform its intended function. So, any management scheme must begin by taking into account the ***absolute*** level at which the design and control complexities are to be equalized.

In terms of general control strategies, there are two complementary approaches. One is to develop bifurcation-free and bifurcation-generating ***feedback*** policies. As has been noted elsewhere, feedback laws have the effect of changing the internal structure of the system they regulate. Of course, in our context this means that any feedback policy has the potential to change the design complexity of the controlled system. Some illustrations of how this can be done are discussed in Casti (1980), although from the somewhat different perspective of optimal control theory, not the more general setting discussed here.

From a management point of view, there are some disadvantages to using feedback policies, the principle one being that any error-actuated feedback law does not even begin to act until the system is already out of control; that is, if there is no error, the system is not being regulated at all. For many engineering systems this situation is quite satisfactory, but in social and behavioral processes we cannot usually be so sanguine about error-actuated control. Generally, in such systems we would like to ***anticipate*** difficulties and take action now to avoid projected malfunctions later. In human systems, we cannot afford the luxury of waiting for the system to fail before we take remedial action. This basic principle leads to the idea of ***anticipatory*** control and ***feedforward*** policies.

The most important feature of anticipatory control systems is that the manager must have a model of the system to be regulated, and his or her actions are dictated by the ***regularities*** between the behavior of the system, as predicted by the model (which is run on a time-scale faster than real-time), and the actual, observed system behavior at the future time of the model prediction. The prediction and observation are then correlated and the model recalibrated, leading to the idea of adaptive control. Surprisingly, there seems to have been very little study of such processes, although some recent work by Rosen (1979, 1984) promises to redress this imbalance of knowledge between feedback and feedforward regulators.

From the above, the broad outline of a research program for complexity management begins to emerge, and consists of the following major components:

(1) *A Theory of Models*. There is a need for development of a sufficiently rich theoretical framework for mathematically representing processes in the social, behavioral, and cultural environment. This theory must of necessity include methods for identifying relevant observables, state spaces, and equations of state, as well as provide a basis for formally incorporating the complexity, adaptation, hierarchy, and emergence concepts discussed above.

(2) *Anticipatory Control.* A deep investigation into the nature of feedforward policies as opposed to feedback is needed, in order to provide the means for balancing complexity levels between the manager/decision-maker/observer and the system under consideration. Such an investigation will include studies of adaptive mechanisms, as well as the role of anticipatory policies in reducing/generating bifurcations in the managed system descriptions.

Each of these points need considerable elaboration before they can constitute a plan for a truly creative research program. But already it is clear that creative research is what is needed if any progress at all is to be made in the complexity management problem. And here the emphasis is on the word creative: no pedestrian, pull-the-pieces-off-the-shelf-and-put-them-together type of program will suffice. New ideas and new approaches are the only currency of this realm. It seems appropriate to close by stating a few general features that serve to identify what we mean by creative research, as opposed to the pedestrian. Our advice to anyone contemplating creative research is to:

- Avoid the research literature.
- Avoid practitioner's problems.
- Never put high hopes on any study for any useful information.
- Never plan – especially not in the long term.
- Never apply for a research grant.
- Never give up if everyone thinks you are wrong.
- Give up immediately when they think you are right.

As Nietzsche said, "that which needs to be proved cannot be worth much", so in today's world I won't hold my breath waiting for any putative "research" organizations to adopt even one of the foregoing principles as part of their official posture and manifesto. Nonetheless, the closer an individual researcher comes to adherence to these guidelines, the closer he or she will be to a position from which to crack the nut of system complexity and its management.

Notes and references

Complexity and simplicity

A detailed consideration of the contention that system complexity necessarily relates to the interaction of a given system with its observer/describer/controller is found in

Phillips, W. and Thorson, S. (1975) Complexity and policy planning, in *Systems Thinking and the Quality of Life, Proc. Soc. for General Systems Research Annual Meeting.*

This paper is notable for its review of various concepts of complexity in the field of social system management and for its conclusion that " ... no adequate characterization of the complexity of a system can be given without specifying the class of observers dealing with the system, as well as the specific purposes of the observers". The author's arguments supporting this view of complexity culminate in the contention that "whichever approach we take to modeling the outer environment – the policy problem – the complexity characteristic of the system is contingent upon our description of the relations between the inner environment and the outer. It is a function of the theories we bring to bear upon problems and the way we view the environment".

Some further views along the same lines were expressed by one of the cybernetics pioneers, W. Ross Ashby, in

Ashby, W.R. (1973) Some peculiarities of complex systems. *Cybernetics Medicine* 9: 1–8.

In this paper, Ashby remarked that "a system's complexity is purely relative to a given observer; I reject the attempt to measure an absolute, or intrinsic, complexity; but this acceptance of complexity as something in the eye of the beholder is, in my opinion, the only workable way of measuring complexity".

Basically, the same points have been emphasized in the philosophy of science literature from a somewhat more fundamental perspective; see, for example,

Quine, W.v.O. (1964) On simple theories of a complex world, in J. Gregg and F. Harris (Eds) *Form and Strategy in Science* (Dordrecht: Reidel),

Wimsatt, W. (1972) Complexity and organization, in K. Schaffner and R. Cohen (Eds) *Studies in the Philosophy of Sciences, Vol. XX* (Reidel, Boston),

and the classic paper

Simon, H., (1969) The architecture of complexity, in *Sciences of the Artificial* (Cambridge, MA: MIT Press).

Management of the complex

The concepts of design and control complexity were introduced by Gottinger in the somewhat different context of an automata–theoretic treatment of complexity. For a recent account of his ideas see

Gottinger, H. (1983) *Coping with Complexity* (Dordrecht: Reidel).

This work represents an approach to the problem of system complexity originally initiated by John Rhodes in

Rhodes, J. (1971) *Application of Automata Theory and Algebra* (Berkeley, CA: Lecture Notes, Department of Mathematics, University of California).

The importance of the symmetry of the interaction between the system and its observer/controller has been particularly emphasized in

Rosen, R. (1984) *Anticipatory Systems* (London: Pergamon),

and

Rosen, R. (1978) *Fundamentals of Measurement and Representation of Natural Systems* (New York: Elsevier).

For a discussion of some of the important matters arising from the interactions present in the political process see

Kirby, M.J.L. (1980) *Reflections on Management of Government Within a Democratic Society in the 1980s, Parts I & II.* (Ottawa: Plaunt Lectures, Carlton University).

Works emphasizing similar aspects of complexity in social and behavioral areas include

Winthrop, H. (1972) Social systems and social complexity in relation to interdisciplinary policymaking and planning. *Policy Sciences* 3: 405–420,

Winham, G. (1976) Complexity in international negotiation, in D. Druckman (Ed) *Negotiations* (Beverly Hills: Sage Publ. Co.),

as well as the Phillips and Thorson article cited earlier.

Systems, observables, and models

A thorough exposition of the ideas surrounding observables, abstractions, and equations of state is found in the Rosen books cited earlier.

The fast–slow distinction as a means of inducing causality is a special case of hierarchical ordering, but in time rather than space. For a discussion of this crucial point, see the book

Fraser, J.T. (1978) *Time as Conflict* (Basel: Birkhäuser).

Additional discussion of the macro–micro problem is found in

Allen, T.F.H. and Starr, T. (1982) *Hierarchy* (Chicago: University of Chicago Press).

Use of an evolutionary metaphor to characterize human systems is far from new, dating back at least to Herbert Spencer and the social Darwinists. A modern attempt to mimic biology as a guide to social development is

Corning, P. (1983) *The Synergism Hypothesis* (New York: McGraw-Hill).

In the economic area, the evolutionary metaphor has been quite well-developed in

Nelson, R. and Winter, S. (1982) *An Evolutionary Theory of Economic Change* (Cambridge, MA: Harvard University Press),

Boulding, K. (1981) *Evolutionary Economics* (Beverly Hills: Sage Publ.).

Singularity theory is treated from a mathematical point of view in

Golubitsky, M. and Guillemin, V. (1973) *Stable Mappings and their Singularities* (New York: Springer),

Lu, Y.C. (1976) *Singularity Theory* (New York: Springer),

Gibson, C. (1979) *Singular Points of Smooth Mappings* (London: Pitman).

The connection between these mathematical results and the theory of equivalent systems is made in

Casti, J. (1984) *System Similarity and Laws of Nature* IIASA WP-84-1 (Laxenburg, Austria: International Institute for Applied Systems Analysis).

The emergence of complexity

For a discussion of the interrelationship between the idea of system complexity as presented here, and the concepts of system error and entropy, see Chapter 5 in Rosen (1978), cited earlier.

Many attempts have been made to define the complexity of a system in terms of properties of the system alone, such as number of components, density of internal interactions, and so forth. Some machine–theoretic efforts along these lines are

Bremermann, H. (1974) Complexity of automata, brains and behavior, in S. Levin (Ed) *Lecture Notes in Biomathematics*, Vol. 4 (Berlin: Springer),

Bremermann, H. (1974) Algorithms, complexity, transcomputability, and the analysis of systems, in W. Reidel, W. Händler, and M. Spreng (Eds) *Proc. Fifth Congress of the Deutsche Gesellschaft für Kybernetik* (Munich: Oldenbourg),

Gaines, B. (1976) On the complexity of causal models. *IEEE Tran. Syst. Man & Cyber*. SMC-6: 56–59,

George, L. (1977) Tests for system complexity. *Int. J. Gen. Syst.* 3: 253–258.

In addition to missing the crucial point that complexity depends upon the *interaction* of a system with another rather than upon the system itself, an annoying aspect of such studies is the way in which the extremely useful term complexity has been usurped by the computer-orientation of such authors and taken to mean something very specific in the context of machines and algorithms. This situation is by no means new, as the computer industry has a long and deplorable history of taking useful terms and concepts, such as information, system, and systems analyst, and then warping the terms to such an extent that their original meanings are totally lost. Normally this distorting process could be dismissed with a casual shrug, as is done in mathematics, for instance, but for the fact that the computer-industry propaganda machines effectively promote their new meaning of these terms to the general public, thereby creating considerable confusion as to the more general, and far more useful interpretations of these important concepts.

A fascinating article involving the use of complexity in assessing aesthetic experience is

Goguen, J. (1977) Complexity of hierarchically organized systems and the structure of musical experiences. *Int. J. Gen. Syst.* 3: 233–251.

This article introduces the concept of *conditional* complexity, based upon past experiences and expectations, and then applies the idea to develop a theory of surprise for musical compositions. For purposes of aesthetic satisfaction, the author concludes that if the conditional complexity of a piece is too low, then our expectations are too easily and too often fulfilled to maintain our interest, whereas if the conditional complexity is too high, our expectations are too often frustrated to permit much listening satisfaction. This argument leads to an aesthetic law of the mean for musical complexity.

The evolution of complexity

System complexity depends upon whether the system is regarded as an object or as a description, a theme explored in detail in

Löfgren, L. (1977) Complexity of systems: a foundational study. *Int. J. Gen. Syst.* 3: 197–214.

The stability and evolutionary potential of self-describing complex systems depends also upon the complementary relation between the dynamic (structural) and linguistic (functional) modes of system description. This relationship is inextricably intertwined with the epistemological problem of measurement. For a detailed consideration of these matters, see

Pattee, H. (1977) Dynamic and linguistic modes of complex systems. *Int. J. Gen. Syst.* 3: 259–266.

A discussion of the several types of evolutionary strategies is found in

Buckley, W. (1977) Sociocultural systems and the challenge of sociobiology, in H. Haken (Ed) *Synergetics: a Workshop* (Berlin: Springer).

Complex systems: adaptation, hierarchy, and bifurcation

A detailed exploration of biological adaptation as a metaphor for human systems is given in

Rosen, R. (1975) Biological systems as paradigms for adaptation, in R. Day (Ed) *Adaptive Economic Models* (New York: Academic Press).

Rather thorough expositions of the nature of adaptive mechanisms in both engineering and living systems are found in the works

Holland, J. (1975) *Adaptation in Natural and Artificial Systems* (Ann Arbor: University of Michigan Press),

Conrad, M. (1983) *Adaptability: The Significance of Variability from Molecule to Ecosystem* (New York: Plenum).

The initial steps toward a theory of anticipatory control involving feedforward loops are outlined in

Rosen, R. (1978) On anticipatory systems: I & II. *J. Social & Biol. Structures* 1: 155–180.

It is worthwhile to note that the formalism for anticipatory control is in the same spirit as the so-called bounded rationality models in economics. See, for example,

Day, R.H. (1985) Disequilibrium economic dynamics: a post-schumpeterian contribution. *J. Econ. Behavior and Org.* (to be published in 1985),

and

Simon, H.A. (1981) *The Sciences of the Artificial* (2nd edn) (Cambridge, MA: MIT Press).

The appearance of hierarchical organizational structures in natural, as well as man-made systems is discussed from several viewpoints in

H. Pattee (Ed) (1973) *Hierarchy Theory* (New York: Braziller).

See also

Jantsch, E. (1980) *The Self-Organizing Universe* (Oxford: Pergamon),

as well as the Allen and Starr book cited earlier.

The emergence of new structures and behavioral modes through parameter fluctuations and environmental variability is discussed in some detail in

Prigogine, I., Allen, P., and Herman, R. (1977) Long term trends and the evolution of complexity, in E. Laszlo and J. Bierman (Eds) *Goals in a Global Community* (New York: Pergamon),

Prigogine, I. (1980) *From Being to Becoming: Time and Complexity in the Physical Sciences* (San Francisco: Freeman).

The concept of surprise as a system bifurcation is explored in

Casti, J. (1982) Topological methods for social and behavioral sciences. *Int. J. Gen. Syst.* 8: 187–210.

A nontechnical consideration of the same circle of ideas and their applied significance is considered in

Holling, C.S. (Fall 1983) Surprise? *IIASA Options* (Laxenburg, Austria: International Institute for Applied Systems Analysis).

A formal theory of surprises, using ideas from algebraic topology, is put forth in

Atkin, R.H. (1981) A theory of surprises. *Environment & Planning B*, 8: 359–365.

While Atkin's theory does not explicitly employ the idea of a system bifurcation, the concept is implicit in his work and a mathematical unification of the two approaches would be a valuable exercise, shedding additonal light on the essential aspects of a nonprobabilistic theory of surprises.

Models, complexity, and management

The question of complexity management is hardly a new one. A nontechnical introduction to some of the important managerial issues that arise is

Beer, S. (1970) Managing modern complexity. *Futures* 2: 245–257.

It is often held that the objective of system management is to stabilize a process in the face of a fluctuating environment and, in this context, that stability and complexity are positively correlated. Discussions of the pros and cons of this dubious argument are found in

Chadwick, G.F. (1977) The limits of the plannable: stability and complexity in planning and planned systems. *Environment and Planning A* 9: 1189–1192,

Pimm, S. (1984) The complexity and stability of ecosystems. *Nature* 307: 321–326.

The question of bifurcation-free feedback control laws is taken up in

Casti, J. (1980) Bifurcations, catastrophes and optimal control, *IEEE Tran. Auto. Control*, AC-25: 1008–1011.

For a discussion of how linear *feedback* control laws alter internal system structure, see

Casti, J. (1977) *Dynamical Systems and their Application: Linear Theory* (New York: Academic Press).

The connection between feedback and feedforward control laws and the effect that each type has on the alteration of system structure is pursued in

Kalman, R. (1971) Kronecker invariants and feedback, in L. Weiss (Ed) *Ordinary Differential Equations* (New York: Academic Press).

The problems of anticipatory control are developed in

Rosen, R. (1979) Anticipatory systems in retrospect and prospect. *General Systems* 24: 11–23.

See also the Rosen works cited earlier.

CHAPTER 7

On Information and Complexity

Robert Rosen

Introduction

We introduce the rather wide-ranging considerations which follow with a discussion of the concept of *information* and its role in scientific discourse. Ever since Shannon began to talk of information theory (by which he meant a probabilistic analysis of the deleterious effects of propagating signals through channels; cf. Shannon and Weaver, 1949), the concept has been relentlessly analyzed and reanalyzed. The time and effort expended on these analyses must surely rank as one of the most unprofitable investments in modern scientific history; not only has there been no profit, but also the currency itself has been debased to worthlessness. Yet, in biology, for example, the terminology of information intrudes itself insistently at every level; code, signal, computation, recognition. It may be that these informational terms are simply not scientific at all; that they are a temporary anthropomorphic expedient; a *facon de parler* which merely reflects the immaturity of biology as a science, to be replaced at the earliest opportunity by the more rigorous terminology of force, energy, and potential which are the province of more mature sciences (i.e. physics), in which information is never mentioned. Or, it may be that the informational terminology which seems to force itself upon us bespeaks something fundamental; something that is missing from physics as we now understand it. We take this latter viewpoint, and see where it leads us.

In human terms, information is easy to define; it is anything that is or can be the answer to a question. Therefore, we preface our more formal considerations with a brief discussion of the status of interrogatives, in logic and in science.

The amazing fact is that interrogation is not ever a part of formal logic, including mathematics. The symbol "?" is not a logical symbol, as, for instance, are "$\vee$", "$\wedge$", "$\exists$", or "$\forall$"; nor is it a mathematical symbol. It belongs entirely to informal discourse and, as far as I know, the purely logical or formal character of interrogation has not been investigated. Thus, if information is indeed connected in an intimate fashion with interrogation, it is not surprising that it has not been formally characterized in any real sense. There is simply no existing basis on which to do so.

I do not intend to go deeply here into the problem of extending formal logic (always including mathematics in this domain) so as to include interrogatives. What I want to suggest here is a relation between our informal notions of interrogation and the familiar logical operation "$\Longrightarrow$"; the conditional, or the implication, operation. Colloquially, this operation can be rendered in the form "If A, then B". My argument involves two steps. First, that ***every*** interrogative can be put into a kind of conditional form:

If A, then B ?

(where B can be an indefinite pronoun like who, what, etc., as well as a definite proposition); and second, and most important, that every interrogative can be expressed in a more special conditional form, which can be described as follows. Suppose I know that some proposition of the form

If A, then B

is true. Suppose I now change or vary A; that is, replace A by a new expression, δA. The result is an interrogative, which I can express as

If δA, then δB ?

Roughly, I am treating the true proposition "If A, then B", as a reference, and I am asking what happens to this proposition if I replace the reference expression A by the new expression δA. I could, of course, do the same thing with B in the reference proposition; replace it by a new proposition δB and ask what happens to A. I assert that every interrogative can be expressed this way, in what I call a *variational form*.

The importance of these notions for us lies in their relation to the external world; most particularly in their relation to the concept of ***measurement***, and to the notions of causality to which they become connected when a formal or logical system is employed to represent what is happening in the external world; that is, to describe some physical or biological system or situation.

Before discussing this, I want to motivate the two assertions made above, regarding the expression of arbitrary interrogatives in a kind of conditional form. I do this by considering a few typical examples, and leave the rest to the reader for the moment.

Suppose I consider the question

"Did it rain yesterday?"

First, I write it as

"If (yesterday), then (rain)?"

which is the first kind of conditional form described above. To find the variational form, I presume I know that some proposition like

"If (today), then (sunny)"

is true. The general variational form of this proposition is

"If δ(today), then δ(sunny)?"

Then, if I put

δ(today) = (yesterday),

δ(sunny) = (rain)

I have, indeed, expressed my original question in the variational form. A little experimentation with interrogatives of various kinds taken from informal discourse (of great interest are questions of classification, including existence and universality) should serve to make manifest the generality of the relation between interrogation and the implicative forms described above; of course, this cannot be *proved* in any logical sense since, as noted above, interrogation remains outside logic.

It is clear that the notions of observation and experiment are closely related to the concept of interrogation. That is why the results of observation and experiment (i.e. data) are so generally regarded as being information. In a formal sense, simple observation can be regarded as a special case of experimentation; intuitively, an observer simply determines what *is*, while an experimenter systematically perturbs what is, and then observes the effects of his or her perturbation. In the conditional form, an observer is asking a question which can generally be expressed as

"If (initial conditions), then (meter readings)?"

In the variational form, this question may be formulated as follows: assuming the proposition

"If (initial conditions = 0), then (meter readings = 0)"

is true (this establishes the reference, and corresponds to calibrating the meters), we ask

"If δ(initial conditions = 0), then δ(meter readings = 0)?"

where, simply

δ(initial conditions = 0) = (initial conditions)

and

δ(meter readings = 0) = (meter readings).

The experimentalist, essentially, takes the results of observation as the reference and asks, in variational form, simply

"If δ(initial conditions), then δ(meter readings)?"

The theoretical scientist, on the other hand, deals with a different class of question; namely, those that arise from assuming a δB (which may be B itself) and asking for the corresponding δA. These are questions that an experimentalist cannot approach directly, not even in principle. It is the difference between the two kinds of questions which distinguishes between experiment and theory, as well as the difference between the explanatory and predictive roles of theory itself; clearly, if we give δA and ask for the consequent δB, we are predicting, whereas if we assume δB and ask for the antecedent δA, we are explaining.

It should be noted that exactly the same duality arises in mathematics and logic themselves; that is, in purely formal systems. Thus, a mathematician can ask (*informally*): If (I make certain assumptions), then (what follows)? Or, the mathematician can start with a conjecture, and ask: If (Fermat's Last Theorem is true), then (what initial conditions must I assume to construct explicitly a proof)? The former is analogous to prediction, the latter to explanation.

When formal systems (i.e. logic and mathematics) are used to construct images of what occurs in the world, then interrogations and implications become associated with ideas of causality. Indeed, the whole concept of natural law depends precisely on the idea that causal processes in natural systems can be made to correspond with implication in some appropriate, descriptive inferential system (e.g. Rosen, 1984, where this theme is developed at great length).

But the concept of causality is itself a complicated one; a fact largely overlooked in modern scientific discourse, to its cost. That causality is complicated has already been pointed out by Aristotle, for whom all science was animated by a specific interrogative: Why? He said explicitly that the business of science was to concern itself with "the why of things". In our language, these are just the questions of *theoretical* science: If (B), then (what A)? and hence we can say B *because* A. Or, in the variational form, δB *because* δA.

However, Aristotle argued that there were four distinct categories of causation; four ways of answering the question *why*. These categories, which he called *material cause, formal cause, efficient cause*, and *final cause*, are not interchangeable. If this is so (and I argue below that, indeed, it is), then there are correspondingly *different kinds of information*, associated with different causal categories. These different kinds of information have been confused, mainly because we are in the habit of using the same mathematical language to describe each of them; it is from these inherent confusions that much of the ambiguity and murkiness of the concept of information ultimately arises. Indeed, we can say more than this: the very fact that the same mathematical language does not (in fact, cannot) distinguish between essentially distinct categories of causation means that the mathematical language we have been using is, in itself, somehow fundamentally deficient, and that it must be extended by means of supplementary structures to eliminate those deficiencies.

The Paradigm of Mechanics

The appearance of Newton's *Principia* toward the end of the seventeenth century was surely an epochal event. Though nominally the theory of physical systems of mass points, it was much more. In practical terms, by showing how the mysteries of the heavens could be understood on the basis of a few simple, universal laws, it set the standards for explanation and prediction which have been accepted ever since. It unleashed a feeling of optimism almost unimaginable today; it was the culmination of the entire Renaissance. More than that: in addition to providing a universal explanation for specific physical events, it also provided a language and a way of thinking about systems which has persisted, essentially unchanged, to the present time; what has changed has only been the technical manifestation of the language and its interpretation. In this language, the word information does not appear in any formal, technical sense; we have only words like energy, force, potential, work, and the like.

It is important to recognize the twin roles played by Newtonian mechanics in science: as a reductionistic ultimate and as a paradigm for representing systems not yet reduced to arrangements of interacting particles. The essential feature of this paradigm is the employment of a mathematical language with an inherent duality, which we may express as the distinction between ***internal states*** and ***dynamical laws***. In Newtonian mechanics, the internal states are represented by points in some appropriate manifold of phases, and the dynamical laws represent the internal or impressed forces. The resulting mathematical image is thus what is called nowadays a ***dynamical system***. However, the dynamical systems arising in mechanics are mathematically rather special ones, because of the way phases are defined (they possess a symplectic structure). Through the work of people like Poincaré, Birkhoff, Lotka, and many others over the years, however, this dynamical system paradigm, or its numerous variants, has come to be regarded as the universal vehicle for the representation of systems which could not, technically, be described in terms of mechanics; systems of interacting chemicals, organisms, ecosystems, and many others. Even the most radical changes occurring within physics itself, like relativity and quantum theory, manifest this framework; in quantum theory, for instance, there was the most fundamental modification of what constitutes a ***state***, and how it is connected to what we can observe and measure; but otherwise, the basic partition between states and dynamical laws is relentlessly maintained. Roughly, this partition embodies a distinction between what is inside or intrinsic (the states) and what is outside (the dynamical laws, which are formal generalizations of the mechanical concept of impressed force).

This, then, is our inherited ***mechanical paradigm***, which in its many technical variants or interpretations has been regarded as a universal language for describing systems and their effects. The variants take many forms; automata theory, control theory, and the like, but they all conform to the same basic framework first exhibited in the ***Principia***.

Among other things, this framework is regarded as epitomizing the concept of causality. We examine this closely here, because it is important when we consider the concept of information within this framework.

Mathematically, a dynamical system can be regarded simply as a vector field on a manifold of states; to each state, there is an assigned velocity vector (in mechanics it is, in fact, an acceleration vector). A given state (representing what the system is intrinsically like at an instant) together with its associated tangent vector (which represents what the effect of the external world on the system is like at an instant) uniquely determine how the system will change state, or move in time. This translation of environmental effects into a unique tangent vector is already a causal statement, in some sense; it translates into a more perspicuous form through a process of ***integration***, which amounts to solving the equations of motion. More precisely, if a dynamical system is expressed in the familiar form

$$dx_i/dt = f_i(x_1,\ldots,x_n) \quad i = 1,\ldots,n \tag{7.1}$$

in which time does not generally appear as an explicit variable (but only implicitly through its differential or derivation, dt), the process of integration manifests the explicit dependence of the state variables $x_i = x_i(t)$ on time,

$$x_i(t) = \int_{t_0}^{t} f_i[x_1(\tau),\ldots,x_n(\tau)]d\tau + x_i(t_0) \ . \tag{7.2}$$

This is a more traditional kind of causal statement, in which the state at time t is treated as an *effect*, and the right-hand side of equation (7.2) contains the *causes* on which this effect depends.

Before going further, let us take a look at the integrands in equation (7.2), which are the velocities or rates of change of the state variables. The mathematical character of the entire system is determined solely by the *form* of these functions. Hence, we can ask: What is it that expresses this form (i.e. what determines whether our functions are polynomials, or exponentials, or of some other form)? And given the general form (polynomial, say), what is it that picks out a specific function and distinguishes it from all others of that form?

The answer, in a nutshell, is *parameters*. As I have written the system (7.1) above, no such parameters are explicitly visible, but they are at least tacit in the very writing of the symbol f_i. Mathematically, these parameters serve as coordinates for function spaces; just as any other coordinate, they label or identify the individual members of such spaces. They thus play a very different role to the state variables, which constitute the arguments or domains of the functions that they identify.

Here we find the first blurring. For the parameters which specify the form of the functions f_i can, *mathematically*, be thrown in as arguments of the functions f_i themselves; thus, we could (and in fact always do) write

$$f_i = f_i(x_1, \ldots, x_n, a_1, \ldots, a_r) \tag{7.3}$$

where a_i are *parameters*. We could even extend the dynamical equations (7.1) by writing $da_i/dt = 0$ (if the a_i are indeed independent of time); thus, mathematically we can entirely eradicate any distinction between the parameters and the state variables.

There is still one further distinction to be made. We pointed out above that the parameters a_i represent the effects of the outside world on the intrinsic system states. These effects involve *both* the system *and* the outside world. Thus, some of the parameters must be interpreted as intrinsic too (the so-called *constitutive* parameters), while others describe the state of the outside world. These latter obey their own laws, not incorporated in equation (7.1), so they are, from the standpoint of equation (7.1), simply regarded as *functions of time* and must be posited independently. They constitute what are variously called *inputs, controls*, or *forcings*. Indeed, if we regard the states $[x_i(t)]$, or any mathematical functions of them, as corresponding *outputs* (that is, output as a function of input rather than just of time) we pass directly to the world of control theory.

So let us review our position. Dividing the world into state variables plus dynamical laws amounts to dividing the world into state variables plus parameters, where the role of the parameters is to determine the *form* of the functions, which in turn define the dynamical laws. The state variables are the arguments of these functions, while the parameters are coordinates in function spaces. Further, we must partition the parameters themselves into two classes; those which are *intrinsic* (the constitutive parameters) and those which are *extrinsic*; that is, which reflect the nature of the environment. The intrinsic parameters are intuitively closely connected with the *system identity*; that is, with the specific nature or character of the system itself. The values they assume might, for example, tell us whether we are dealing with oxygen, carbon dioxide, or any other chemical *species*, and, therefore, cannot change without our perceiving that a *change of species* has occurred. The environmental parameters, as well as the state variables, however, can change without affecting the species of the system.

These distinctions cannot be accommodated with the simple language of vector fields on manifolds; that language is too abstract. We can only recapture these distinctions by (a) superimposing an informal layer of *interpretation* on the formal language, as we have done above, or (b) changing the language itself, to render it less abstract. Let us examine how this can be done.

In order to have names for the various concepts involved, I call the constitutive parameters, which specify the *forms* of the dynamical laws, and hence the species of system with which we are dealing, the system *genome*; the remaining parameters, which reflect the nature of the external world, I call the system *environment*, and the state variables themselves I call *phenotypes*. This rather provocative terminology is chosen to deliberately reflect corresponding biological situations; in particular, I have argued (cf. Rosen, 1978) that, viewed in this light, the genotype–phenotype dualism which is regarded as so characteristically biological has actually a far more universal currency.

The mathematical structure appropriate to reflect the distinctions we have made is that of genome-parameterized mappings from a space of environments to a space of phenotypes; that is, mappings of the form

$$f_g : E \to P$$

specified in such a way that given any initial phenotype, environment-plus-genome determines a corresponding trajectory. Thus, we have no longer a simple manifold of states, but rather a fiber-space structure in which the basic distinctions between genome, environment, and phenotype are embodied from the beginning. Some of the consequences of this scenario are examined in Rosen (1978, 1983); we cannot pause to explore them here.

Now we are in a position to discuss the actual relation between the Newtonian paradigm and the categories of causation described earlier. In brief, if we regard the phenotype of the system at time t as *effect*, then

(1) Initial phenotype *is* material cause.
(2) Genome g *is* formal cause.
(3) $f_g(a)$, as an operator on the initial phenotype, *is* efficient cause.

Thus, the distinctions we have made between genome, environment, and phenotype are directly related to the old Aristotelian categories of causation. As we shall soon discover, that is why these distinctions are so important.

Note that one of the Aristotelian categories is missing from the above; there is no *final cause*. Ultimately, this is the reason why final cause has been banished from science; the Newtonian paradigm simply has no room for it. Indeed, it is evident that any attempt to superimpose a category of final causation upon the Newtonian world would effectively destroy the other categories within it.

In a deep sense, the Newtonian paradigm has led us to the notion that we may effectively *segregate the categories of causation* in our system descriptions. Indeed, the very concept of system state segregates the notion of material cause from other categories of causation, and tells us that it is correct to deal with all aspects of material causation independent of other categories: likewise with the concepts of genome and environment. I, in fact, claim that *this very segregation into independent categories of causation is the heart of the Newtonian paradigm.* When stated in this way, however, the universality of the paradigm perhaps no longer appears so self-evident.

Information

We said above that information is, or can be, the answer to a question, and that a question can generally be put in the variational form: If δA, then δB?. This serves as the connecting bridge between information and the Newtonian paradigm. In fact, it has played an essential role in the historical development of Newtonian mechanics and its variants, under the rubric of *virtual displacements.*

In mechanics, a virtual displacement is a small, imaginary change imposed on the *configuration* of a mechanical system, while the impressed forces are kept fixed. The animating question is: If such a virtural displacement is made under given circumstances, then what happens? The answer, in mechanics, is the well-known *Principle of Virtual Work*: if a mechanical system is in equilibrium, then the virtual work done by the impressed forces as a result of the virtual displacement must vanish. This is a static (equilibrium) principle, but it can readily be extended from statics to dynamics, where it is known as *D'Alembert's Principle.* In the dynamical case, it leads directly to the differential equations of motion of a mechanical system when the impressed forces are known. Details can be found in any text on classical mechanics.

In what follows, we explore the effect of such virtual displacements on the apparently more general class of dynamical systems of the form

$$dx_i/dt = f_i(x_1,\dots,x_n) \quad i = 1,\dots,n \ . \tag{7.4}$$

There is, however, a close relationship between the general dynamical systems (7.4) and those of Newtonian mechanics; indeed, the former systems can be regarded as arising out of the latter by the imposition of a sufficient number of nonholonomic constraints.[1]

As we have already noted, the language of dynamical systems, like that of Newtonian mechanics, does not include the word information; the study of such systems revolves around the various concepts of *stability.* However, in one of his analyses of oscillations in chemical systems, Higgins (1967) drew attention to the quantities

$$u_{ij}(x_1,\dots,x_n) = \partial/\partial x_j(dx_i/dt) \ .$$

These quantities, which he called cross-couplings if $i \neq j$ and self-couplings if $i = j$, arise fundamentally from the conditions which govern the existence of oscillatory solutions to equations (7.4). It turns out that it is not so much the magnitudes as the signs of these quantities that are important. In order to have a convenient expression for the signs of these quantities, he proposed that we call the jth state variable, x_j, an *activator* of the ith, in the state $(x_1^0,\dots,x_n^0)$, whenever the quantity

$$u_{ij}(x_1^0,\dots,x_n^0) = \frac{\partial}{\partial x_j}\left(\frac{dx_i}{dt}\right)_{(x_1^0,\dots,x_n^0)} > 0$$

and an *inhibitor* whenever

$$u_{ij}(x_1^0,\dots,x_n^0) < 0 \ .$$

Now, activation and inhibition are *informational* terms. Thus, Higgins' terminology provides an initial hint as to how dynamical language might be related to informational language, through the Rosetta stone of stability.

Now let us examine what Higgins' terminology implies. If x_j activates x_i in a particular state, then a (virtual) increase in x_j increases the *rate of change* of x_i or, alternatively, a (virtual) decrease of x_j decreases the rate of change of x_i. It is, intuitively, eminently reasonable that this is the role of an activator. Conversely, if x_j inhibits x_i, it means that an increase in x_j decreases the rate of change of x_i, etc.

Thus, the n^2 functions, $u_{ij}(x_1,\dots,x_n)$; $i, j = 1,\dots,n$, constitute a form of informational description for the dynamical system (7.4), which I have elsewhere (Rosen, 1979) called an *activation–inhibition pattern*. As we have noted, such a pattern concisely represents the answers to the variational questions: If we make a virtual change in x_j, what happens to the rate of production of x_i?.

There is no reason to consider only the quantities u_{ij}. We can, for instance, go one step further, and consider the quantities

$$u_{ijk}(x_1,\dots,x_n) = \partial/\partial x_k[\partial/\partial x_j(\mathrm{d}x_i/\mathrm{d}t)] \ .$$

Intuitively, these quantities measure the effect of a (virtual) change in x_k on the *extent* to which x_j activates or inhibits x_i. If such a quantity is positive in any particular state, it is reasonable to call x_k an *agonist* of x_j with respect to x_i; if negative, an *antagonist*. That is, if u_{ijk} is positive, a (virtual) increase in x_k increases or facilitates the activation of x_i by x_j, etc. The quantities u_{ijk} thus define another layer of informational interaction, which we may call an *agonist–antagonist pattern*.

We can iterate this process, in fact to infinity, to produce at each state r a family of n^r functions, $u_{ij\dots r}(x_1,\dots,x_n)$. Each layer in this increasing sequence describes how a (virtual) change of a variable at that level modulates the properties of the preceding level.

So far we have considered only the effects of virtual changes in state variables, x_j, on the velocities, $\mathrm{d}x_i/\mathrm{d}t$, at various informational levels. We could similarly consider the effects of virtual displacements at these various levels on the second derivatives, $\mathrm{d}^2x_i/\mathrm{d}t^2$ (i.e. on the *accelerations* of x_i), the third derivatives $\mathrm{d}^3x_i/\mathrm{d}t^3$, and so on. Thus, we have a doubly infinite web of informational interactions, defined by the functions

$$u_{ijk\dots r}^m(x_1,\dots,x_n) = \frac{\partial}{\partial x_r}\left[\cdots\ \frac{\partial}{\partial x_j}\left(\frac{\mathrm{d}x_i^m}{\mathrm{d}t^m}\right)\cdots\right]$$

If we start from the dynamical equations (7.4), then nothing new is learned from these circumlocutions beyond, perhaps, a deeper insight into the relations between dynamical and informational ideas. Indeed, given any layer of informational structure, we can proceed to succeeding layers by mere differentiation, and to antecedent layers by mere integration. Thus, knowledge of any layer in this infinite array of layers determines all of them and, in particular, the dynamical equations themselves. If we know, for instance, the activation–inhibition pattern

$u_{ij}(x_1,..., x_n)$, we can reconstruct the dynamical equations (7.4) through the relationship

$$df_i = \sum_{j=1}^{n} u_{ij} dx_j \tag{7.5}$$

(note in particular that the differential form on the right-hand side resembles a generalized *work*), and then set the function $f_i(x_1,..., x_n)$ so determined equal to the rate of change, dx_i / dt, of the ith state variable.

However, *our ability to do all this depends fundamentally on the exactness of the differential forms which arise at every level of our web of informational interaction,* and which relate each level to its neighbors. If the forms in equation (7.5) are not exact, there are no functions $f_i(x_1,..., x_n)$ whose differentials are given by it, and hence *no rate equations of the form (7.4).* In such a situation, the simple relationship between the levels in our web breaks down completely; the levels become independent of each other, and must be posited separately. So two systems could have the same activation–inhibition patterns, but vastly different agonist–antagonist patterns, and hence manifest entirely different behaviors.

To establish firmly these ideas, let us examine what is implied by the requirement that the differential forms

$$\sum_{j=1}^{n} u_{ij} dx_j$$

defined by the activation–inhibition pattern be exact. The familiar, necessary conditions for exactness here take the form

$$\frac{\partial}{\partial x_k} u_{ij} = \frac{\partial}{\partial x_j} u_{ik}$$

for all $i, j, k = 1,..., n$. Intuitively, these conditions mean that *the relations of agonism and activation are entirely symmetrical* (commutative); that x_k as an agonist of the activator x_j is exactly the same as x_j as an agonist of the activator x_k; and similarly for all other levels.

Clearly, such situations are extremely degenerate in informational terms. They are so because the requirement of exactness is highly nongeneric for differential forms. Thus, these very simple considerations suggest a most radical conclusion: that *the Newtonian paradigm, with its emphasis on dynamical laws, restricts us from the outset to an extremely special class of systems, and that the most elementary informational considerations force us out of that class.* We explore some of the implications of this situation in the following section.

Meanwhile, let us consider some of the ramifications of these informational ideas that hold even within the confines of the Newtonian paradigm. These concern the distinctions made in the preceding section between environment, phenotype, and genome; the relations of these distinctions to different categories of causation; and the correspondingly different categories of information which these causal categories determine.

First, let us recall that according to the Newtonian paradigm, every relation between physical magnitudes (i.e. every equation of state) can be represented as a genome-parameterized family of mappings

$$f_g : E \rightarrow P$$

from environments to phenotypes. It is worth noting specifically that every dynamical law or equation of motion is of this form, as is shown by

$$d\boldsymbol{x}/dt = f_g(\boldsymbol{x}, \boldsymbol{a}) \ . \tag{7.6}$$

Here, in traditional language, $\boldsymbol{x}$ is a vector of states, $\boldsymbol{a}$ is a vector of external controls (which together with states constitutes *environment*), and the phenotype is the tangent vector $d\boldsymbol{x}/dt$ attached to the state $\boldsymbol{x}$.[2] In this case, then, the tangent vector or phenotype constitutes *effect*; the genome g is identified with formal cause, state x with material cause, and the operator $f_g(\ldots, \boldsymbol{a})$ with efficient cause.

By analogy with the activation–inhibition networks and their associated informational structures, described above, we can consider formal quantities of the form

$$\frac{\partial}{\partial(\text{cause})}\left[\frac{d}{dt}(\text{effect})\right] \tag{7.7}$$

As always, such a formal quantity represents an answer to a question: If (cause is varied), then (what happens to effect)? This is the same question as we asked in connection with the definition of activation–inhibition networks and their correlates, but now set in the wider context to which our analysis of the Newtonian paradigm has led us. That is, we may now virtually displace *any* magnitude which affects the relation (7.6), whether it be a genomic magnitude, an environmental magnitude, or a state variable. In a precise sense, the effect of such a virtual displacement is measured by the quantity (7.7).

It follows that there are indeed different *kinds* of information. What kind of information we are dealing with depends on whether we apply the virtual displacement to a genomic magnitude (associated with formal cause), an environmental magnitude (efficient cause), or a state variable (material cause). Formally, we can now distinguish at least the following three cases:

(1) Genomic information,

$$\frac{\partial}{\partial(\text{genome})}\left[\frac{d}{dt}(\text{effect})\right] .$$

(2) Phenotypic information,

$$\frac{\partial}{\partial(\text{state})}\left[\frac{d}{dt}(\text{effect})\right] .$$

(3) Environmental information,

$$\frac{\partial}{\partial(\text{control})}\left[\frac{d}{dt}(\text{effect})\right] .$$

We confine ourselves herein to these three, which generalize only the activation–inhibition patterns described above.

We now examine an important idea; namely, ***the three categories defined above are not equivalent***. Before justifying this assertion, we must briefly discuss what is meant by equivalent. In general, the mathematical assessment of the effects of perturbations (i.e. of real or virtual displacements) is the province of

stability. For example, the effect on subsequent dynamical behavior of modifying or perturbing a system state is the province of Lyapunov stability of dynamical systems; that of perturbing a control is part of control theory; and that of perturbing a genome relates to structural stability. To establish this firmly, let us consider genomic perturbations, or ***mutations***. A virtual displacement applied to a genome g replaces the initial mapping f_g determined by g with a new mapping $f_{g'}$. Mathematically, we say that the two mappings, f_g and $f_{g'}$, are equivalent, or similar, or conjugate, if there exist appropriate transformations

$$\alpha : E \to E \ ,$$

$$\beta : P \to P \ ,$$

such that the diagram

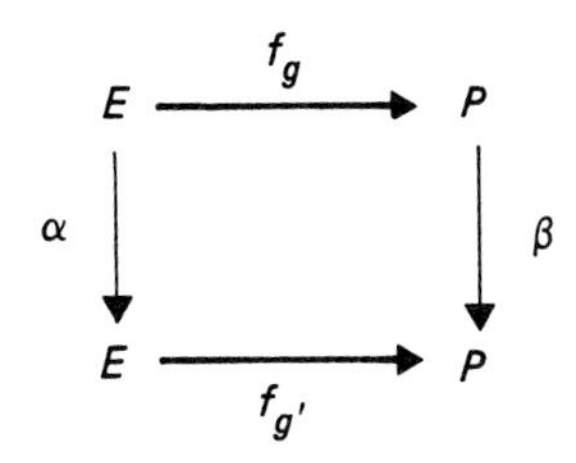

commutes; that is, if

$$\beta[f_g(e)] = f_{g'}[\alpha(e)]$$

for every e in E. Intuitively, this means that a mutation $g \mapsto g'$ can be counterbalanced, or nullified, by imposing suitable ***coordinate transformations*** on the environments and phenotypes. Stated yet another way, a virtual displacement of genome can always be counteracted by corresponding displacements of environment and phenotype so that the resultant variation on effect vanishes.

We have elsewhere (Rosen, 1978) shown at great length that this commutativity may not always obtain; that is, that there may exist genomes which are bifurcation points. In any neighborhood of a bifurcating genome g, there exist genomes g' for which f_g and $f_{g'}$ fail to be conjugate.

With this background, we return to the question of whether the three kinds of information (genomic, phenotypic, and environmental) defined above are equivalent. Intuitively, equivalence would mean that the effect of a virtual displacement δg of genome, supposing all else is fixed, could equally well be produced by a virtual displacement of environment, δa, or of phenotype, δp. Or stated another way, the effect of a virtual displacement δg of genome can be nullified by virtual displacements $-\delta a$ and $-\delta p$ of environment and phenotype, respectively. This is simply a restatement of the definition of conjugacy or similarity of mappings.

If all forms of information are equivalent, it follows that there could be no bifurcating genomes. We note in passing that the assumption of equivalence of the three kinds of information defined above thus creates terrible ambiguities when it comes to ***explanation*** of particular effects. We do not consider that aspect here, except to say that it is perhaps very fortunate that, as we have seen, they are not equivalent.

Let us examine one immediate consequence of the nonequivalence of genomic, environmental, and phenotypic information, and of the considerations which culminate in that conclusion. Long ago (cf. von Neumann, 1951; Burks, 1966) von Neumann proposed an influential model for a self-reproducing automaton, and subsequently, for automata which grow and develop. This model was based on the famous theorem of Turing (1936), which established the existence of a universal computer (universal Turing machine). From the existence of such a universal computer, von Neumann asserted that there must also exist a universal constructor. Basically, he argued that computation (i.e. following a program) and construction (following a blueprint) are both algorithmic processes, and that anything holding for one class of algorithmic processes necessarily holds for any other class. This universal constructor formed the central ingredient of the self-reproducing automaton.

Now, a computer acts, in the language we have developed above, through the manipulation of efficient cause. A constructor, if the term is to bear any resemblance to its intuitive meaning, must essentially manipulate material cause. The inequivalence of the two categories of causality, in particular manifested by the nonequivalence of environmental and phenotypic information, means that we cannot blithely extrapolate from results pertaining to efficient causation into the realm of material causation. Indeed, in addition to invalidating von Neumann's specific argument, we learn that great care must be exercised in general when arguing from purely logical models (i.e. from models pertaining to efficient cause) to any kind of physical realization, such as developmental or evolutionary biology (which pertain to material cause).

Thus, we realize how significant are the impacts of informational ideas, even within the confines of the Newtonian paradigm, in which the categories of causation are essentially segregated into separate packages. We now consider what happens when we vacate the comforting confines of the Newtonian paradigm.

An Introduction to Complex Systems

Herein, I call any natural system for which the Newtonian paradigm is completely valid a *simple system*, or *mechanism*. Accordingly, a *complex system* is one which, for one reason or another, resides outside this paradigm. We have already seen a hint of such systems in the preceding section; for example, systems whose activation–inhibition patterns u_{ij} do not give rise to exact differentials $\sum u_{ij} dx_j$. However, some further words of motivation must precede a conclusion that such systems are truly complex (i.e. reside fundamentally outside the Newtonian paradigm). We must also justify our very usage of the term complex in this context.

What I have been calling the Newtonian paradigm ultimately devolves upon *the class of distinct mathematical descriptions* which a system can have, and the relations which exist between these descriptions. As noted earlier, the basis of system description arising in this paradigm is the fundamental dualism between states and dynamical laws. Thus, the mathematical objects which can describe natural systems comprise a category which may be called general dynamical systems. In a formal sense, it appears that any mathematical object resides in this category, because the Newtonian partition between states and dynamical laws exactly parallels the partition between propositions and production rules (rules of inference) which presently characterize all logical systems and logical theories. However, we argue that, although this category of general dynamical systems is large, it is not everything, and, indeed, it is far from large enough.

The Newtonian paradigm asserts much more than simply that every image of a natural system must belong to a given category. It asserts certain relationships between such images. In particular (and this is the reductionistic content of the paradigm), it asserts that among these images there is the universal one, which effectively maps on all the others. Intuitively, this is the master description or ultimate description, in which every shred of physical reality has an exact mathematical counterpart; in category–theoretic terms, it is much like a free object (a generalization of the concept of free semigroup, free group, etc.).[3]

There is still more. The ingredients of this ultimate description, by their very nature, are themselves devoid of internal structure; their only changeable aspects are their relative positions and velocities. Given the forces acting between them, as Laplace noted long ago, everything that happens in the external world is in principle predictable and understandable. From this perspective, everything is determined; there are no mysteries, no surprises, no errors, no questions, and no information. This is as much true for quantum theory as for classical; only the nature of state description has changed. And it applies to everything, from atoms to organisms to galaxies.

How does this universal picture manifest itself in biology? First, from the standpoint of the physicist, biology is concerned with a rather small class of extremely special (indeed, inordinately special) systems. In the theoretical physicist's quest for general and universal laws, there is thus not much contact with organisms. As far as he or she is concerned, what makes organisms special is not that they transcend the physicist's paradigms, but rather that their specification within the paradigm requires a plethora of special constraints and conditions, which must be superimposed on the universal canons of system description and reduction. The determination of these special conditions is an empirical task; essentially someone else's business. But it is not doubted that the relationship between physics and biology is the relationship between the general and the particular.

The modern biologist, in general, avidly embraces this perspective.[4] Historically, biology has only recently caught up with the Newtonian revolution which swept the rest of natural philosophy in the seventeenth century. The three-century lag arose because biology has no analog of the solar system; no way to make immediate and meaningful contact with the Newtonian paradigm. Not until physics and chemistry had elaborated the technical means to probe microscopic properties of matter (including organic matter) was the idea of molecular biology even thinkable. And this did not happen until the 1930s.

At present, there is still no single inferential chain which links any important effect in physics to any important effect in biology. This is a fact; a datum; a piece of information. How are we to understand it? There are various possibilities. Kant, long ago, argued that organisms could only be properly understood in terms of final causes or intentionality; hence, from the outset he suggested that organisms fall completely outside the canons of Newtonian science, which are applicable to everything else. Indeed, the essential telic nature of organisms precluded even the possibility that a "Newton of the grassblade" would come along, and do for biology what Newton did for physics. Another possibility is the one we have already mentioned; we have simply not yet characterized all those special conditions which are necessary to bring biology fully within the scope of universal physical principles. Yet a third possibility has developed within biology itself, as a consequence of theories of evolution; it is that much of biology is the result of *accidents* which are *in principle* unpredictable and hence governed by no laws at all.[5] In this view biology is as much a branch of history as of science. At present, this last hypothesis lies in a sort of doublethink relation with reductionism; the two are quite inconsistent, but do allow modern biologists to enjoy the benefits of vitalism and mechanism together.

Yet a fourth view was expressed by Albert Einstein, who wrote in a letter to Leo Szilard: "One can best appreciate, from a study of living things, how primitive physics still is".

So, the present prevailing view in biology is that the Newtonian canons are indeed universal, and we are lacking only knowledge of the special conditions and constraints which distinguish organisms from other natural systems within those canons. One way of describing this with a single word is to assert that organisms are *complex*. This word is not well defined, but it does connote several things. One of these is that complexity is a system property, no different from any other property. Another is that the *degree* to which a system is complex can be specified by a number, or set of numbers. These numbers may be interpreted variously as the dimensionality of a state space, or the length of an algorithm, or as a cost in time or energy incurred in solving system equations.

On a more empirical level, however, complexity is recognized differently, and characterized differently. If a system surprises us, or does something we have not predicted, or responds in a way we have not anticipated; if it makes errors; if it exhibits emergence of unexpected novelties of behavior, we also say that the system is complex. In short, complex systems are those which behave counter-intuitively.

Sometimes, of course, surprising behavior is simply the result of incomplete characterization; we can then hunt for what is missing, and incorporate it into our system description. In this way, the planet Neptune was located from unexplained deviations of Uranus from its expected trajectory. But sometimes this is not the case; in the apparently analogous case of the anomalies of the trajectory of the planet Mercury, for instance, no amount of fiddling within the classical scenario succeeded and only a massive readjustment of the paradigm itself (via general relativity) availed.

From these few words of introduction, we can conclude that the identification of complexity with situations where the Newtonian paradigm fails is in accord with the intuitive connotation of the term, and is an alternative to regarding as complex any situation which merely is technically difficult within the paradigm.

Now let us see where information fits into these considerations. We recall that information is the actual or potential response to an interrogative, and that every interrogative can be put into the variational form: If δA, then δB? The Newtonian paradigm asserts, among other things, that the answers to such interrogatives follow from dynamical laws superimposed on manifolds of states. In their turn, these dynamical laws are special cases of *equations of state*, which link or relate the values of system observables. Indeed, the concept of an observable was the point of departure for our entire treatment of system description and representation (cf. Rosen, 1978); it was the connecting link between the world of natural phenomena and the entirely different world of formal systems which we use to describe and explain.

However, the considerations we have developed above suggest that this world is not enough. We require also a world of variations, increments, and differentials of observables. It is true that every linkage between observables implies a corresponding linkage between differentials, but as we have seen, the converse is not true. We are thus drawn to the notions that a differential relation is a generalized linkage and that a differential form is a type of generalized observable. A differential form which is not the differential of an observable is thus an entity which assumes no definite numerical value (as an observable does), but which can be incremented.

If we do think of differential forms as generalized observables, then we must correspondingly generalize the notion of equation of state. A generalized equation of state thus becomes a linkage or relation between ordinary observables and differentials or generalized observables. Such generalized equations of state are the vehicles which answer questions of our variational form: If δA, then δB?

But as we have repeatedly noted, such generalized equations of state do not usually follow from systems of dynamical equations, as they do in the Newtonian paradigm. Thus, we must find some alternative way of characterizing a system of this kind. Here is where the informational language introduced above comes to the fore. Let us recall, for instance, how we defined the activation–inhibition network. We found a family of functions u_{ij} (i.e. of observables) which could be thought of in the dynamical context as modulating the effect of an increment dx_j on that of another increment df_i. That is, the values of each observable, u_{ij}, measure precisely the extent of activation or inhibition which x_j exerts on the rate at which x_i is changing.

In this language, a system falling outside the Newtonian paradigm (i.e. a complex system) can have an activation–inhibition pattern, just as a dynamical (i.e. simple) system does. Such patterns are still families of functions (observables), u_{ij}, and the pattern itself is manifested by the differential forms

$$\omega_i = \sum u_{ij} dx_j$$

But in this case, there is no global velocity observable, f_i, that can be interpreted as the rate of change of x_i; there is only a velocity *increment*. It should be noted explicitly that u_{ij}, which define the activation–inhibition pattern, need not be functions of x_i alone, or even functions of them at all. Thus, the differential forms which arise in this context are different from those with which mathematicians generally deal, and which can always be regarded as cross sections of the cotangent bundle of a definite manifold of states.

The next level of information is the agonist–antagonist pattern, u_{ijk}. In the category of dynamical systems, this is completely determined by the activation–inhibition pattern, and can be obtained from the latter by differentiation:

$$u_{ijk} = \frac{\partial}{\partial x_k} u_{ij} \ .$$

In our world of generalized observables and linkages, u_{ijk} are independent of u_{ij}, and must be posited separately; in other words, complex (non-Newtonian) systems can have identical activation–inhibition patterns, but quite different agonist–antagonist patterns.

Exactly the same considerations can also be applied to every subsequent layer of the informational hierarchy; each is now independent of the others, and so must be posited separately. Hence a complex system requires an *infinite* mathematical object for its description.

We cannot examine herein the mathematical details of the considerations sketched so briefly above. Suffice it to say that a complex system, defined by a hierarchy of informational levels of the type described, is quite a different object to a dynamical system. For one, it is quite clear that there is no such thing as a set of *states*, assignable to such a system once and for all. From this alone, we might expect that the nature of causality in such systems is vastly different to what it is in the Newtonian paradigm; we come to this in a moment.

The totality of mathematical structures of the type we have defined above forms a category. In this category the class of general dynamical systems constitutes a very small subcategory. We are suggesting that the former provides a suitable framework for the mathematical imaging of complex systems, while the latter, by definition, can only image simple systems or mechanisms. If these considerations are valid (and I believe they are), then the entire epistemology of our approach to natural systems is radically altered, and it is the basic notions of information which provide the natural ingredients.

There is, however, a profound relationship between the category of general dynamical (i.e. Newtonian) systems, and the larger category in which it is embedded. This can only be indicated here, but it is important indeed. Namely, there is a precise sense in which an informational hierarchy can be *approximated*, locally and temporarily, by a general dynamical system. With this notion of approximation there is an associated notion of *limit*, and hence of topology. Using these ideas, it can be shown that what we call the category of complex systems is the completion, or limiting set, of the category of simple (i.e. dynamical) systems.

The fact that complex systems can be approximated (albeit locally and temporarily) by simple ones is crucial. It explains precisely why the Newtonian paradigm has been so successful, and why, to this day, it represents the only effective procedure for dealing with system behavior. But in general, it is apparent that it can usually supply *only* approximations, and in the universe of complex systems this amounts to replacing a *complex* system with a *simple subsystem*. Some of the profound consequences are considered in detail in Rosen (1978).

This relationship between complex systems and simple ones is, by its very nature, without a reductionistic counterpart. Indeed, what we presently understand as physics is seen in this light as *the science of simple systems*. The relation between physics and biology is thus not at all the relation of general to particular; in fact, quite the contrary. It is not biology, but physics, which is too special. We can see from this perspective that biology and physics (i.e. contemporary physics) develop as two divergent branches from a *theory of complex systems*, which as yet can be glimpsed only very imperfectly.

The category of simple systems is, however, still the only one that we know how to use. But to study complex systems by means of approximating simple systems resembles the position of early cartographers, who were attempting to map a sphere while armed only with pieces of planes. Locally, and temporarily, they could do very well, but globally, the effects of the topology of the sphere become progressively important. So it is with complexity; over short times and only a few informational levels, we can always make do with a simple (i.e. dynamical) picture. Otherwise, we cannot; we must continually replace our approximating dynamics with others as the old ones fail. Hence another characteristic feature of complex systems; they appear to possess a multitude of partial dynamical descriptions, which cannot be combined into one single complete description. Indeed, in earlier work (Rosen, 1977), we took this as the defining feature of complexity.

I add a brief word about the status of causality in complex systems, and about the practical problem of determining the functions which specify their informational levels. Complex systems do not possess anything like a state set which is fixed once and for all. Also, the categories of causality become intertwined in a way which is not possible within the Newtonian paradigm. Intuitively, this follows from the independence of the infinite array of informational layers which constitutes the mathematical image of a complex system. Variation of any particular magnitude connected with such a system typically manifests itself independently in many of these layers, and thus reflects itself partly as material cause, partly as efficient cause, and even partly as formal cause in the resultant variation of other magnitudes. We feel that it is, at least for the most part, this involvement of magnitudes simultaneously in each of the causal categories which makes biological systems so refractory to the Newtonian paradigm.

Also, this intertwining of the categories of causation in complex systems makes the direct interpretation of experimental results of the form: If δA, then δB, extremely difficult. If we are correct so far, such an observational result as it stands is far too coarse to have any clear-cut meaning. In order to be meaningful, an experimental proposition of this form must isolate the effect of a variation δA on a single informational level, keeping the others clamped. As might be appreciated, this will in general not be an easy task. In other words, the experimental study of complex systems cannot be pursued with the same tools and ideas that are appropriate for simple systems.

One final conceptual remark is also needed. As mentioned earlier, the Newtonian paradigm has no room for the category of final causation. This category is closely linked to the notion of anticipation, which in turn is linked to the ability of systems to possess internal predictive models of themselves and their environments, which can be utilized for the control of present actions. We have argued at great length elsewhere (cf. Rosen, 1984) that anticipatory control is indeed a distinguishing feature of the organic world, and have described some of the unique features of such anticipatory systems. Herein we have shown that for a system to be anticipatory, it must be complex. Thus, our entire treatment of anticipatory systems becomes a corollary of complexity. In other words, complex systems can admit the category of final causation in a perfectly rigorous, scientifically acceptable way. Perhaps this alone is sufficient recompense for abandoning the comforting confines of the Newtonian paradigm, which has served so well over the centuries. It will continue to serve us well, provided we recognize its restrictions and limitations, as well as its strengths.

Notes

[1] Newton's original ***particle mechanics***, or ***vectorial mechanics***, is hard to apply to many practical problems, and was early on (through the work of people like Euler and Lagrange) transmuted into another form, generally called ***analytical mechanics***. This latter form is usually used to deal with extended matter (e.g. rigid bodies). In particle mechanics, the rigidity of a macroscopic body is a consequence of interparticle forces, which must be explicitly taken into account in describing the system. Thus, if there are N particles in the system (however large N may be) there is a phase space of $6N$ dimensions, and a set of dynamical equations which expresses for each particle the resultant of ***all*** forces experienced by that particle. In analytical mechanics, on the other hand, any rigid body can be completely described by giving only six configurational coordinates (e.g. the coordinates of the center of mass, and three angles of rotation about the center of mass), however many particles it contains. From the particulate approach the internal forces which generate rigidity are replaced by ***constraints***; supplementary conditions on the configuration space which must be identically satisfied. Thus, the passage from particle mechanics to analytical mechanics involves a partition of the forces in an extended system into two classes: (a) the internal or ***reactive*** forces, which hold the system together, and (b) the ***impressed*** forces, which push the system around. The former are represented in analytical mechanics by algebraic constraints, the latter by differential equations in the configuration variables (six for a rigid body).

A system in analytical mechanics may have additional constraints imposed upon it by specific circumstances; for example, a ball may roll on a table top. It was recognized long ago that these additional constraints (which, like all constraints, are regarded as expressing the operation of reactive forces) can be of two types, which were called by Hertz ***holonomic*** and ***nonholonomic***. Both kinds of constraints can be expressed locally, in infinitesimal form, as

$$\sum_{i=1}^{n} u_i(x_1,\ldots,x_n)\mathrm{d}x_i = 0$$

where $x_1,\ldots,x_n$ are the configuration coordinates of the system. For a holonomic constraint, the above differential form is exact; that is, the differential of some global function $\varphi(x_1,\ldots,x_n)$ is defined over the whole configuration space. Thus, the holonomic constraint translates into a global relation

$$\varphi(x_1,\ldots,x_n) = \text{constant} \ .$$

This means that the configurational variables are no longer independent, and that one of them can be expressed as a function of the others. The constraint thus reduces the dimension of the configuration space by *one*, and therefore reduces the dimension of the phase space by *two*.

A nonholonomic constraint, on the other hand, does not allow us to eliminate a configurational variable in this fashion. However, since it represents a relation between the configuration variables and their differentials, it does allow us to eliminate a coordinate of *velocity*, while leaving the dimension of the configuration space unaltered. That is, a nonholonomic constraint serves to eliminate one degree of freedom of the system. It thus also eliminates one dimension from the space of impressed forces which can be imposed on the system without violating the constraint.

Similarly, if we impose r independent nonholonomic constraints on our system, we (a) keep the original dimension of the configuration space; (b) eliminate r coordinates of velocity, and thus reduce the dimensionality of the phase space by r; and (c) similarly, reduce by r the dimensionality of the set of impressed forces which can be imposed on the system.

Let us express these facts mathematically. A nonholonomic constraint can be expressed locally in the general form

$$\varphi\left(x_1,\ldots,x_n,\frac{dx_1}{dt},\ldots,\frac{dx_n}{dt}\right)=0$$

which can (locally) be solved for one of the velocity coordinates (dx_1/dt, say). Thus, it can be written in the form

$$\frac{dx_1}{dt}=\psi\left(x_1,x_2,\ldots,x_n,\frac{dx_2}{dt},\ldots,\frac{dx_n}{dt}\right)$$
$$=\Psi(x_1,\mathbf{a})$$

where we have written $\mathbf{a}=(x_2,\ldots,dx_n/dt)$. [At this point the reader is invited to compare this relation with equation (7.6) in the main text.]

Likewise, if there are r nonholonomic constraints, these can be expressed locally by r equations

$$dx_i/dt=\psi_i(x_1,\ldots,x_r,\mathbf{a}) \quad i=1,\ldots,r$$

where now $\mathbf{a}$ is the vector $(x_{r+1},\ldots,x_n,dx_{r+1}/dt,\ldots,dx_n/dt)$. These equations of constraint, which intuitively arise from the *reactive* forces holding the system together, now become more and more clearly the type of equations we always use to describe general dynamical or control systems.

Now what happens if $r=n$? In this case, the constraints leave us only *one degree of freedom; they determine a vector field on the configuration space.* There is in effect only one *impressed* force that can be imposed on such a system, and its only effect is to move the system; once moving, the motion is determined entirely by the *reactive* forces, and not by the *impressed* force. Mathematically, the situation is that of an autonomous dynamical system, whose manifold of states is the *configuration* space of the original mechanical system.

This relationship between dynamics and mechanics is quite different from the usual one, in which the manifold of states is thought of as generalizing the mechanical notion of *phase*, and the equations of motion as generalizing the *impressed* force. In the above interpretation, however, it is quite different; the manifold of states correspond now to mechanical *configurations*, and the equations of motion come from the *reactive* forces.

[2] The reader should be most careful not to confuse two kinds of propositions, which are equivalent mathematically but completely different epistemologically and causally. On the one hand, we have a statement like

$$d\mathbf{x}/dt=f_g(\mathbf{x},\mathbf{a}) .$$

This is a local proposition, linking a tangent vector or velocity $d\mathbf{x}/dt$ to a state $\mathbf{x}$, a genome g, and a control $\mathbf{a}$. Each of these quantities is derived from observables assuming definite numerical values at any instant of time, and it is *their values at a common instant* which are related by this proposition.

On the other hand, the integrated form of these dynamical relations is

$$\mathbf{x}(t)=\int_{t_0}^{t} f_g[x,a(\tau)]d\tau .$$

This relationship involves time *explicitly* and links the values of observables *at one instant* with values (assumed by these and other observables) *at other instants*.

Each of these epistemically different propositions has its own causal structure. In the first, we treat the tangent vector dx/dt as effect and define its causal antecedents as we have done. In the integrated form, on the other hand, we take $\mathbf{x}(t)$ as effect and find a correspondingly different causal structure. In general, the mathematical or logical equivalence of two expressions of linkage or relationship in physical systems does not at all connote that their causal structures are identical. This is merely a manifestation of what was discussed earlier, that the mathematical language we use to represent physical reality has abstracted away the very basis on which such causal discriminations can be made.

[3] It should be recognized that this reductionistic part of the Newtonian paradigm can fail for purely mathematical reasons. If it should happen that there is no way to effectively map the master description onto some partial description, then this is enough to defeat a reductionistic approach to those system behaviors with which the partial description deals. This is quite a different matter from the one we are considering here, in which no Newtonian master description *exists*, and the program fails for *epistemological* reasons, rather than mathematical ones.

[4] This statement is not simply my subjective assessment. In 1970 there appeared a volume entitled *Biology and the Future of Man*, edited by Philip Handler (1970), then President of the National Academy of Sciences of the USA. The book went to great lengths to assure the reader that it spoke for biology as a science; that in it biologists spoke with essentially one voice. At the outset, it emphasized that the volume was not prepared as a (mere) academic exercise, but for serious pragmatic purposes:

> Some years ago, the Committee on Science and Public Policy of the National Academy of Sciences embarked on a series of 'surveys' of the scientific disciplines. Each survey was to commence with an appraisal of the 'state of the art'.... In addition, the survey was to assess the nature and strength of our national apparatus for continuing attack on those major problems, e.g., the numbers and types of laboratories, the number of scientists in the field, the number of students, the funds available and their sources, and the major equipment being utilized. Finally, each survey was to undertake a projection of future needs for the national support of the discipline in question to assure that our national effort in this regard is optimally productive....

To address these serious matters, the Academy proceeded as follows:

> ....Panels of distinguished scientists were assigned subjects.... Each panel was given a general charge...as follows:
>
> The prime task of each Panel is to provide a pithy summary of the status of the specific sub-field of science which has been assigned. This should be a clear statement of the prime scientific problems and the major questions currently confronting investigators in the field. Included should be an indication of the manner in which these problems are being attacked and how these approaches may change within the foreseeable future. What trends can be visualized for tomorrow? What lines of investigation are likely to subside? Which may be expected to advance and assume greater importance?... Are the questions themselves...likely to change significantly?.... Having stated the major questions and problems, how close are we to the answers? The sum of these discussions, panel by panel, should constitute the equivalent of a complete overview of the highlights of current understanding of the Life Sciences.

There were twenty-one such Panels established, spanning the complete gamut of biological sciences and the biotechnologies. The recruitment for these Panels consisted of well over 100 eminent and influential biologists, mostly members of the Academy. How the panelists themselves were chosen is not indicated, but there is no doubt that they constituted an authoritative group.

In due course, the Panels presented their reports. How they were dealt with is described in colorful terms:

> .... In a gruelling one week session of the Survey Committee...each report was *mercilessly* exposed to the criticism of all the other members.... Each report was then rewritten and subjected to the *searching, sometimes scathing*, criticisms of the members of the parent Committee on Science and Public

Policy. The reports were again revised in the light of this exercise. Finally, the Chairman of the Survey Committee...devoted the summer of 1968 to the final editing and revising of the final work.

Thus we have good grounds for regarding the contents of this volume as constituting a truly authoritative consensus, at least, as of 1970. There are no minority reports; no demurrals; biology does indeed seem guaranteed here to speak with one voice.

What does that voice say? Here are a few characteristic excerpts:

The theme of this presentation is that life can be understood in terms of the laws that govern and the phenomena that characterize the inanimate, physical universe and, indeed, that at its essence life can be understood *only* in the language of chemistry. [emphasis added]

A little further along, we find this:

Until the laws of physics and chemistry had been elucidated, *it was not possible even to formulate* the important, penetrating questions concerning the nature of life... . The endeavors of thousands of life scientists...have gone far to document the thesis...(that) living phenomena are indeed intelligible in physical terms. And although much remains to be learned and understood, and the details of many processes remain elusive, those engaged in such studies hold *no doubt* that answers will be forthcoming in the reasonably near future. Indeed, *only two major questions* remain enshrouded in a cloak of *not quite* fathomable mystery: (1) the origin of life...and (2) the mind–body problem...yet (the extent to which biology is understood) even now constitutes a satisfying and exciting tale. [emphases added]

Still further along, we find things like this:

While *glorying* in how far we have come, these chapters also reveal how large is the task that lies ahead... . If (molecular biology) is exploited with vigor and understanding...a shining, hopeful future lies ahead. [emphasis added]

And this:

Molecular biology provides the closest insight man has yet obtained of the nature of life – and therefore, of himself.

And this:

It will be evident that the huge intellectual triumph of the past decade will, in all likelihood, be surpassed tomorrow – and to the everlasting benefit of mankind.

It is clear from such rhapsodies that the consensus reported in this volume is not only or even mainly a scientific one; it is an emotional and aesthetic one. And indeed, anyone familiar with the writings of Newton's contemporaries and successors will recognize them.

The volume to which we have alluded was published in 1970. But it is most significant that nothing fundamental has changed since then.

[5] In the inimitable words of Jacques Monod (1971, pp 42–43):

We can assert today that a universal theory, however completely successful in other domains, could never encompass the biosphere, its structure and its evolution as phenomena deducible from first principles... . The thesis I shall present...is that the biosphere does not contain a predictable class of objects or events but constitutes a particular occurrence, compatible with first principles but not deducible from these principles, and therefore *essentially unpredictable*. [emphasis added]

References

Burks, A. (1966) *Theory of Self-Reproducing Automata* (Urbana, IL: University of Illinois Press).

Handler, P. (Ed) (1970) *Biology and the Future of Man* (Oxford: Oxford University Press).

Higgins, J. (1967) Oscillating chemical reactions. *J. Ind. & Eng. Chem.* 59: 18–62.

Monod, J. (1971) *Chance and Necessity* (New York: Alfred A. Knopf).

von Neumann, J. (1951) The general and logical theory of automata, in L.A. Jeffress (Ed) *Cerebral Mechanisms in Behavior* (New York: John Wiley & Sons) pp 1–41.

Rosen, R. (1977) Complexity as a system property. *Int. J. General Systems* 3: 227–32.

Rosen, R. (1978) *Fundamentals of Measurement and Representation of Natural Systems* (New York: Elsevier).

Rosen, R. (1979) Some comments on activation and inhibition. *Bull. Math. Biophysics* 41: 427–45.

Rosen, R. (1983) The role of similarity principles in data extrapolation. *Am. J. Physiol.* 244: R591–9.

Rosen, R. (1985) *Anticipatory Systems* (London: Pergamon Press) in press.

Shannon, C. and Weaver, W. (1949) *The Mathematical Theory of Communication* (Urbana, IL: University of Illinois Press).

Turing, A.M. (1936) On computable numbers. *Proc. London Math. Soc. Ser.* 2, 42: 230–65.

CHAPTER 8

Organs and Tools: A Common Theory of Morphogenesis

René Thom

Introduction: Toward a Comprehensive Biological Theory

At the beginning of the sixteenth century, people began to anatomize dead bodies and discovered organs for which the putative function had to be found. The simplest method of establishing this was to associate the organ with a tool, which apparently naive procedure led, nevertheless, to striking successes. Within that century, Harvey showed the heart to be a pump that sent blood through natural pipes, the blood vessels. The skeleton (bones, joints, and muscles) provided obvious mechanical interpretations (a member acting as a lever, for instance); and the lungs were compared to a pair of bellows (with the obvious omission of the fundamental physiological function of gas exchange between air and blood). All these mechanical analogies led to Descartes' theory of the animal machine. It was only with our noblest organ, the brain (the seat of the soul), that these analogical explanations met with difficulties: How could consciousness and thinking be generated inside this apparently amorphous gray or white substance? But the mechanical imagery was to achieve, around 1950, its most notable success. The almost simultaneous appearance, in the middle of the twentieth century, of computers and molecular biology developed the idea that the genetic material, DNA, was the analogue of a computer program for the development of an adult organism from an egg. This interpretation offered a new, important breakthrough: previously the mechanical analogy had constantly raised the problem of biological finality. How could all these organs, so beautifully adapted to their function and of such a huge efficiency, be formed, apparently by themselves, during embryological development? Was it not necessary to postulate the assistance of a "genius", a demiurge who had to direct and control the whole process of epigenesis? The proposition that the complete structure of the organism might be encoded in the genomic nucleotide sequence of DNA solved the problem: one had only to admit that DNA, playing the role of the demiurge, directed the full development of the embryo, exactly as an engineer dictates orders to his or her subordinates in a manufacturing plant. Thus, teleology could be rendered more acceptable under the new term of teleonomy. This idea gained acceptance as soon as it was discovered that the traduction mechanism, DNA → protein, was effectively a code, in the technical sense of the word; to any triple of nucleotides there corresponds only one amino acid. But, in the initial case of metazoa embryonic development, the situation is entirely different: one has to understand how the genetic information supposedly included in DNA can render itself in the three-dimensional (3-D) organic structure of the embryo (and later of the adult). (Here I am referring to the Metazoa; the case of Procaryotes is somewhat different.) As a result, all modern biological thought has been trapped in the fallacious homonym associated with the phrase "genetic code", and this abuse of language has resulted in a state of conceptual sterility, out from which there is little hope of escape.

To achieve some progress in solving this difficulty, only a major theoretical jump will be of any help. And theory cannot exist – in biology no less than in any other discipline – without introducing imaginary entities. After all, life itself, in the usual reductionist view, cannot be anything other than an imaginary concept. Thus, in the organ–tool conceptualization, it may be more sensible to reverse the sense of the explanation; instead of explaining the organ by the tool, could we not explain the tool by the organ? This has already been suggested by many vitalist philosophers, such as Bergson. Could the intuitive imagination which led *Homo faber* to build his extraordinarily efficient tools – centuries before the appearance of modern science – be none other than the manifestation through phylogenetic evolution of some biological unconsciousness? A simple idea can be taken here as evidence: the tool, generally, extends the action, and the action, in its essential motor structure, is genetically inherited. We have here to reconsider the Lamarckian axiom: function creates the organ. Not, as is often trivially stated, that organs are created and develop as the result of the frequent performance of a function, but in a more abstract, platonic sense. All the regulatory properties of an organic structure rely on some geometric properties of a "figure of regulation" which lies in some abstract space of metabolic activities. A function is then a regulatory apparatus of a formal, dynamic nature insuring homeostasis of some physiologically important character (or parameter), such as chemical energy content, oxygen content, organic waste content, etc. The performance of such a function may involve a wide variety of physicochemical agents as well as the most diverse organs. When all these functions perform correctly, they insure canalization of the metabolic state of the system around a specific attractor (the figure of regulation).

The aim of theoretical biology is to describe (with the utmost accuracy possible) this geometric object. We are certainly still very far from this, but as an approximate procedure, we may try to give local descriptions associated with a specific function, using the data of partial models of the following type. If the figure of regulation is described as an invariant, closed object of flow X in a space of very high dimension Ω, a function may be given by an auxiliary dynamic $(M; Y)$, where Y is a flow in a phase space M of low dimension, and we have a smooth map $F(\Omega_x) \to M$ such that for any $x \in \Omega$, with $y = F(x)$, the image vector $\mathbf{F}(X; x)$ is very near $Y(y)$. Moreover, the image dynamics $(M; Y)$ have to be algebraically simple (as in catastrophe theoretic modeling). The local coordinates in Y have biological meaning; they insure that the semiotic character of life is the most important value to be defended against the potentially deadly threats of external stresses. [Here we have the axiologic character of these imaginary entities, frequently described mathematically by a potential function, following Kergosien's terminology (1984).] Of course, the use of such imaginary entities produces difficult philosophical and methodological problems. In fundamental physics, most basic notions (such as mass, force, fields, etc.) are also imaginary, but their existence is made legitimate by their implication in highly accurate quantitative laws. In biology we cannot expect such a justification; but, at least, we may introduce our imaginary entities in to ***qualitatively stable*** models. In the same way as physical laws are frequently the results of symmetry constraints, we may try to impose as rigorously as possible constraints for the spatiotemporal propagation of our entities. Practically, all these constraints arise as the consequences of a single principle, ***the principle of locality***, which can be stated as follows: any local process inside a living being has to be explained as a result of only local deterministic processes. Action at a distance is prohibited (e.g., pure magic).

In its most general formulation, the principle of locality could be stated not only with respect to the usual space–time dimensions, but also to all semiotic (imaginary) spaces in which our imaginary entities are embodied. However, this would exclude any kind of qualitative discontinuity in the behavior of such entities; such a restriction is too stringent and one has to allow that at exceptional loci discontinuities of a qualitative character may occur (catastrophes), in the same way as sound propagation has to accept the presence of shock waves. But, in such cases, the discontinuities themselves have to be explained in terms of local deterministic processes (this justifies the importance of catastrophe theoretic formalism).

Salient Forms and Pregnances

The concepts used in such analyses are of two kinds:

(1) *Salient* forms, that is forms which are defined by a sharp boundary in their background space. Molecules, cells, organs, and organisms are salient forms, with well-defined spatiotemporal localization. Most of these forms are clearly individualized and in space any two have to be disjoint (impenetrability).

(2) *Pregnances* denote all field-like entities; they are not localized so two distinct pregnances may be present at the same point of space–time without interacting. Pregnances propagate in space, according to local deterministic principles [for instance, partial differential equations (PDE) for physical fields]. Any pregnance involves some energy content and, as such, any propagating pregnance is also an energy flux. (In some sense, the concept of energy is the most general physical pregnance.) Biological pregnances are sometimes subjectively defined (as life, fear, etc.); objectively, they propagate in space–time via material support or appropriate salient forms (the support of these forms may be material or field-like; olfactory signals are of the first kind, visual and auditory of the second).

Interactions between salient forms and pregnances may be described as follows (cf. Thom, 1983). Pregnances emanate from salient forms called source forms. They may invest other salient forms, in which they create a change of internal state with, sometimes, perceptible effects (the so-called figurative effect); generally, invested salient forms become secondary, induced source forms for the same pregnance. Subjective pregnances propagate from salient form to salient form by two modes of action; action by contiguity (contact) and action by similarity. Objective pregnances propagate only by contact (the axiom of locality). Using this ontology of salience and pregnances, we may establish the following classification of local processes:

(1) *A salient form invested by a pregnance*. This may be subjective; for example, a subject under the influence of fear communicated by nearby subjects. Or it may be objective, such as excitation of an atom (salient form) by an electromagnetic field (a physical pregnance); microbial contagion of an animal is another example, the contagious disease being here an objective pregnance. The fact that the pregnance is carried by specific salient forms (the bacteria) is no reason to exclude a contagious disease from the pregnance concept.
(2) *Emission of a pregnance by a salient form*. An infested individual may propagate a contagious disease. In fact, it should be realized that in order for a form to be salient, it has to be carried to an observer by a physical field (such as light or sound).
(3) *Interaction of a pregnance with a salient form*. Here we are basically concerned with physical flows that satisfy, locally, the invariance of the kinetic momentum. If we immerse a solid (s) in such a flow, then the topology of the flow may be drastically changed according to the position of s inside the flow. This type of interaction is the basis of the notion of ***preprogram***, developed below.
(4) *Pregnance–pregnance interactions*. Little is known in general about such interactions, except the case where the pregnances are described by potential functions on the same space, where the catastrophe theoretic formalism may be valid. In particular, a pregnance may interact with itself in such a way as to create shocks (that is salient forms), but these rarely materialize as independent entities.
(5) *Interactions between salient forms*. I believe that such interactions can always be described as an exchange of pregnances between the forms.

Interactions of type (5) are the only ones recognized in modern biology (for instance, communications between cells). The logicist ideal is to eliminate pregnances and reduce them to contiguity processes that involve only contact, absorption, or emission of material bodies. But physical fields are not matter and biochemical interpretation of the great pregnances (the effects) is still very far from being achieved.

As said earlier, any physiological function has, in practice, a large number of biochemical carriers (even inside the same individual); this renders the biochemical interpretation of pregnances very difficult, although a complete theory should also explain the variations of biochemical support of a given function within, for instance, the animal kingdom. Thus, a comparative evolutionary physiology is needed, which should parallel, and complete, the comparative evolutionary embryology. Here, quite certainly, historical events played an important role, and accounting for their importance would require much painful study.

Among the imaginary entities that must be introduced are the classical epigenetic gradients of embryology. Nobody knows the biochemical basis for an animal's cephalocaudal gradient, but nobody can doubt its importance. On this basis alone, the theoretical introduction of pregnances seems to me in need of no further justification. But, of course, in each specific case one has to find constraints to the propagation of pregnances. Do they need specific salient forms to invest? What kind of partial differential equation, if any, must they satisfy to produce smooth propagation? If the physicochemical basis of the pregnance is known (molecular diffusion, electromagnetic fields, etc.) such constraints may be established. If the basis is unknown, one may have to rely only on a locality axiom, acting in a purely qualitative way. In such cases, it may be impossible to build a model in the usual quantitative sense of the word, but we may, nevertheless, arrive at a metaphor which, while not allowing strict control of the situation, could bring about a better understanding of the phenomena.

The Notion of Preprogram

In my book *Structural Stability and Morphogenesis* (Thom, 1972) I proposed that the major accidents of early embryology (gastrulation, etc.) could be explained by mechanisms of type (4), where initially we have only epigenetic gradients unfolding some singularity of a potential that is of metabolic origin. Such a concept is acceptable only in a static situation, in which the inner fluxes of energy within the embryo may be neglected. As soon as some inner circulation arises, then kinetic variations of these fluxes have to be taken into consideration; this corresponds to the somewhat mysterious process denoted ***internalization of an external variable*** in my previous work (Thom 1972). This is why any coherent theory of morphogenesis has to account for how a propagative flux of energy can be modified in its topology under a variation of the boundary constraints. This is the object of the classical obstacle problem in PDE theory (see, for instance, Arnol'd's book, *Catastrophe Theory*, 1984). I propose to summarize all that we need in the following metatheorem.

We consider a domain U in some Euclidean space $R^{n+1}(x; y)$; with x a specific, real coordinate, and $y \in R^n$. Suppose that the domain U meets the slice $0 \leq x \leq 1$ along some set B and suppose that U is crossed by an energy (or material) flow which propagates in the sense of increasing x. This flow emanates from a source situated distantly in the negative x half-space. We suppose that the boundary ∂B can be subjected to a deformation

$$G: B \times I \to R^{n+1}, \quad 0 < x < 1,$$

depending on the parameter $s \in I$, leaving the edges fixed, $B \cap [x = 0,1]$. Let $m(s; y)$, $y \in R^n$, be the asymptotic density of the energy flow for a value s of the deforming parameter [if s is time, then the deformation $G(s)$ is supposed to be infinitely slow with respect to the speed of the flow]. For any s, the boundary $G(s)$ is impenetrable to the flow, which must always be tangent to the boundary. Let C be the set of critical values of the smooth, real-valued function $m(s; y)$ in the plane $(s; m)$. Then, by the genericity assumption, the set C is a smooth curve with normal crossings as the only singularities, and its projection on the s-axis admits only fold points (f) as critical points (Figure 8.1). Proof of this metatheorem requires showing that the associated map F which transforms the function space of the deformation $G(s)$ into the function space of the density $m(s; y)$ is sufficiently surjective to permit transversality of the m functions which are not generic. I do not know to what extent theorems of this type are found in the current literature. It should be observed, also, that the spatial framework in which this theorem was stated may allow any kind of interpretation, whatever the propagating process we began with.

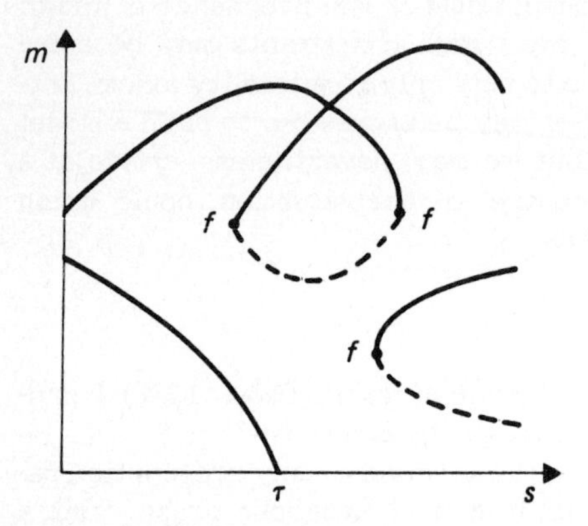

Figure 8.1 Arcs of maxima (—); arcs of index $n-1$(---); τ = stopping point.

In applications to morphogenesis, U is the usual domain of 3-D Euclidean space. We consider, on the curve C, those arcs J associated with the maxima of the density $m(s;y)$, and we restrict ourselves to only fairly sharp maxima, with very low intervening thresholds. The standard fold point of C is the junction of a J-arc with an arc of index one in the density $m(s;y)$. If the parameter s denotes time, then for a value s^0 for which a maximum $\mu(s)$ disappears, we have at the disparition point two possible situations:

(1) The disparition point $(s^0;Y)$ is a fold point of C; there the maximum μ of m coincides with a saddle of index one, according to a local model of the type

$$m(s;Y) = m(s^0;Y) + (s-s^0)y_1 + y_1^3 + \sum_{1<j<n-1} (y_j)^2$$

for local coordinates y_1, y_j in Y, and $m(s^0;Y) \neq 0$. Then the fold point in the $(s;m)$ plane is a *flex point* (with horizontal tangent) of the graph of the function $m(s^0;y_1)$. Hence it belongs to the basin of some other maximum τ of $m(s^0;y_1;y_j)$, whose local flow *captures* the disappearing flow (Figure 8.2).

(2) At the disparition point $(s^0;Y)$, the local intensity $m(s^0;Y)$ vanishes. This is, strictly speaking, not a generic situation, but it is of utmost practical importance; it corresponds to the task of *stopping* entirely the local flow.

These two situations can be symbolized by the two graphs 1A and 1B in Figure 8.3. By reversing the sense of time, we obtain the two symmetric morphologies 2A and 2B in Figure 8.3. These four morphologies play a fundamental role in classifying organs and tools according to their main functions. In the technical world, one of the most frequent preprograms occurs when a solid (s) is immersed in a material flow; here the control space U may be the full 6-D space describing all positions of U. If we consider k-D families of the positions of s we may obtain in such a control space higher singularities than those described above. For instance, in a 2-D U we

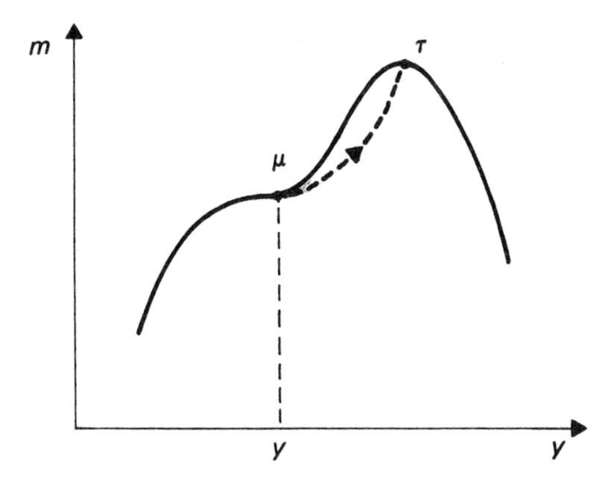

Figure 8.2 Capture of a disappearing maximum μ by a higher maximum τ.

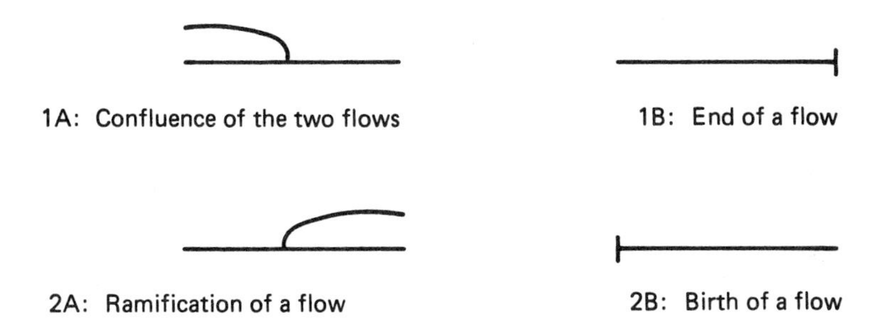

Figure 8.3 Morphologies 1A, 1B, 2A, and 2B.

may have points where a flow may possess a triple ramification (stably in U), but usually the solid s is constrained in a low-dimensional family of positions.

This notion of preprogram appears as a dynamical metaphor for the classic notion of gatekeeper due to Kurt Lewin (1951), the founder of topological psychology; that is, an individual whose position in a society allows him or her to modify deeply the structure and the intensity of some particular economic (or communicational) flow. The basic aim of this chapter is to show the ubiquity of preprograms in biological organization; in that respect, DNA – in my view – is no longer the program directing the whole development and regulation of the organism. It is a preprogram – with, undoubtedly, some central character – but a preprogram among many, as practically any organ has the same preprogramming capacity.

Organs and Tools as Preprograms: A Taxonomy

Let us return to the four canonical morphologies in Figure 8.3. The morphologies 1A and 1B are of a concentrating, converging character; morphologies 2A and 2B are, on the contrary, of a ramifying, diverging character. Generally, one may say that for life and technical needs, convergence is more important than divergence. In fact, as we shall see, even in the ramifying case 2A at least one of the created flows (usually the "upper" one) is strictly canalized.

Morphology 1B: End of a flow

We start with the simplest of the morphologies, 1B (Figure 8.3). If the s parameter is spatial, not temporal, then morphology 1B has the character of spatially limiting a flow. This means that the flow runs across a spatial domain U limited by a compact surface W such that the density $m(s; y)$ of the flow is zero outside W. In simpler terms, the surface W is a *wall* containing the flow in its interior. If, as earlier, we denote Ox as the direction of the flow, then, for almost all values of x, the projection of the wall W onto Ox is regular, and the flow has a product structure $H \times I$ defined by this projection on any (sufficiently small) interval I around the regular value x_0. The flow is encapsulated in a tube of section H and is said to be *canalized*. In most cases, the 2-D section is a disk and the wall W appears as a tubular neighborhood of some central trajectory; we call such a situation a *simple canalization*.

Most flows that occur in living or inanimate nature almost always exhibit simple canalization; the same is *a fortiori* true of the flows used in human technology (pipes, electric conductors, etc.). The only exceptions are the global fluid flows such as the oceanic currents (e.g., Gulf stream) or atmospheric wind.

Canalization – the underlying concept of Waddington's (1957) notion of "chreod" – may be given a metaphoric description in terms of a potential well. A 3-D potential well limited by two vertical walls is a good representation of the bed of a simple canalized flow (for a section $x = cst$). It is sometimes convenient to treat the potential well as a smooth, parabolic well (Figure 8.4). This amounts to replacing the original density function $m(x; y)$ (y is the transverse coordinate in the section) by a potential function $V(y; x)$, which may be roughly taken as the opposite; that is $-m(x; y) = V(y; x)$ (Figure 8.5). Also, to describe generic singularities of simple canalized flows, we can use the catastrophe theoretic formalism on the family of potential functions $V(y; x)$, where y is the internal variable and x the control variable. This leads to the morphologies 1A, 1B, 2A, and 2B given earlier (Figure 8.3).

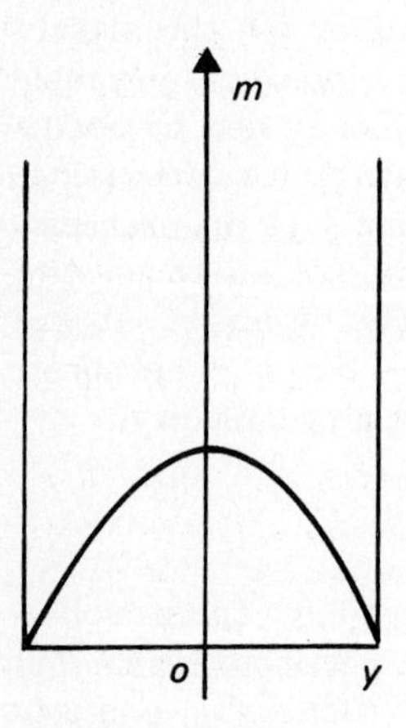

Figure 8.4 Simple canalized flow.

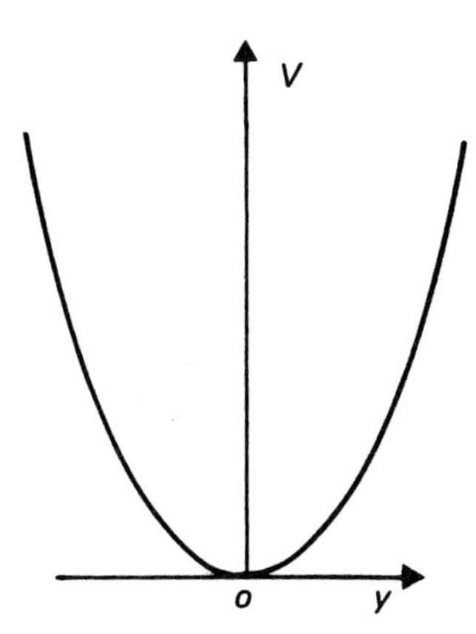

Figure 8.5 Equivalent parabolic well for the flow in Figure 8.4.

Canalized morphology 1B

Let us then consider morphology 1B for simple canalized flows; we take here the *s* parameter as being Ox, a spatial coordinate. The spatial end of a flow provides a paradox, as matter and energy cannot disappear at the end point. The flow may disappear as the result of a phase transition (as for rivers that flow from Moroccan Atlas and disappear by evaporation in the hot sands of the Sahara). It may disappear as a canalized flow, like those sewers near the Riviera beaches at which ends the waste is left to diffuse freely in the ambient water; a similar situation is that of industrial chimneys, where the smoke diffuses freely in the atmosphere. (Here the flow really ends in a generalized ramifying morphology, cf. Figure 8.7 and the discussion of morphology 2A below.) If the flow is not permanent, but starts (and finishes) at definite times, then another solution may exist: the tubular neighborhood of the canalization may extend itself as a tubular neighborhood of the end point singularity. This leads to a partial expansion of the end point, which bounds a spatial domain in which the matter carried by the flow may accumulate (Figure 8.6). This is the origin of the ***container morphology***, of which we can cite a large number of examples, both biological and technological. A ***vase***, for instance, is the material realization of a potential well in which solid or liquid may be stored; if the flow is gaseous, then the container has to be closed (the gas not being sufficiently subject to gravity). Sometimes closure of a pipe occurs by fixing a specific solid obstacle, such as the cork on a bottle neck; here elasticity of the cork symbolizes the expanded end point. In organic morphology, for example, the stomach and the bladder are such containing organs; the cell itself, inside its membrane, can be considered as such a receptacle. Nothing is more fundamental, both in life and technology, than this requirement of canalization and containment; if the technique of nuclear fusion is still ineffective, it is because we do not know how to *contain* a plasma. [1]

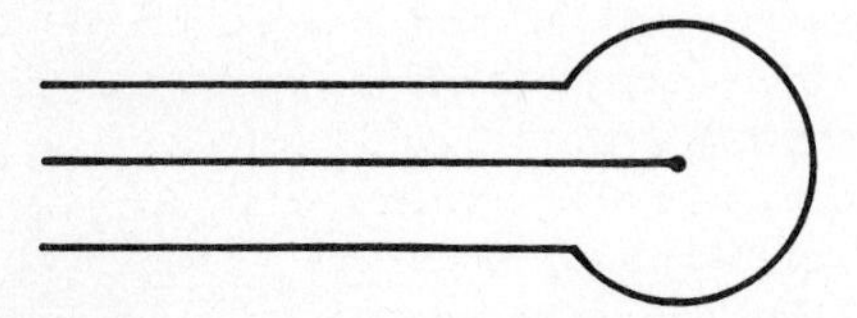

Figure 8.6 Origin of a container as an enlarged end point of a canalized flow.

A very interesting case of morphology 1B occurs in blood coagulation; here an accidental gap in a canalizing wall is repaired by the transported flow itself. An example of this being partially achieved in technology is the *valve*. Manufacturers of tubeless tires have also been able, at least partially, to mimic this process.

Morphology 2B: birth of a flow

According to the well-known maxim, *ex nihilo nihil*, a flow cannot be born out of nothing. A natural example of morphology 2B is the source of a spring; liquid water, previously circulating below ground, suddenly appears and creates a spring. Here we have, before the water surfaces, a subterranean concentration of small streams converging towards the source point. This is the opposite morphology to the diffusion morphology that occurs when canalization discontinues (a convergent, ramifying process as opposed to the diverging ramification of diffusion, Figure 8.7).

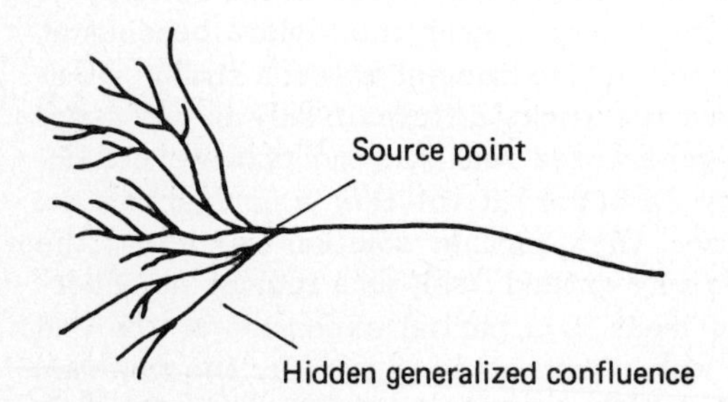

Figure 8.7 Birth of a flow (morphology 2B).

The case of phase transition described for morphology 1B occurs also for 2B. For instance, an electric bulb is a light source in which the light field is produced by thermal energy due to the Joule effect in the filament. Sun, as a light source, emits energy produced by nuclear reactions and gravitational collapse.

Examples of concentration processes providing sources are frequent in biology, such as capillaries anastomosing into veins; the *funnel* is a typical preprogram transforming a noncanalized flow into a simple canalized one. The *umbrella* is also a preprogram creating a canalization within a noncanalized flow (rain), but here the inside of the canalized domain has zero flow (in principle). An important case is that in which the liquid of the flow emanates from a reservoir where it has been stored. The usual appliance for creating alternately the morphologies 2B and 1B is the *cock*. This reversibility constraint is expressed in the Hamiltonian character of the control dynamics in the U control space of the cock; the cock is rotated, and the coupling of the Hamiltonian dynamics with the irreversibility of the effect (opening or closing) appears in the helicoidal nature of the screw thread. The ***door*** is a similar preprogram in which the control dynamics is also a rotation. Biologically, the function of the cock is taken over by ring-like muscles known as *sphincters*; here we have in the wall surface a reversible enlarging of an end point. In terms of physical fields, such as light, we think of the light source; the light flow exhibits a kind of inner canalization due to the ray structure, so any opaque object imbedded in the field acts as a preprogram. Placing a screen across a ray-bundle stops the light flow; removing the screen allows the light to flow again. The concentration of a ray-bundle by a converging lens at the focal point is also a form of sharp canalization (but of limited duration). In general, the theory of a total apparent contour, viewing an object along all possible directions, can be considered as the preprogramming capability of any opaque object in a fixed bundle of parallel light rays. When we move this object according to the full rotation group $SO(3)$, we obtain, in the 3-D control space, $U = SO(3)$, a bifurcation set with known generic singularities, which Kergosien has aptly named the ***obturation set***. Some of these singularities (of codimension one) correspond to stopping (or creating) the flow of light through some hole in the body. The difference in morphologies 2B and 1B corresponds to the classic opposition found in physiology between inhibition and excitation. The energetic fluxes which are excited or inhibited may have the most varied organic or biophysical supports, but the metaphor of the cock remains in each case valid.

Morphology 1A: Confluence of two flows

For noncanalized flows, the dichotomic junction of 1A (Figure 8.3) does not present a problem; it amounts to mixing two flows. Joining two pipes with a junction piece (such as the French *culotte*) is the simplest example of morphology 1A (Figure 8.8). But such simplicity is a little fallacious; if we have strict canalization, then owing to the presence of a potential well the genericity assumption renders the situation asymmetrical. Generally, one of the flows that arrives at the confluency point is in a metastable flex point of the potential well and so is captured by the lower potential flow. This occurs in the geosphere, for instance, when the tributary river reaches the main valley, U-shaped by glaciation, to produce the water falls of a suspended valley (often, nowadays, a hydroelectric plant is built below the falls). Hence, generically, morphology 1A is energetically favorable (and perhaps always entropically favorable).

There are numerous examples of confluences of flows in the biosphere (veins anastomoses, confluency of the bile duct with the intestine, etc.); at a higher level, we have phagocytosis between cells and predation among animals. But here, again, there is an asymmetrical situation where one of the individuals, the predator, is endowed with intentionality. Among superior animals predation becomes a conflict and hence frequently dramatic (the agôn).

Between two solid bodies, the junction morphology may be realized by using

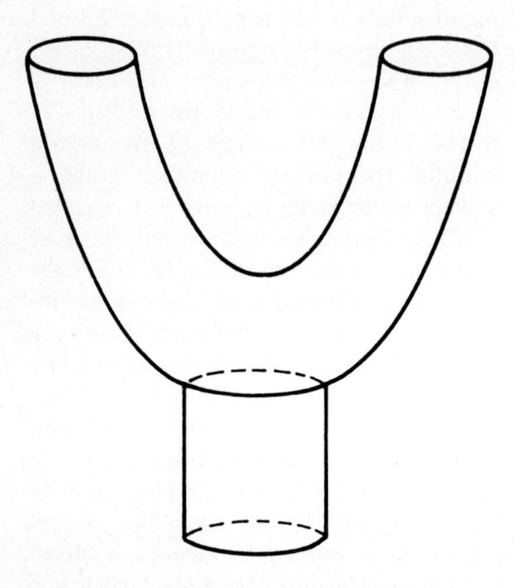

Figure 8.8 Confluence of two pipes (morphology 1A).

an auxiliary object (such as binding string) or a chemical (e.g. glue). But, in general at least, a partial congruence of some parts of the bodies' surfaces is required.

Morphology 2A: dichotomic ramification

Reversing the junction morphology of two pipes can be done, but it requires energy to work efficiently (for example, higher fluid pressure is required upstream). A natural stream, like a river, is generally canalized by its own flow, the river-bed being limited by the higher banks (and valley slopes). This canalization is strong inasmuch as the flow is fast and there is a large downhill slope, which results in strong erosive power. But for simple canalized flows, ramification is energetically unfavorable because it requires lifting a part of the flow to an upper minimum (Figure 8.9). This is demonstrated by the fact that, generally, natural drainage systems usually develop confluences and only rarely ramifications. When the latter do occur, as in delta-heads, this expresses the fact that the river has lost its erosive, hence its canalizing, power. In flat areas, the canalization is frequently man-made, with artificial dams. In fact, ramification of a stream is often obtained by building a dam (provided with sluices) across the river, and forcing part of the flow into an artificial canal (as frequently occurs for irrigation).

It is very important that the ramifying morphology 2A, requiring energy, is more difficult to realize than the confluency 1A; we discuss later some philosophical implications of this. For the present, it suffices to say that the required energy for ramification may be furnished immediately by the flow itself, as in explosive processes (e.g. the electric spark), but then the ramification itself is not strictly controlled and may repeat itself, possibly giving rise to a divergent, infinitely ramifying process (Figure 8.7). If we require a ramification of type 2A

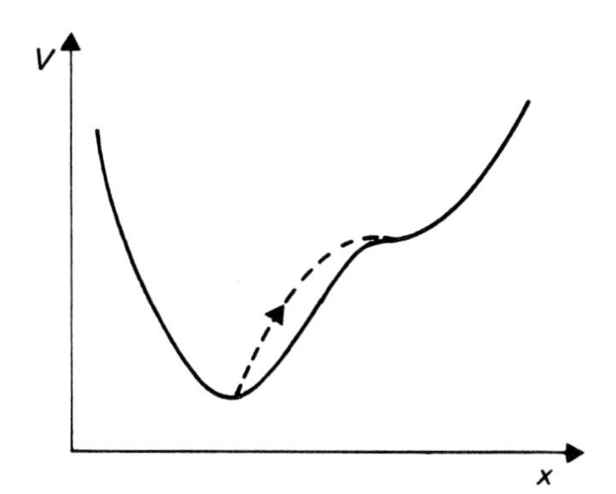

Figure 8.9 Creating an auxiliary flow from the main flow (ramifying morphology 2A).

that is strictly controlled, then we must use a preprogram which enables some part of the energy carried by the main flow to force the derived flow into the alternative canal (Figure 8.10). If this diverted flow must be raised above the level of the main flow, we can no longer use a fixed preprogram (like the dam above), and have to use a *moving* preprogram. The standard example here is the *Noria*, the millwheel used in the Middle East to lift the water of a river into a higher irrigation canal (the Archimedes screw is another device for the same purpose).

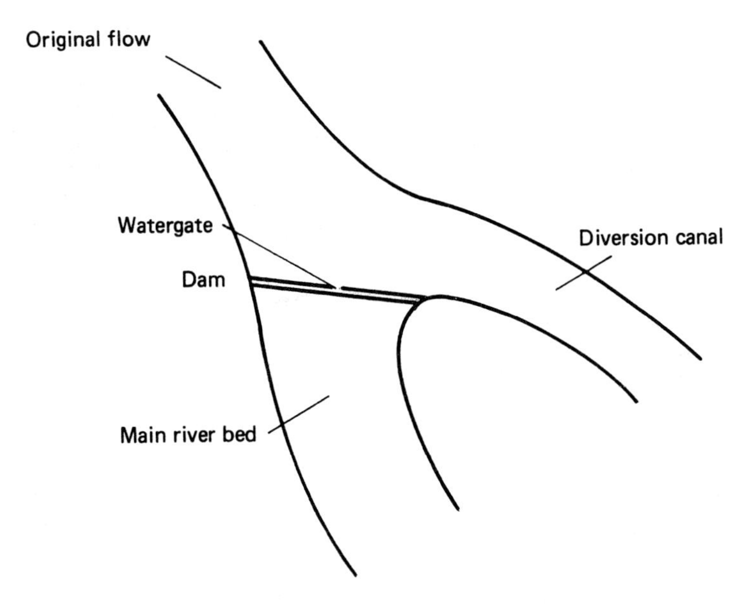

Figure 8.10 Creating morphology 2A.

Moving Preprograms

In the preceding section, we considered a fixed solid s within a stationary flow, in which it is obvious that s is subjected to forces emanating from the current, but the control is such that s remains at rest. Suppose we allow s to move inside some space U; then under the pressure of the fluid, s moves and develops a coupling with the fluid. It is predictable that this interaction will achieve an asymptotic regime, described by an attractor for the motion of s in U. The simplest case (for a nonpoint attractor) is a closed trajectory (the only nontrivial attractor if U is 1-D). An example is when a millwheel is moved by a current; we may consider that the kinetic energy of the wheel (and its axis) is a ramified branch of the full energy flow of the current. Conversely, if we give to s a periodic motion, then the fluid may take a stationary regime. This is the principle behind the propulsion of boats by paddle-wheels or propellers. The *pump* is such a moving preprogram, which couples the push–pull periodic motion of the piston with the irreversible character of the valves, and so transfers this motion to a periodic pushing of the fluid.

A priori, the formation of such moving preprograms seems to require human participation, but geological examples show that such preprograms may occur quite naturally. A river erodes its banks and sometimes a detached rock may be trapped by the ambient current in a circular movement (forming a pothole). (We discuss later the theory of periodic, oscillatory movements entrained by a continuous flow.) Moving preprograms are among the most ancient tools used by mankind. For example, the ***knife blade*** (or ***ax***) is a tool used to cut or split a solid into two parts. In catastrophe theoretic formalism, the blade can be considered as a dual cusp associated with the potential

$$V = x^4/4 + ux^2/2 + vx$$

as shown in Figure 8.11. Its interaction with the potential well canalizing the body creates a scission of the potential well. For the ***sieve*** (or ***net***) we have a material flow moving across the mesh with particles separated according to their size relative to that of the threshold magnitude, the mesh of the sieve.

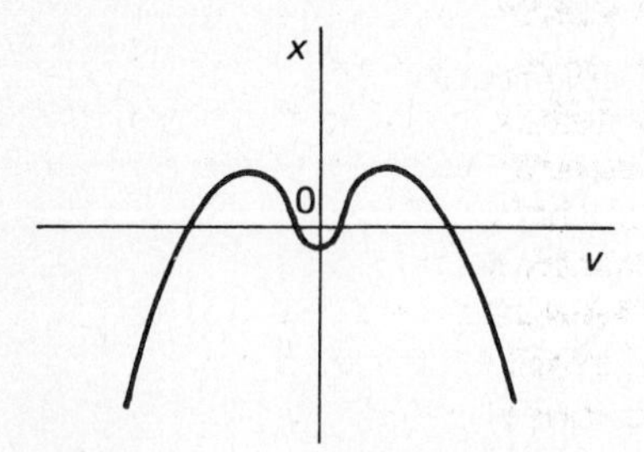

Figure 8.11 Potential associated on the Maxwell line of the dual cusp.

For many normal tools, in fact, we implicitly use the principle of relativity of movement; instead of placing a fixed obstacle inside a moving flow, we have a moving preprogram thrown against the treated body at rest. Which seems to show that the relativity principle was innately used much earlier than Galileo or Einstein...

Continuing these general considerations, I suggest that due to this fundamental asymmetry between morphologies 2A and 1A, the symbolism of the movement direction of 1A might be a cultural universal. The transition of 2 units → 1 unit is always easier than the reverse, 1 → 2. For instance, we may place in a basket different objects with no difficulty at all, but if we want to extract from the basket a specific object, we have to *choose* this object from the whole set of contained objects. This requires the use of some gestaltist, perceptive criterion which is transformed into a motor preprogram by the mind's activity. Even without choosing, we have, in order to achieve morphology 2A, to lift the basket and turn it in such a way that the contents fall to the ground. Here we meet with, perhaps, two of the most difficult questions in physics: the origin of time irreversibility and the nature of the second principle of thermodynamics. Remember that for the origin of irreversibility in Boltzmann's H-theorem, we have the *Ansatz* that two molecules are freely independent before colliding, but after collision they form a single system united by correlations; that is, morphology 1A is preferred to 2A. In the same spirit, Gibbs considered the following paradox. Consider two balloons joined by a pipe (Figure 8.12), with a cock to allow the contents to mix. If, initially, the contents are chemically different gases, mixing produces an increase in entropy. If the two balloons contain the same gas at the same temperature, mixing gives no entropy increase. To explain the result, Schrödinger had to invoke the quantum principle of indiscernability of particles! (For in the second case, quantum-wise, nothing happens and so the mixing is imaginary...) From the viewpoint of qualitative dynamics, the superiority of 1A over 2A is obvious. For if we allow some coupling between two differential systems $(M_1;X_1)$, $(M_2;X_2)$ (notation of p 198) so that the product structure $(M_1;X_1) \times (M_2;X_2)$ is (in general) unstable; it breaks under the weakest form of coupling (at the slightest resonance). The split property of the system and the split systems form a set of infinite codimensions in the function space of flows on $M_1 \times M_2$. One cannot avoid the general feeling that if we explain the spatial individuality of a system by some form of canalizing dynamics, then placing all these systems inside the same huge potential well strongly impairs each individual dynamics. To extract a specific system from the melting pot requires restoring its canalizing dynamics, which cannot be done without time, cunning, and energy. In this metaphor, extracting the chosen system requires the use of qualitative properties of the system, which (a) are specific to it and (b) resist the perturbations due to mixing. Hence the use of a *sieve*: we rediscover here Maxwell's demon as an antidote to the second principle.

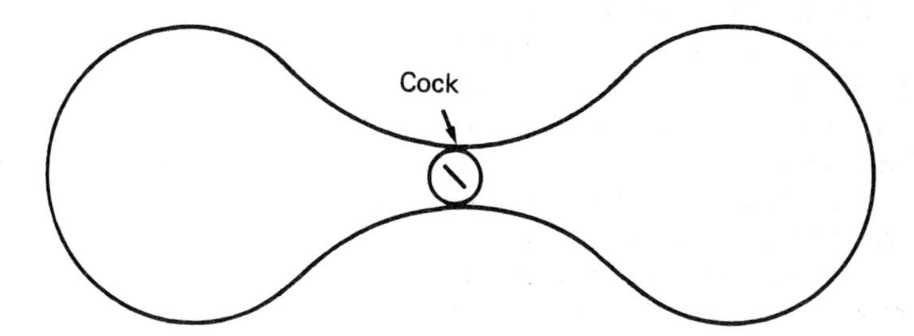

Figure 8.12 Mixing of two gases from two balloons (Gibbs' paradox).

Entraining an Oscillator by a Continuous Flow: The Dynamical Origin of the Cybernetic Loop

The millwheel

The motion of a millwheel in a steady current is apparently smooth; but a more detailed analysis can be achieved if we consider a wheel with one single blade. Then the rotation of the system can be decomposed into two periods:

(1) The blade enters the water, receives energy from the current, and delivers it to the system; this is the *entraining* phase of the period.

(2) The blade enters the air, the wheel rotates by inertia (invariance of angular momentum), but as the system has to overcome friction or other consuming demands, it loses some energy (not enough to prohibit it from reentering the current); this is the *retroflux* phase (since the blade is moving in the opposite direction to the current).

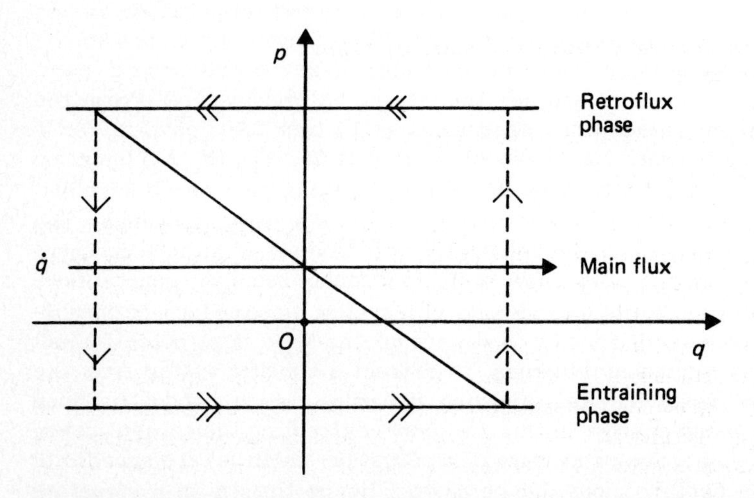

Figure 8.13 Fully anharmonic entrainment of an oscillator by a continuous flow: q, spatial coordinate; p, associate momentum. The total area, A, of the hysteresis loop divided by the total length, T, of the period is equal to the gained energy, E.

So the complete cycle of the blade is divided into the two phases, entraining and retroflux, with short, catastrophic transitional periods in between (Figure 8.13). (Here the discontinuities affect second-order derivatives.) As a millwheel usually has many blades breaking the rotational symmetry, the total movement is a superposition of all the cycles; hence the final movement appears smooth. Note also that the retroflux phase is the one in which the system delivers useful energy (cf., the example of the *Noria*, quoted above). If this useful energy is directed to a preprogram acting up-stream on the entraining flow itself, then we have the classical cybernetic loop of retroaction, as exemplified by Watt's regulator for steam engines.

The violin string under the bow

When the string moves in the same direction as the bow, the relative velocity is less and the friction coefficient greater; this is the entraining phase. When the string moves in the opposite direction, the relative velocity is larger and the friction coefficient smaller; this is the retroflux phase. Overall, the string receives kinetic energy from the bow, dissipated through friction; due to the catastrophic transitions between the two phases, the oscillation is strongly anharmonic, which ensures the richness of the timbre of the emitted sound.

The clock's escapement

Here again a detailed analysis of the escapement mechanism reveals an entraining period, where the pendulum travels in the same direction as the cogwheel moved by the falling weight, and a phase of retroflux, where the pendulum moves in the opposite direction.

Harmonicity Versus Anharmonicity of Oscillations: The van der Pol Theory

Consider the standard potential of the cusp singularity

$$V = x^4/4 - ux^2/2 + vx \ .$$

The first derivative in x, $V_x = x^3 - ux + v$, defines in the $(v;x)$ plane (here the x axis is vertical!) the smooth curve C of equations $V_x = 0$, $v = ux - x^3$. For positive values of u, this curve has the well-known S-shape wiggle which defines the cusp catastrophe: it has fold points of coordinates $x = \pm 1/\sqrt{3}$, $v = \pm 2/3\sqrt{3}$, respectively. The upper and lower branches of C ($x > 1/\sqrt{3}, x < -1/\sqrt{3}$, respectively) correspond to minima of potential V and so are stable regimes for the dynamics defined by $-\text{grad}_x V$. The middle arc $(-1/3 < x < 1/3)$ corresponds to an unstable regime (V maximum). For u going to 0 and becoming negative, the curve $C(u)$ unfolds its wiggle and becomes a simple curve of negative slope, dx/dv. The limiting case is $u = 0$, where $C(0)$ has a flex at 0 with a vertical tangent (the wiggle disappears).

We consider now the dynamics defined by $-\text{grad}V$ with respect to the hyperbolic metric $ds^2 = dx^2 - k\,dv^2$, with k positive. The components of this gradient are:

$$\dot{x} = x^3 - ux + v \qquad \dot{v} = -x/k$$

For k going to $+\infty$, this dynamics tends towards a constrained dynamics, which admits the stable arcs of the curve $C(u)$ as a slow manifold. Set $u = (k-1)/k$, then for any positive k, the flow admits the origin 0 as the only singularity. The linear parts of the flow at 0 ($\dot{x} = -ux + v$ and $\dot{v} = -x/k$) define the (2,2) matrix

$$\begin{vmatrix} -u & 1 \\ -1/k & 0 \end{vmatrix}$$

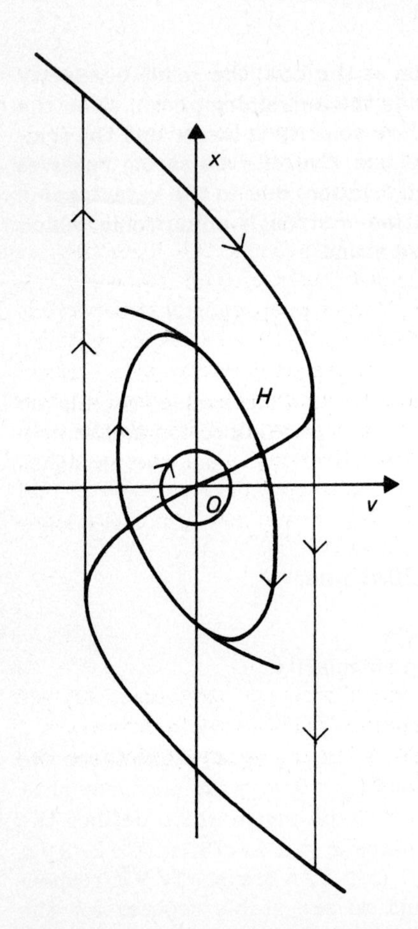

Figure 8.14 Continuous deformation of an attracting cycle born by Hopf bifurcation into an hysteresis loop (van der Pol theory). Intermediate cycles are provided by "rivers" flowing into the cycle and prefigurating the stable branches of the associated characteristic.

with the characteristic polynomial in s, $s^2 + us + 1/k$. For negative u and small $|u|$, with $k < 1$, grad$v$ has 0 as a repelling focus; hence for $-$grad$V$ there is an attracting focus. As u passes through 0 and k through +1, we obtain the standard Hopf bifurcation, which transforms this attracting focus into a repelling one and creates an attracting cycle $G(k)$. A more detailed study shows that for k increasing to $+\infty$, this cycle develops into the hysteresis loop H associated with the wiggle $C(1)$ (Figure 8.14). We conclude from this brief study that an hysteresis loop of type H can be deformed (in a continuous family of easy definition, depending on the parameter k) into an interior smooth cycle which may later concentrate into an attracting focus; that is, a return towards the organizing center ($u = v = 0$, $V = x^4/4$) of the cusp catastrophe.

Morphological versus cybernetical approach

The general metatheorem stated on pp 201–203 gives some credence to the apparently common belief that any regulatory procedure can be described by a diagram involving only excitations and inhibitions of specific operations (i.e., opening and closing of apertures). For any path in a control space may be given a slight deformation to the characteristic of a preprogram, in such a way that it admits only the four singularities 1A, 1B, 2A, and 2B. But the interpretation of confluences 1A or ramifications 2A may lead to difficulties if the branches are qualitatively distinct or if the relative intensities of fluxes play an important role later. In this respect, the problem of concatenation of preprograms is important; it occurs each time a branch arising from a ramification 2A is so directed as to act by local diffeomorphism in the control space of a subsequent preprogram. In such cases, genericity assumptions may be kept valid by applying the composition of generic smooth maps. But this would require having converging morphologies only, and not diverging ones [Dufour theorem (1977)]; it is a fact that in most ramifying processes 2A only one branch is of interest (it bears some pregnance, some value). Moreover, as we shall see in the "budding process" described later, optimality constraints may impair the transversality requirement.

Biological Finality and the Dynamics of Life

In this section, we discuss what may be considered as profound analogies between primitive forms of teleological construction and the basic processes of biological replication, at the cellular and organic levels. We first describe this construction with an example of a technical nature.

Millwheels and diversion canals

The first millwheels were originally built on river banks and were rotated only by the natural current. But people quickly became aware that they could obtain a much higher efficiency if the mill was built on a steeper river gradient. Hence the idea of creating an artificial fall, by diverting water further up-stream into a canal of very low gradient, and allowing it to enter the main stream via the blades of the millwheel. It should be noted that this idea was developed with practically no scientific or technical knowledge. (It occurred certainly in the early Middle Ages, perhaps even in Antiquity ... after all, beavers also know how to build dams.)

To build the diversion canal, the river has to be dammed with varying aperture levels (sluices) to allow the excess of water to flow through the original river bed; this is a case of the ramifying morphology 2A. Between the dam and the millwheel(s) the stream is in a state of bimodality. Where the canal water reenters the river bed, a junction of morphology 1A occurs. In considering the bimodality of the gravitational potential energy, it is natural to use a hysteresis loop, h_1

(Figure 8.15), which can be obtained by folding into an S-shape (cliff) the original, smooth, linear slope of the stream. This can be done at the cost of introducing, for the side *SD* of the loop h_1, a fictive retroflux, which precisely describes the containing effect of the dam and the ensuing lack of kinetic energy in the main stream.

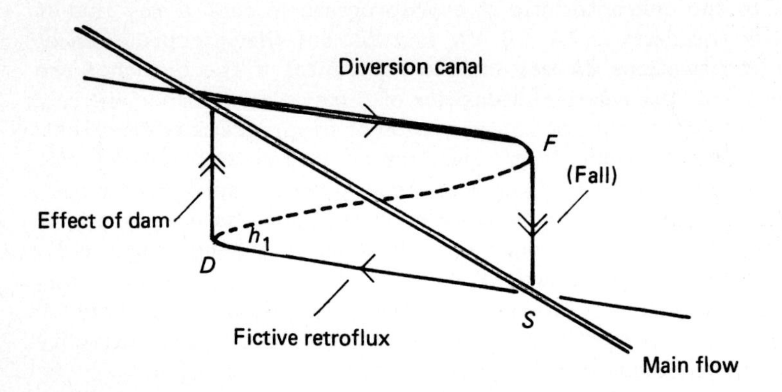

Figure 8.15 Hysteris loop h_1 associated with a diversion canal.

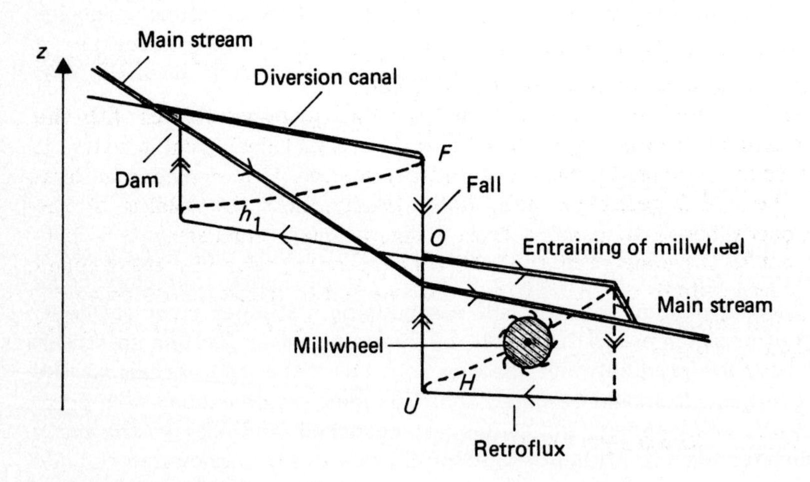

Figure 8.16 Millwheel entrained by an artificial water fall.

As explained earlier, the millwheel has itself an hysteresis loop *H*. Since the internal energy of the wheel is taken from the potential energy of the river, its energy is considered as negative in terms of the axis *z*. Figure 8.16, so constructed, may be considered as resulting from the concatenation of the two hysteresis loops h_1 and *H*, ***with coincidence of the vertical movements*** associated with the transitions (a) direct flux to retroflux for h_1 and (b) retroflux to direct flux for *H*. The two cofold lines meet at *O*, but have to spatially coincide. The energy of falling water in the segment *FO* is transferred to the wing of the wheel

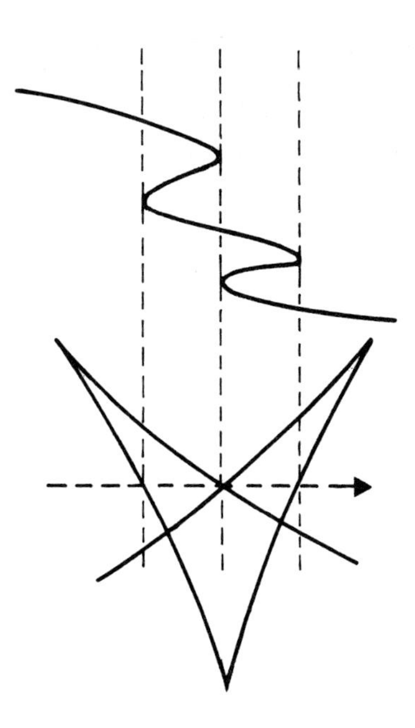

Figure 8.17 Section of butterfly catastrophe corresponding to Figure 1.16.

entering the fall along UO. Note that between the two loops h_1 and H, there is a saddle connection at O, common point of the two loops.

The process by which a ramification of type 2A occurs up-stream with the creation of an auxiliary loop is a typical, primitive form of teleological activity. It is also, in some sense, a primitive model of cell replication if we consider an hysteresis loop to be a 2-D geometric cell. Note, finally, that the folding of the characteristic corresponds to passing from a simple cusp singularity to a butterfly one, subject to the constraint that the corresponding path in control space has to be of the type which expresses that the two-fold extreme points coincide. In a standard plane section of the butterfly bifurcation locus this path meets the central double point (Figure 8.17).

Dynamics of Life: A Bird's Eye View

The metabolism of living things is – at a given instant – a kind of stationary metabolic regime, represented in a space of biochemical parameters as an attractor for a flow; this metabolic activity is entrained by an energetic flux of dissipative nature. For Earth, energy comes from solar light and finally degradates into thermal energy; it is believed that in the initial, primitive soup energy sources were of a chemical and radiative nature. This flux had to be canalized; how are we to imagine the walls of this canalization? A general answer can be provided as follows. The chemical reactions involved in biochemistry are fundamentally of two very different types: first, fast, stoichiometric reactions involving small metabolites colliding freely in an aqueous medium; second, nonstoichiometric reactions involving macromolecules. These latter polymerizations, or degradations and depolymerizations, are much slower than those of the first type. Spatial regions in which macromolecules are more densely concentrated have a much slower metabolism than those free regions in which free collision dominates. Owing to the adherence of small molecules to macromolecules, free collisions are rarer in the macromolecular regions. The turbulent attractor, of large dimension, which characterizes the free region degenerates in macromolecular regions to a quasi-point attractor (a stationary state). Of course, even in macromolecular regions small metabolites do interact; but owing to their adherence to macromolecules, which in general exhibit an ordered configuration, the free collisions become spatially guided, hence canalized. There are well known examples of this in organelles such as chloroplasts and mitochondria. As a result, we may consider the energy flux in living matter as a kind of stream canalized between banks provided by metabolic states of macromolecular regions. Moreover, life is itself a ***spatially canalized phenomenon***. Of course, the formation of cells different from the surrounding medium may be purely due to a phase transition, as when two liquids cease to be miscible (formation of lipid bilayers as membranes); but such nuclei have to be metrically controlled and replicate regularly by schizogenesis. Hence the very likely conjecture that biochemical canalization (by macromolecules) and spatial canalization (by membranes) had to be strongly coupled initially; which means that membranes have to be spanned by regularly ordered macromolecules. As the limiting membrane is relatively fixed (neglecting such phenomena as pseudopodia, pinocytosis, etc.), it must be colonized by macromolecules such as those which form a cytoskeleton. Although this limiting cytoskeleton has to increase in interphase by polymerization, we may suppose that the growth is a relatively ordered process, allowing us to define for any point of the membrane its temporal trajectory (mathematically, the cytoskeleton defines a connection). Hence the membrane can be considered as the fixed bank of the energetic flux due to the metabolism of small molecules. We have to consider the total cytoskeleton as a macromolecular system susceptible to biochemical vibrations; the period of such a vibration – entrained by the ambient energy flux – is the cell replication cycle, in which one cell becomes two (morphology 2A). If the membrane is considered to be at rest, some parts of the cytoskeleton must still be internal to the cell and exhibit variation. Here the analogy with the millwheel becomes relevant. The axis of the millwheel must be bound to the fixed banks, although, at the same time, it has to rotate; and it carries blades which must fully interact with the current. These properties strongly suggest that the axis of the millwheel is realized – in our metaphor – by genomic DNA. We justify this analogy by developing the metaphor; that is, by describing more precise relationships between biochemical and spatial canalization.

In a biphasic entrained system, we distinguish between the entrainment phase and the retroflux phase. In classic descriptions of the mitotic cycle, the G_1 and S phase have to be considered as the entraining phase (synthesis and polymerization of macromolecules). The G_2 phase is the retroflux period, with progressive deceleration of the metabolism. Mitosis proper (i.e. spatial scission) is the catastrophe that enables the passage from retroflux to entraining. But the main problem for life is to transform the chemical, dissipative current into a spatial current. Suppose the available energy exists outside the cell in the form of specific molecules (e.g. glucose); then capturing this energy would be relatively easy if these molecules happened to belong to a current hitting the cell wall. In this case, it would suffice to have in the wall a macromolecular system of intertwined strings constituting a net, such that the pores formed act as a *sieve* for the required molecules. (Some marine animals actually use this method to feed on plankton.)

In general, however, for a cell unable to move by itself such a current does not exist; hence it must be created within the cell by internal cellular mechanisms. For example, there may be on the outer wall itself predator organelles (macromolecular systems) able to capture the prey molecules and transfer them within, where they are treated and their available free energy undergoes a long sequence of transformations involving a series of chemical carriers, resulting in, for instance, the ATP–ADP system. As, in metabolism, most of the energy is of a chemical nature (neglecting kinetic or electrical energy), we may suppose that this energy has continuous trajectories within the cell and defines there a flow. For instance, some energy carried by intermediate molecules of the sequence must be transferred back to the outer wall to provide the predator organelles with the energy they require to capture and transfer. Hence this energy flow must be somewhat complex within the cell. Let us idealize our living cell as a 3-D Euclidean ball, defined in Euclidean three-space $Oxyz$ by

$$x^2 + y^2 + z^2 \leq R^2 \ ,$$

where R is the radius. Let $\mathbf{x}$ be a vector field defining this energy flow in B. We call $\mathbf{x}_T$ the flow obtained when the normal component vanishes on the boundary sphere by local smoothing. We may admit that for a generic time t in interphase, the flow $\mathbf{x}_T$ is conservative (condition of stationariness of the asymptotic regime). But such a flow $\mathbf{x}_T$, tangent to the boundary sphere, must have on the boundary an even number of singular points, and as $\mathbf{x}_T$ inherits from the original flow $\mathbf{x}$ a gradient structure with respect to time (every trajectory of $\mathbf{x}$ entering B has to leave after a finite, fixed time), then any trajectory of $\mathbf{x}_T$ leaving a singular point on ∂B must reach ∂B at another singular point. As for a flow on R^2, not all minimal sets can consist of saddle connections and there should exist at least one center; there has to be a connection between centers for the flow $\mathbf{x}_T$, which has to exhibit internal vortices.

The simplest example of such a flow is that defined by a rotation of the ball B around the z axis (Figure 8.18). The two centers are then the two poles ($x = y = 0$; $z = \pm 1$) and there is a gradient structure on the z axis from, for example, the North Pole to the South Pole. This is thus the axis of invariant points akin to the genome.

Here we remark on the dynamics of cell replication. If we consider an equatorial section of the ball B, we obtain a 2-D disk D having the origin O as center. There exists a flow in D, with O as a singular point (center) and tangent (nonvanishing) to the boundary ∂D. How are we to cut this cell into two? The most obvious approach is to join the two opposite points JK [Figure 8.18(b)], of coordinates $x = 0$, $y = \pm 1$, and to collapse by continuous deformation the segment JK to the origin O. We thus obtain a figure of eight, formed by two disks meeting at O, a common point which could be a saddle-point of the inherited flow. But this requires an anterior duplication of the singularity of the flow, hence the constraint of duplicating the genome before the cell itself.

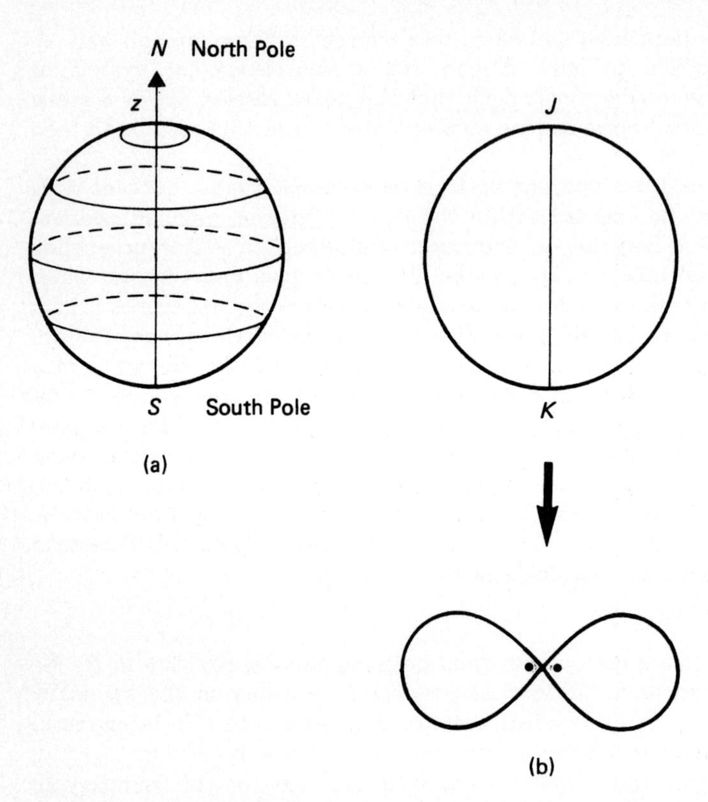

Figure 8.18 (a) Rotation of a ball around the z-axis; (b) breaking a cycle into two requires preliminary duplication of the central singular points.

How can the singularity O duplicate itself? Here the mathematician's goal is to replace the cybernetic execution of a program by sheer analytic continuation. The simplest approach is to consider the initial flow in D (the one of concentric circles centered at O) as given by the linear pencil of equation

$$x^2 + y^2 - r^2 t \ , \qquad t = 0 \ ,$$

defining the line at infinity. Now let this line at infinity approach and pass through O as the axis Oy. Biologically, this means that the genome moves to the boundary wall and that the wall, extending to infinity as a straight line, must necessarily break (Figure 8.19). The flow is then given by the *meromorphic* potential

$$V = \frac{x^2 + y^2}{2y}$$

the flow lines of which are given by a very similar equation

$$H = \frac{x^2 + y^2}{2x} = cst$$

Then the mirror axis Oy assumes the role of the mitotic equatorial plane and both genomes separate to create two distinct cells. Of course, the idea that replication

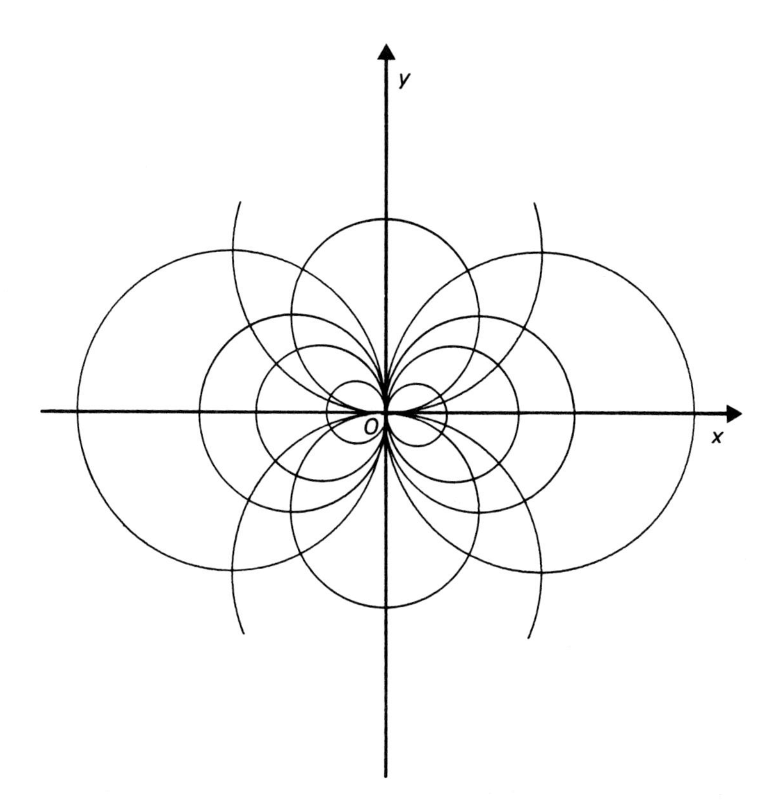

Figure 8.19 Two pencils of orthogonal circles.

occurs through some kind of analytic continuation, where an inducer structure, being in contact with a competent medium, extends itself into a similar structure is difficult to imagine (although the centriole is not very far from achieving this). We know how life has solved the problem; it has anticipated the splitting of the genome along Oy by replacing the single molecule O with a dipole system of axes parallel to Oy (note that a dipole flow is very similar to the flow defined by the Hamiltonian H). By some reflection principle, this dipole creates across the mirroring axis Oy a new dipole homologous to the preceding one, thus producing a quadrupole (Figure 8.20). Now this ***ménage à quatre*** reorganizes as two dipoles of axes parallel to Ox, thus allowing the dividing line Ox to be different from the mirroring line Oy; the reader has probably recognized here the fork-like nature of the DNA semi-conservative mechanism of replication. [2] As for breaking the envelope, this phenomenon is well known for the nuclear wall in Eukaryotes (although it occurs very posteriorly to DNA duplication). Obviously the genome has very different structures and functions in Prokaryotes as opposed to Eukaryotes.

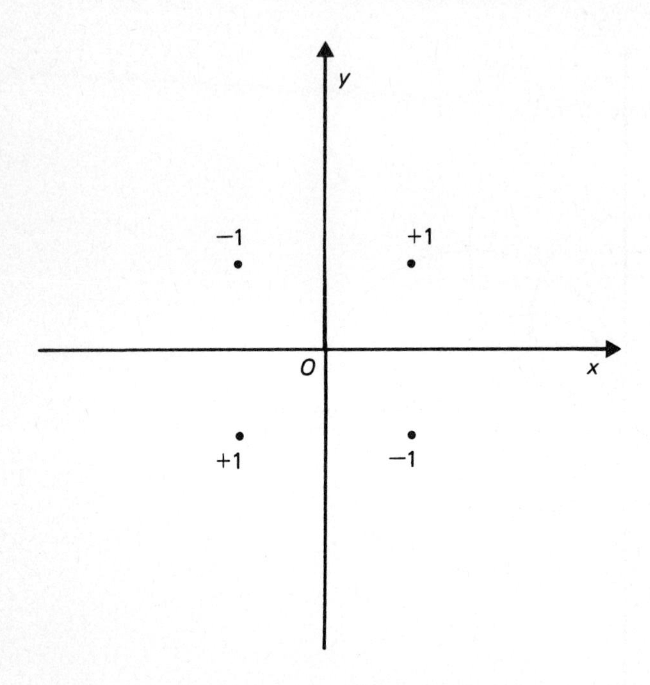

Figure 8.20 Quadrupole composed of two dipoles.

The Genetic Code

If we believe the metaphor that direct flux represents the dissipative energy flow of small metabolites and retroflux the polymerization of macromolecules, then there should be in the cell a catastrophe locus where the local regime shifts from entraining to retroflux. It is likely that the ribosomes are those organelles in which such a change happens; polymerization occurs also along the DNA string, in the synthesis of Messenger RNA. I propose that the locus of trajectories of the Messenger RNA strings, when they retreat from the DNA chain, can be considered as a surface having the genome as boundary line (perhaps the genome in its activated state can be considered as an edge of regression of a singular surface: one sheet carrying the necessary precursors, the other the synthesized strings). In the spherical model B, the RNA strings move along a helicoidal surface with, as the director curve, a spiral drawn on a cone with Oz as the axis and S (South Pole) as the vertex. In extending itself, the RNA string finally contacts ribosomes which, on this conical surface, are regularly situated at generator lines (Figure 8.21). The conical surface can be considered as a conveyor belt bringing precursor material to the growing wall. Passing through the ribosomes, of course, the RNA string is translated into an amino acid chain.

The dynamics of the system may be more intelligible if we remember the toys that children used to make when domestic heating was provided by coal stoves. The children cut from cardboard a spiral, which they attached to the top of a knitting needle stuck in a cork placed on the stove (Figure 8.22). The spiral was held in a delicate equilibrium, its center on the top of the needle; the cardboard

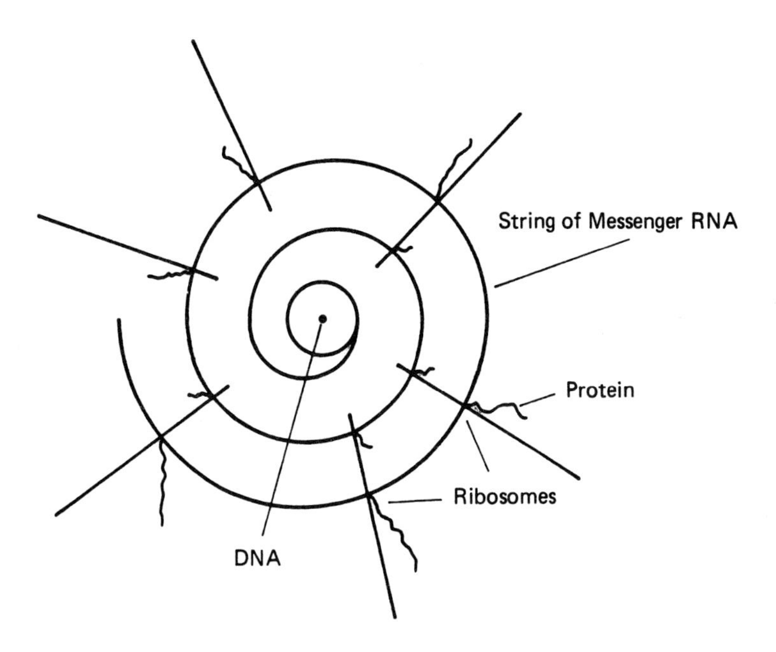

Figure 8.21 Helicoidal surface on which RNA strings move.

spiral comprises two opposite gradients which, along with its own weight and the ascending hot air, cause it to rotate around the needle.

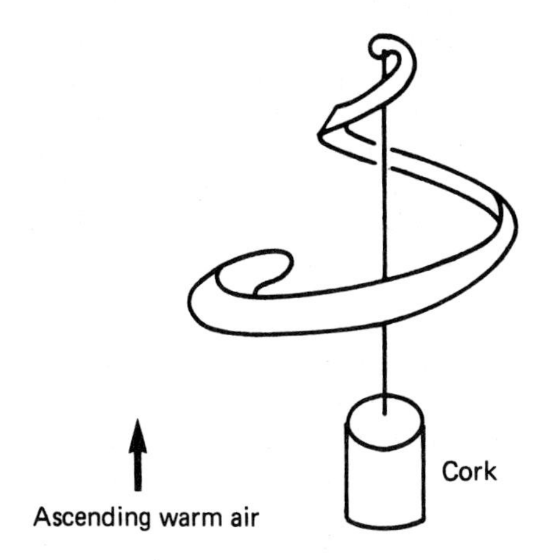

Figure 8.22 Illustration of the dynamics of Figure 8.21.

Of course, biochemists say that all the reactions involving DNA → RNA → proteins may be realized *in vitro*, in the presence of the necessary precursors, ATP, ribosomes, and the relevant enzymes. This proves, perhaps, only that the enzymes are able to direct locally the energy flow in such a way as to provide in a micro-environment the necessary energetic configuration. Anyway, the kinetics *in vivo* are considerably faster than *in vitro*.

Perhaps to understand the origin of the genetic code we should consider the ribosomal reading mechanism as a ***modulated sieve***: in a flow of amino acid precursors the ribosome is a funnel constructed such that the shape of the bottom hole is determined by the triplet structure of the Messenger RNA. One may well imagine proto-biotes living in a flow of useful molecules and able to modify the topology of the capturing sieve according to the nature of the incident molecules. Initially, we can imagine that a specific molecular flow creates a puffing of a specific part of the chromosome, leading to exploitation of this flow or triggering, by a convenient sieve-like preprogram, a regulatory synthesis of proteins. One has to imagine a phonemic ritualization of such a process involving "puffs" only three nucleotides long. Such a number is, in fact, the least number of elements able to delineate a planar hole, the bottom of the ribosome funnel. But here, of course, we entertain yet more speculation...

Dynamics of Development in Metazoa

Here, we are interested in the embryology of triploblastic animals – vertebrates in particular, recalling the basic scheme expounded in Thom (1972). The first differentiation splits entoderm from mesectoderm; it is interpreted as the result of a cusp catastrophe, in which the lower regime represents the subject, the predator (entoderm), and the upper the object, the prey (ectoderm). The mesoderm's function is to capture the prey; that is, to push it to the fold point K (the mouth; Figure 8.23). This is the result of performing a hysteresis loop between the upper and lower regimes [pumping energy from the entodermal reserves (liver) to catch the prey]. But the presence of bilateral symmetry quickly complicates the morphology; axial mesoderms (notochords) and paraaxial mesoderms (somites) appear which split into three parts (dermatome, clerotome, and myotome) and later form vertebrae around the neural tube created by neurulation; and finally lateral mesoderms appear (kidneys, gonads, and the two lateral sheets, somatopheure and splanchnopleure, enclosing the coelom). The first cycle (the dorsal one) essentially develops the voluntary muscles and bones; the second (the ventral one) the involuntary muscles and vascular system (in particular, the heart).

The central theme is that the main functional hysteresis loop of Figure 8.23 splits into two according to Figure 8.24. A dorsal cycle involves chasing and ingesting the prey; a ventral cycle involves digesting the prey, storing the chemical energy produced by digestion, and delivering this energy to the motor organs (essentially, blood circulation). I suspect that, at least for the higher vertebrates from fishes to mammals, this splitting process takes place *via* the budding of a loop described earlier. The fundamental hysteresis loop $OAKL$ of the mesoderm then becomes a smooth loop [by decreasing the k coefficient of van der Pol's theory we may produce a continuous family of concentric cycles, a conservative dynamics that is within the rectangle of the hysteresis loop, Figure 8.25(a)]. Now the straight line $t = 0$ of the model is an internal image of the symmetry axis; it occupies the direction $v - x = cst$, near to the upper left corner A [Figure

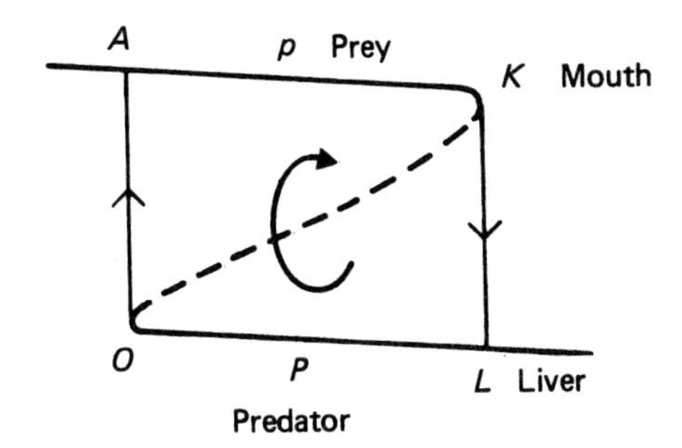

Figure 8.23 Fundamental cycle of the mesoderm.

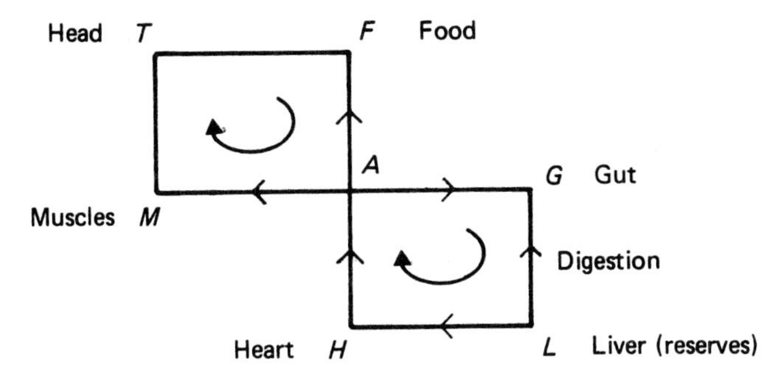

Figure 8.24 Splitting the mesodermal fundamental cycle into two: the dorsal (*FAMT*) and ventral (*AGLH*).

8.25(b)]; the stable regime within the little triangle *BAC* describes the notochord (singularity of the elliptic umbilic)

$$V = xy(x + y - c)$$

10.2In the paraaxial mesoderm (somites), the constant c vanishes, and the line $v = x - x_A$ acts as a mirroring axis for the tangent dynamics of A; the centers of the concentric cycles are pushed towards the boundary, at the corner A. As a result, a symmetric dynamics is formed by reflection on the symmetric angle [Figure 8.25(c)]; the centers then move away from A (as in mitosis) and the dorsal cycle is created [and also, symmetrically, the ventral cycle, Figure 8.25(d)]. This describes the classic process of neural induction, where the mesodermal plate induces by contact the formation of neural tissue in the overlying ectoderm. The cell budding process occurs both in the induced ectoderm and in the inducer mesoderm (but in opposite directions). In the latter tissue, after the centers have moved away, one may obtain attracting cycles both in the dorsal and in the ventral oscillator, thus realizing a torus [Figure 8.25(d)]. A further degeneracy toward 1-D resonance may explain, by breaking the quotient circle periodically, the metameric splitting into somites. And by unfolding each somite along the cephalocaudal

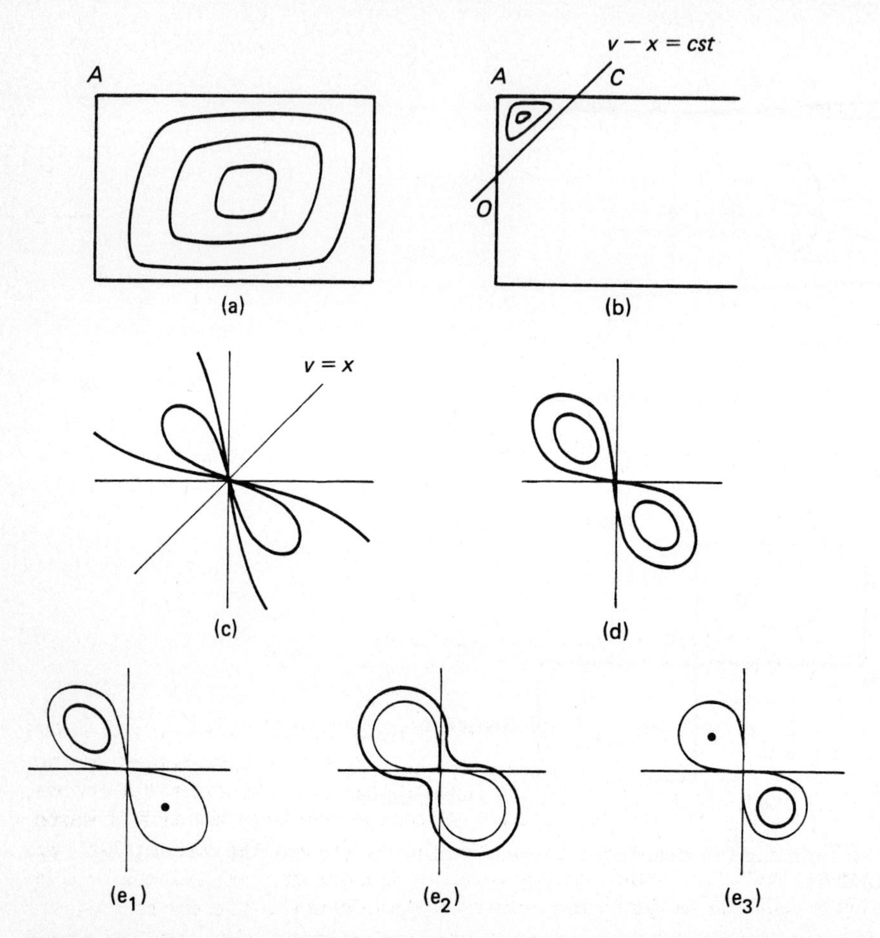

Figure 8.25 (a) Continuous family of concentric cycles within the hysteresis loop; (b) the notochord; (c) axial dynamics (general); (d) paraaxial dynamics (somites); (e_i) sclerotome, (e_{ii}) dermatome, (e_{iii}) myotome.

gradient, we may find an interpretation of the ternary splitting (sclerotome, dermatome, and myotome) arising later in somites [Figure 8.25(e_i–e_{iii})]. Later in the development, the fusion of sclerotome and myotome is a suggestive image for the canonical vertebrae [Figure 8.25(f)], with the neural arch enclosing the neural tube, and the haemal arch enclosing the dorsal aorta. The internal regimes of the somtatopleure [Figure 8.25(h)] and the splanchopleure [Figure 8.25(i)] are purely dorsal and ventral, respectively (here the conservative character of the flow disappears). The last regime, in its most cephalic part, keeps some oscillatory character (attracting cycle) which manifests itself in the cardiac cells by elementary contractions (note that the periods of this last cycle may have nothing in common with the previously considered cycles).

To explain the intermediate development of lateral mesoderm (nephrotomes, gonads) one has to postulate that some further budding of the system of loops occurs. For instance, at the vertex L of the ventral rectangle (Figure 8.26, lower

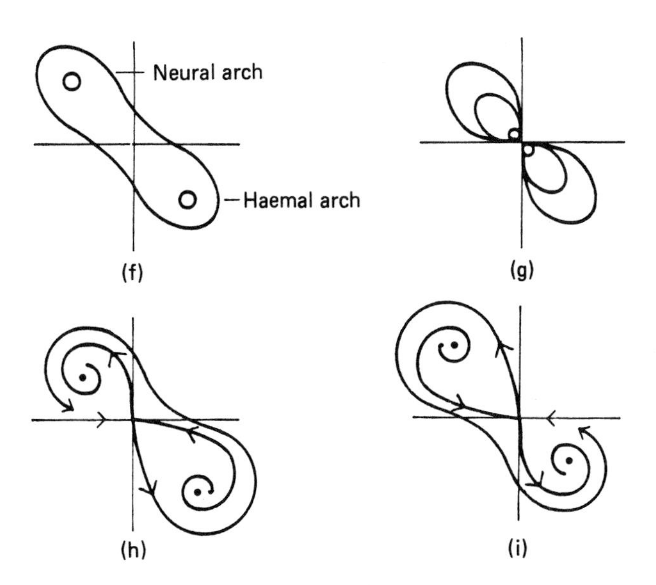

Figure 8.25 (*cont.*) (f) schema of vertebrae; (g) germinal dynamics; (h) somatopleure; and (i) splanchnopleure.

right) a new loop may be born by further budding (*LKPO*); *LK* represents production of waste material and *OP* symbolizes the external world. The edge *KP* is the (catastrophic) excretion of waste into the external world. The retroflux arrows *PQ* and *QL* represent, respectively, the loss of energy due to production of waste and the necessary injection of inorganic material (such as water and oxygen) for producing waste material (urine, carbon dioxide). Gonads are also related to this repelling cycle; but the germinal line cells emanate in general from the splanchnopleure, very near the origin A. Here again, one may conjecture that the return to germinal dynamics occurs via the straightening of the fundamental S-characteristic of the first catastrophe (ectoderm–entoderm), leading to the $V = x^4/4$ organizing center, and then allowing the two roots emanating from zero to develop Hopf bifurcations, in such a way as to develop for each pair of symmetrical roots a figure of eight [Figure 8.25(g)].

Another very important budding process is the ***cephalic*** loop extending from the upper left vertex T into *TBIS* (Figure 8.26, upper left). The edge *TB* represents tracking the prey; the retroarrow *IB* represents the projection emanating from the eye to the external object (the active pregnance of seeing); *IS* is the progressive localization of the prey, and *ST* its spatial localization by sensory inputs. This model invokes the concept that underlies the whole embryology; there is a kind of symbolic ***physiological blastula***, the cells of which are hysteresis loops, causally related by the identity of the same energetic flow. This blastula forms by successive mitoses (budding) of an original "cell" the fundamental hysteresis loop of the mesoderm. As the basic physiological constraints are the same for the whole animal kingdom, one may suspect that this physiological blastula is a universal structure within animal embryology (and, perhaps, also its temporal method of formation). But the coupling with spatial unfoldings (ill-defined,

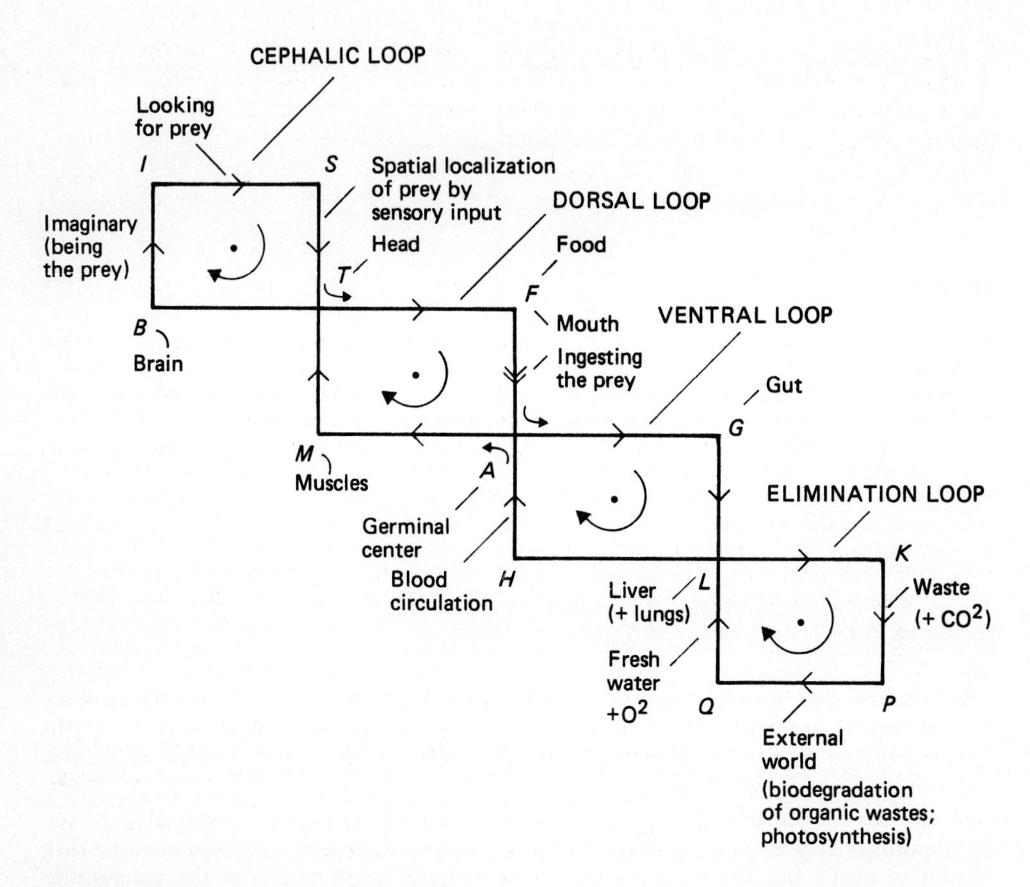

Figure 8.26 The physiological blastula.

both qualitatively and quantitatively) creates the variety of *Bauplans* observed in nature. To take a single – striking – example, let us consider the starfish, which develops from a bilaterally symmetric larva into an organism of five-fold symmetry. In vertebrates, the cephalocaudal axis is mapped on the vertex A of the physiological blastula, with a tendency to unfold along the diagonal TAL (mouth in the head, T; anus at the caudal extremity, L). When the starfish undergoes metamorphosis, its cephalocaudal axis is spatially vertical (the mouth–anus axis); it is mapped along the ventral diagonal AL of the symbolic blastula. But the dorsal diagonal AT (devoted to prehension activity) is mapped in the adult along the horizontal circle dual (perpendicular) to the central vertical axis, and it is there broken into a five-order symmetry. Such huge geometrical remodelings of the body obviously necessitate an almost total lysis of the organic structure during metamorphosis.

Concluding Remarks

I hope to have shown in this chapter how the consideration of very crude energetic constraints, together with some genericity assumptions, may increase our understanding of tools, organs, and embryology. Of course, orthodox scientists may say that such ideas are not useful, as they do not allow experimental prediction or verification. To that I would like to answer that I do not claim here to pursue scientific data, my only aim being to revive the concept of natural philosophy.

Notes

[1] As pointed out on pp 201–211, complete vanishing of a flow is a nongeneric situation. Hence it is extremely difficult to produce preprograms that allow simultaneous perfect opening and closing of a flow. This is why, in almost all cases, shut off faucets do leak – a truth any owner of an apartment (or country house) can hardly doubt. This necessary dependence of life processes on nongeneric situations such as perfect containment has led some thinkers (e.g., J. Monod in his celebrated *Chance and Necessity*) to appeal to contingence and randomness to explain the origin of life. But we should not forget the constraints which control propagations of specific physical entities; already in optics, total absence of light (full darkness) is relatively easy to obtain in a stable way. A solid body (a crystal for instance) is a closed set in Euclidean space to a very good approximation. Molecular biology is undoubtedly correct to express the importance of these constraints due to molecular bindings, and hence to the underlying spatial geometry.

[2] In this passage from dipole to quadrupole by symmetry, the symmetry is not only spatial, but also internal (the symmetrical equivalent of a +1 charge is a −1 charge). In organic epigenesis such an internal symmetry cannot exist between cells; as a result, the symmetry plane of bilateral embryos is frequently also the separation plane of twin embryos. This appears in double-monster teratology (Siamese twins, spina bifida, etc.). Note that if we have in the $(x\,;v)$ plane the dipole field defined by the potential $V = x/(x^2 + v^2)$ then the fold map $y \to x$ defined by $x = y^2$ induces on the $(y\,;v)$ plane a field defined by the meromorphic potential

$$W = y^2/(y^4 + v^2)$$

This is precisely the type of potential that occurs in our consideration of van der Pol theory, pp 213–214.

References

Arnol'd, V.I. (1984) *Catastrophe Theory* (Berlin: Springer) translated from Russian.

Dufour, J.P. (1977) Sur la stabilité des diagrammes d'applications différentiables, *Ann. Sc. Ecole Norm. Sup.* (4) **10** (2): 153–74.

Kergosien, Y. (1984) Sémiotique de la Nature, *3e Séminaire de Biologie théorique*, (AMTB) Solignac (Paris: Editions du CNRS).

Lewin, K. (1951) Psychological ecology, in *Field Theory in Social Sciences* (New York: Harper and Brothers) pp 174–87.

Schrödinger, E. (1957) *Statistical Thermodynamics* (Cambridge: Cambridge University Press).

Thom, R. (1972) *Stabilité Structurelle et Morphogenèse* (Paris: Intereditions).

Thom, R. (1983) Human psychism *vs* animal psychism, in E. de Grolier (Ed.) *Glossogenetics* (Paris: Harwood Academic Publishers).
Waddington, C.H. (1957) *The Strategy of the Genes* (London: Allen and Unwin).

This text is partly taken from a translation into English of a lecture given in French at the Séminaire André Lichnérowicz, Collège de France Paris, 1983–84, under the title *La notion de programme en Biologie*, to appear in *Collection Interdisciplinaire*, edited by Pierre Delattre, Maloine, Paris.

CHAPTER 9

The Language of Life

David Berlinski

Introduction

In the spring of 1984, I delivered two lectures at IIASA under the title *The Language of Life*. Dianne Goodwin was kind enough to prepare a verbatim transcript of my talks; I have used the months since then to purge the written record of what I said of its incoherence, vagrant inaccuracies, and general slovenliness.

This chapter is at once long and terse – an unhappy combination, and one that makes severe demands of the reader. Many arguments are highly compressed and must be elaborated before they appear convincing. I have not hesitated to make use of mathematical concepts in expressing myself; but I draw no mathematical conclusions. I thus run the risk of alienating the general reader even as I antagonize the mathematician. For these reasons, it may be helpful if in this introduction I endeavor to place this chapter in a somewhat wider personal and intellectual context.

As it stands, *The Language of Life* represents a draft of one-third of a larger work entitled *Language, Life and Logic*. Another part of that more ambitious project was delivered at IIASA two years ago as a set of lectures. The written record of those lectures, which I hope to publish separately as a working paper, is entitled *Classification and its Discontents*.

My aim in *Language, Life and Logic* is to explore a certain complicated common ground that holds between language, on the one hand, and the graphic arts, on the other. These are the classic systems of representation of the human imagination. In both, there is a curious division between the system's syntactic and semantic structures: a theory, for example, consists of a finite set of sentences, the sentences of words; paint and then pigment comprise a painting; and yet, words and sentences, paints and pigments, manage, somehow, to cohere and, then, in a miraculous act of self-transcendence, to make contact with a distinct and different external world. The problems of theoretical biology, it might seem, have nothing much to do with issues that arise in the philosophy of language or the philosophy of art. Not so. A gene comprises a linear array of nucleotides that under certain conditions expresses a protein or set of proteins. The proteins, in turn, are organized to form a structure as complicated as a moose or a mouse. The nucleotides are plainly alphabetic or typographic in character; the organism itself is rich, complex, complete, continuous, unlike an alphabet. How is it, then, that such typographic structures as DNA manage to express so much that is not typographic at all? This is a question quite similar to questions that might be raised about language itself, or works of the graphic arts; and when it is pursued, certain metaphors and quite peculiar images begin drifting from one subject to the other. There is the notion of meaning, of course, which is common to language, art, and life; but also the idea that life is itself a language-like system; or that art is organic. The relations of satisfaction, representation, and expression, while formally distinct, of course, nonetheless display points of contact. In order to explain how it is that a painting may represent a face, for example, one has recourse to the notion of a metaphor, a concept from the philosophy of language and linguistics; to make sense of gene expression, one deals in concepts such as code, codon, information, and regulation. In a general way, a theory, a painting, and a gene belong to the class of interpreted or significant typographic objects. It is for this reason that it has seemed to me profitable to explore some of the concerns of theoretical biology and the philosophy of art and language in a single volume.

Within the context of *this* chapter, my aim is to explore the ramifications of a controlling metaphor: the idea that life comprises a language-like system. I do this against the background of the neo-Darwinian theory of evolution – the most global and comprehensive scheme of thought in theoretical biology. My argument at its most general is constructed as a dilemma: if life is a language-like system, then certain concepts are missing from Darwinian thought; if not, then Darwinian thought is suspicious in the sense that its principles do not naturally apply to cognate disciplines. The intellectual pattern to this chapter is thus one of movement between two unyielding points, a kind of whiplash.

Part One establishes the historical and contemporary background to Darwinian thought; and makes the argument that much of biology cannot be reduced to physics. In Part Two, I consider the confluence of certain concepts: distance in the metric spaces of organisms and of strings, metric spaces in phase, complexity, simplicity, Kolmogorov complexity, the ideas of a weak theory, and a language-like system. Part Three plays off concepts of probability against the hypothesis that molecular biological words are high in Kolmogorov complexity – with results that are inconclusive. In Part Four, I examine evolution or biological change as a process involving paths of proteins. The discussion is set in the mathematical contexts of ergodic theory and information theory. In many respects, the classical concepts of information and entropy are most natural in discussing topics such as the generation of protein paths by means of stochastic devices; but there is a connection between Kolmogorov complexity and entropy in the sense of information theory, which remains to be explored. Almost all of Part Four represents a tentative exploration of concepts that require, and will no doubt receive, a far fuller mathematical treatment.

Many of the points that I make in this paper I first discussed with M. P. Schutzenberger in Paris in 1979 and 1980. Indeed, it was our intention and hope to publish jointly a monograph on theoretical biology. This has not come to pass. Still, to the extent that my ideas are interesting, they are his; to the extent that they are not, they are mine.

John Casti read the penultimate draft of this essay and discovered any number of embarrassing errors. I am grateful for his stern advice, which I have endeavored to heed.

PART ONE

A System of Belief

The natural thought that theoretical biology comprises a kind of intellectual Lapland owes much to the idea that biology itself is somehow a derivative science, an analogue to automotive engineering or dairy management, and, in any case, devoid of those special principles that lend to the physical or chemical sciences their striking mahogany lustre. This is the position for which J.J.C. Smart (1963) provided a classic argument in *Philosophy and Scientific Realism*.[1] Analytic philosophers, for the most part, agree that nothing in the nature of things compels them to learn organic chemistry; Feyerabend, Putnam, and Kuhn have wondered whether *any* discipline can properly be reduced to anything at all; and, then, whether anything is ever scientific, at least in the old-fashioned and honorific sense of that term.[2] Naive physicists – the only kind – are all too happy to hear that among the sciences physics occupies a position of prominence denied, say, to urban affairs or agronomy. The result is ***reductionism from the top down***, a crude but still violently vigorous flower in the philosophy of science. The physicist or philosopher, with his eye fixed on the primacy of physics, thus needs to sense in the other sciences – sociology, neurophysiology, macrame, whatever – intimations of physics, however faint. This is easy enough in the case of biochemistry: chemistry is physics once removed; biochemistry, physics at a double distance. Doing biochemistry, the theoretician is applying merely the principles of chemistry to living systems: like the Pope, his is a reflected radiance.

In 1831, the German chemist Uriel Wohler synthesized urea, purely an organic compound – the chief ingredient in urine, actually – from a handful of chemicals that he took from his stock and a revolting mixture of dried horse blood. It was thus that organic chemistry was created: an inauspicious beginning, but important, nonetheless, if only because so many European chemists were convinced that the attempt to synthesize an organic compound would end inevitably in failure. The daring idea that all of life – I am quoting from James Watson's textbook (1965), *The Molecular Biology of the Gene* – will ultimately be understood in terms of the "coordinative interaction of large and small molecules" is now a commonplace among molecular biologists, a fixed point in the wandering system of their theories and beliefs. The contrary thesis, that living creatures go quite beyond the reach of chemistry, biochemists regard with the alarmed contempt they reserve for ideas they are prepared to dismiss but not discuss. Francis Crick, for example, devotes fully a third of his little monograph, *Of Molecules and Men*, to a denunciation of vitalism almost ecclesiastical in its forthrightness and utter lack of detail.[3] Like other men, molecular biologists evidently derive some satisfaction from imagining that the orthodoxy they espouse is ceaselessly under attack.

Theoretical biologists still cast their limpid and untroubled gaze over a world organized in its largest aspects by Darwinian concepts; and so do high-school instructors in biology – hardly a group one would think much inclined to the idea of the survival of the fittest; but unlike the theory of relativity, which Einstein introduced to a baffled and uncomprehending world in 1905, the Darwinian theory of evolution has never quite achieved canonical status in contemporary thought, however much like a cold wind over water its influence may have been felt in economics, sociology, or political science. Curiously enough, while molecular genetics provides an interpretation for certain Darwinian concepts – those differences between organisms that Darwin observed but could not explain – the Darwinian theory resists reformulation in terms either of chemistry or physics. This is a point apt to engender controversy. Woodger, Hempel, Nagel, and Quine cast reduction as a logical relationship: given two theories, the first may ***directly*** be reduced to the second when a mapping of its descriptive apparatus and domain of interpretation allows the first to be derived from the second. I am ignoring details, now. The standard and, indeed, the sole example of reduction successfully achieved involves the derivation of thermodynamics from statistical mechanics. In recent years, philosophers have come to regard direct reduction with some unhappiness. There are problems in the interpretation of historical terms: the Newtonian concept of mass, for example; and theories that once seemed cut from the same cloth now appear alarmingly incommensurable. Kenneth Schaffner has provided a somewhat more elaborate account of reduction: his definition runs to five points.[4] By a *corrected theory*, he means a theory logically revived to bring it into conformity with current interpretations: Newton upgraded, for example. His general scheme for reduction, then, is this:

(1) All of the terms in the corrected theory must be matched to terms in the reducing theory – a requirement of *completeness*.
(2) The corrected theory must be deducible from the larger theory, given the existence of suitable reduction functions – a requirement of ***derivability***.
(3) The corrected theory must indicate why the original theory was incorrect – a requirement of epistemological *insight*.
(4) The original theory must be explicable in terms of the reducing theory – a requirement of *cogency*.
(5) The original and corrected theories must resemble each other – a requirement of intellectural *symmetry*.

In the case of theoretical biology, to speak crisply of deriving, say, molecular biology from biochemistry is rather like endeavoring to cut steel with butter: there is a certain innocence to the idea that molecular biology has anything like a discernable logical structure. What one actually sees is a mass of descriptive detail, a bewildering plethora of hypothetical mechanisms, much by way of anecdotal evidence, a few tiresome concepts, and an array of metaphors drawn from physics, chemistry, information theory, and cybernetics. The definition of reduction just cited is, in addition, incomplete, its flagrant inapplicability aside. In Mendelian genetics, the concept of a gene is theoretical, and genes figure in that theory as abstract entities. To what should they be pegged in molecular genetics in order to reduce the first theory to the second? DNA, quite plainly, but how much of the stuff counts as a gene? "Just (enough) to act as a unit of function," argues Michael Ruse, a philosopher whose commitment to prevailing orthodoxy is a model of steadfastness.[5] The functions that he has in mind are biochemical: the capacity to generate polypeptides; but to my way of thinking, the reduction achieved thus is illicit. In biochemistry, the notion of a unit of function is otiose, unneeded elsewhere. To the extent that molecular genetics is biochemistry, it does not reflect completely Mendelian genetics; to the extent that it does, it is not biochemistry, but biochemistry beefed-up by extrinsic concepts, a conceptual padded shoulder. What holds in a limited way for molecular genetics holds in a much larger way for molecular biology. Concepts such as code and codon, information, complexity, replication, self-organization, stability, negative entropy (grotesque on any reckoning), transformation, regulation, feedback, and control – the stuff required to make molecular biology work – are scarcely biochemical: the biochemist following some placid metabolic pathway need never appeal to them. Population genetics, to pursue the argument outward toward increasing generality, is a refined and abstract version of Darwin's theory of natural selection, applied directly to an imaginary population of genes: selection pressures act directly on the molecules themselves, a high wind that cuts through the flesh of life to reach its buzzing core. Has one achieved anything like a reduction of Darwinian thought to theories that are *essentially* biochemical, or even vaguely physical? Hardly. The usual Darwinian concepts of fitness and selection appear unvaryingly in place. These are ideas, it goes without saying, that do not figure in standard accounts of biochemistry, which very sensibly treat of valences and bonding angles, enzymes and metabolic pathways, fats and polymers – anything but fitness and natural selection. To Schaffner's list of five, then, I would add a sixth: *no reduction by means of inflation* – a contingent and cautionary restriction that, for the time being at least, enforces a stern separation between biology and mathematical physics.

The Darwinian theory of evolution is the great, global organizing principle of biology, however much molecular biologists may occupy themselves locally in determining nucleotide sequences, synthesizing enzymes, or cloning frogs. Those biologists who look forward to the withering away of biology in favor of biochemistry and then physics are inevitably neo-Darwinians, and the fact that *this* theory – *their* theory – is impervious to reduction they count as an innocent inconsistency. If mathematical physics offers a vision of reality at its most comprehensive, the Darwinian theory of evolution, like psychoanalysis, Marxism, or the Catholic Faith, comprises, instead, a *system of belief*. Like Hell itself, which is said to be protected by walls that are seven miles thick, each such system looks especially sturdy from the inside. Standing at dead center, most people have considerable difficulty in imagining that an outside exists at all.

The Historical Background

Charles Darwin completed his masterpiece, *On the Origin of Species*, in 1859. He was then forty-nine, ten years younger than the century, and not a man inclined to hasty publication. In the early 1830s, he had journeyed to the islands of the South Atlantic as a naturalist aboard *H.M.S. The Beagle*. The stunning diversity of plant and animal life that he saw there impressed him deeply. Prevailing biological thought had held that each species is somehow fixed and unalterable. Looking backward in time along a line of dogs, it is dogs all the way. Five years in the South Atlantic suggested otherwise to Darwin. The great shambling tortoises of the Galapagos, surely the saddest of all sea-going creatures, and countless subspecies of the common finch, seemed to exhibit a pattern in which the spokes of geographic variation all radiated back to a common point of origin. The detailed sketches that Darwin made of the Galapagos Finch, which he later published in *On the Origin of Species*, show what caught his eye. Separated by only a few hundred miles of choppy ocean, each subspecies of the finch belongs to a single family; and yet, Darwin noted, one group of birds had developed a short, stubby beak; another, living northward, a long, pointed, rather Austrian sort of nose. The variations among the finch were hardly arbitrary: birds that needed long noses got them. By 1837, Darwin realized that what held for the finch might hold for the rest of life and this, in turn, suggested the dramatic hypothesis that far from being fixed and frozen, the species that now swarm over the surface of the Earth ***evolved*** from species that had come before in a continuous, phylogenetic, saxophone-like slide.

What Darwin lacked in 1837 was a theory to account for speciation, but the great ideas of fitness and natural selection evidently came to him before 1842, for by 1843 he had prepared a version of his vision, and committed it to print in the event of his death. He then sat on his results in an immensely slow, self-satisfied, thoroughly constipated way until news reached him that A.R. Wallace was about to make known *his* theory of evolution. Wallace, so far as I know, had never traveled to the South Atlantic, sensibly choosing, instead, to collect data in the East Indies, and, yet, considering the same problem that had earlier vexed Darwin, he had hit on precisely Darwin's explanation. The idea that Wallace might hog the glory was too much for the melancholic Darwin: he lumbered into print just months ahead of his rival; but in science, as elsewhere, even seconds count.

The theory that Darwin proposed to account for biological change is a conceptual mechanism of only three parts. It involves, in the first instance, the observation that living creatures vary naturally. Each dog is a member of a common species and thus dog-like to the bone; but every dog is doggish in his own way: some are fast, others slow, some charming, and others bad from the first, suitable only for crime. Darwin wrote before the mechanism of genetic transmission was understood, but he inclined to the view that variations in the plant and animal kingdoms arise by *chance*, and are then passed downward from fathers to sons.

The biological world, Darwin observed, striking now for the second point to his three-part explanation, is arranged so that what is needed for survival is generally in short supply: food, water, space, tenure. Competition thus ensues, with every living thing scrambling to get his share and keep it. The struggle for life favors those organisms whose variations give them a competitive edge. Such is the notion of *fitness*. Fast feet make for fitness among the rabbits, even as a feathery layer of oiled down makes the Siberian swan a fitter foul. At any time, those creatures fitter than others are more likely to survive and reproduce. The winnowing in life effected by competition Darwin termed *natural selection*.

Working backward, Darwin argued that present forms of life, various and wonderful as they are, arose from common ancestors; working forward, that biological change, the transformation of one species to another, is the result of small increments that accumulate, step by inexorable step, across the generations, until natural selection recreates a species entirely. The Darwinian mechanism is both random and determinate. Variations occur without plan or purpose – the luck of the draw; but Nature, like the House, is aggressive; organized to cash in on the odds.

The Central Dogma

Everthing that lives, lives just once. To pass from fathers to sons is to pass from a copy to a copy. This is not quite immortality, even if carried on forever, but it counts for something, as every parent knows. The higher organisms reproduce themselves sexually, of course, and every copy is copied from a double template. Bacteria manage the matter alone, and so do the cells within a complex organism, which often continue to grow and reproduce after their host has perished, unaware, for a brief time, of the gloomy catastrophe taking place around them. It is possible, I suppose, that each bacterial cell contains a tiny copy of itself, with the copy carrying yet another copy; biologists of the early eighteenth century, irritated and baffled by the mystery of it all, actually thought of reproduction in these terms: peering into crude, brass-rimmed microscopes, they persuaded themselves that on the thin, stained glass, they actually saw a homunculus; the more diligent proceeded to sketch what they seemed to see. The theory that emerged had the great virtue of being intellectually repugnant. Much more likely, at least on the grounds of reasonableness and common sense, is the idea that the bacterial cell contains what Erwin Schroedinger called a *code script* – a sort of cellular secretary organizing and recording the gross and microscopic features of the cell. Such a code script would be logically bound to double duty. As the cell divides in two, it, too, would have to divide without remainder, doubling itself to accommodate two bacterial cells where formerly there was only one. Divided, and thus doubled without loss, each code script would require powers sufficient to organize anew the whole of each bacterial cell. The code script that Schroedinger (1945) anticipated in his moving and remarkable book, ***What is Life***? – he wrote in the 1940s – turns out to be DNA, a long and sinewy molecule shaped rather like a spiral in two strands. The strands themselves are made of stiff sugars, and stuck in the sugars, like beads in a sticky string, are certain chemical bases: adenine, cytosine, guanine, and thymine: A, C, T, and G, in the now universal abbreviation of biochemists. It is the alternation of these bases along the backbone of DNA that allows the molecule to store information.

One bacterial cell splits in two: each is a copy of the first. All that physically passes from cell to cell is a strand of DNA: the message that each generation sends faithfully into the future is impalpable, abstract almost, a kind of hidden hum against the coarse wet plops of reproduction, gestation, and birth itself. James Watson and Francis Crick provided the correct description of the chemical structure of DNA in 1952. They knew, as everyone did, that somehow the bacterial cell in replication sends messages to each of its immediate descendents. They did not know how. As it turned out, the chemical structure of DNA, once elaborated, suggests irresistably a mechanism both for self-replication and the transmission of information. In the cell itself, strands of DNA are woven around each other and by an ingenious twist of biochemistry matched antagonistically: A with T, and C with G. At reproduction, the cell splits the double strand of DNA. Each half floats for a time, a gently waving genetic filament; chemical bonds are then repaired as each base fastens to a new antagonist, one simply picked from the ambient broth of the cell and clung to, as in a single's bar. The process complete, there are now two strands of double-stranded DNA where before there was only one.

What this account does not provide is a description of the machinery by which the genetic code actually organizes a pair of new cells. To the biochemist, the bacterial cell appears as a kind of small sac enclosing an actively throbbing biochemical factory; its products are proteins chiefly – long and complex molecules composed, in their turn, of twenty amino acids. The order and composition of the amino acids along a given chain determines which protein is which. The bacterial cell somehow contains a complete record of the right proteins, as well as the instructions required to assemble them directly. The sense of genetic identity that marks *E. Coli* as *E. Coli* and not some other bug must thus be expressed in the amino acids by means of information stored in the nucleotides.

The four nucleotides, we now know, are grouped in a triplet code of 64 codons or operating units. A particular codon is composed of three nucleotides. The amino acids are matched to the codons: C–G–A, for example, to arginnine. In the translation of genetic information from DNA to the proteins, the linear ordering of the codons themselves serves to induce a corresponding linear ordering first onto an intermediary, messenger RNA, and then onto the amino acids themselves – this via yet another messenger, transfer RNA. The sequential arrangement of the amino acids finally fixes the chemical configuration of the cell.

Molecular biologists often allude to the steps so described as *the Central Dogma*, a queer choice of words for a science.

The dour Austrian monk, Gregor Mendel, founded the science of genetics on purely a theoretical notion of a gene, which he likened to a bead on a string. In DNA, one looks on genetics bare: the ultimate unit of genetic information is the nucleotide. All that makes for difference, and hence for charm, in the natural world, and which is not the product of culture, art, artifice, accident, or hard work, all this, which is brilliantly expressed in maleable flesh, is a matter of an ordering of four biochemical letters along two ropey strands of an immemorial acid.

The Central Dogma describes genetic replication; but the concepts that it scouts plainly illuminate Darwinian theory from within. Whether as the result of radiation or chemical accident, letters in the genetic code may be scrambled; one letter shifted for another; entire codons replaced, deleted, or altered. These are genetic mutations: arbitrary, because unpredictable; and yet enduring, because they are variations in the *genetic* message. The theory by which Darwin proposed to account for the origin of species and the nature of biological diversity now admits of expression in a single English sentence. Evolution, or biological change, so the revised, the *neo*-Darwinian theory, runs, is the result of natural selection working on random mutations.

PART TWO

Evolutionary Theories

The popular view of evolution tends to be a tight shot on a tame subject: the dinosaur, who did not make it; the shark, who did; but the maturation of an organism is itself much like the evolution of a species; only our intimate acquaintance with its precise and unhesitating character suggests, misleadingly, I think, that the two processes differ in degree of freedom. Psychology, economics, urban affairs, anthropology, political science, and history also describe processes that begin in a state of satisfying and undemanding simplicity, and end later with everything complex, unfathomable, chaotic. The contrast to physics is sobering.

The dynamics of evolutionary theories are often divided into two conceptual stages. In economics, there are macro- and micro-economic theories, aggregate demand versus the theory of the firm; within linguistics, language at the continuous level of speech, and language some levels below, discrete, a matter of the concatenation of words or morphemes. Biology, too, is double-tiered: above, the organism prances; unseen, below, at a separate level, its life is organized around the alphabetic nucleotides.

Metric Spaces

By a metric space S I mean a space upon which a function

$$d: S \times S \rightarrow R_+$$

has been defined, assigning to each pair of points s, s' in S a nonnegative real number – the *distance* $d(s, s')$ – and satisfying the usual axioms:

$$d(s, s') = 0 \Longleftrightarrow s = s' \ ; \tag{9.1}$$

$$d(s, s') = d(s', s) \ ; \tag{9.2}$$

$$d(s, s') + d(s', s'') \geq d(s, s'') \ . \tag{9.3}$$

Double metrics

The distance between organisms

The disciplines of comparative anatomy and systematic zoology classify creatures into ever-larger sets and sets of sets: individuals (dogs, say), species, genera, families, orders, classes, phyla, taxa, and kingdoms. The classification itself forms an algebraic lattice, with individuals acting as the system's atoms. Comparative anatomists and zoologists bring an exquisitely refined and elaborate intuition to the task of sorting the various biological creatures into appropriate categories: the obvious cases leap to the eye; at the margins of the system, where the whale resides, difficult matters are decided by reference to historical and comparative anatomy, parallel structure, common organization, biological traits, and, often, levels of biological achievement. If the image of a lattice is for the moment taken literally, then each level of the lattice, from the atoms upward, comprises a set or ***ensemble***: of individuals, in the first instance, of sets of individuals, in the second. An ensemble at any distinct level of the lattice, I assume, satisfies equations (9.1)–(9.3), and counts thus as a metric space.

The distance between strings

DNA is a string drawn from a four-letter alphabet; proteins are strings of fixed length composed of 20 amino acids; as such, both strings belong to a wider family of string-like objects: computer programs written in a given language, the sentences of a natural language, formal systems; and acquire by osmosis a distinct conceptual and mathematical structure. It makes little difference whether strings of DNA or strings of amino acids are taken as fundamental; and, in any case, I often alternate between the two. By an ***alphabet*** A I mean a fixed and finite collection of elementary entities called ***words***; by the ***universe of strings*** over a finite alphabet, the set of all finite sequences A^* whose elements lie in A.

The natural distance between words $W = w_1 \ldots w_m$, $V = v_1 \ldots v_n$ $(W, V \in A)$ is $|n| + |n| - 2 \times |k|$, where k is the maximum of the length of a word $U = u_1 \ldots u_l$, which is a subword both of W and V. For example, let W = cadbabbd, V = xcaaba. An appropriate U is U = caab; hence $\hat{D}(W,V) = 8 + 6 - 2 \times k$.

Grantham (1974) has proposed a definition of distance in a Euclidean metric space of proteins based on properties of composition, polarity, and volume; but the theory of evolution suggests that changes in biological strings come about through mutations – random flash points at which letters are scrambled. Some strings may change in a large-hearted way, with whole blocks of letters wheeling and shifting like cavalry horses; but the least mechanism to which these operations may be resolved is the simple one of erasure and substitution – deletion and insertion. The elementary processes of evolution at the molecular level lend to the natural metric a certain simple plausibility in the face of fancy competition. $T = A^*$, then, is a ***typographic metric space***; d_T, its natural distance.

Metric spaces in phase

M and M^*, suppose, are two metric spaces; $g: M \to M^*$ assigns to each point p in M a distinct point P^* in M^*. M and M^* are ***in phase*** under g if g acts roughly to preserve distances: for any $\xi > 0$, there exists a $\varphi > 0$, such that for all p and q in M

$$d_M(p, q) < \varphi \to d_{M^*}[g(p), g(q)] < \xi \ .$$

g is thus ***uniformly continuous*** on M; φ is, of course, a function of ξ. It often happens that a particular mapping between metric spaces is especially natural – for reasons that are not mathematical. The English alphabet, for example, makes for two metric spaces: strings of letters, sets of words. Strings of letters are close if they agree in spelling; words if they agree in meaning. Small typographic changes give rise to large differences in meaning: these metric spaces are not in phase. This observation is often regarded as a paradox in the context of theoretical biology. In an important and influential article, King and Wilson recount evidence showing that chimpanzee and human polypeptide sequences are more than

99 percent identical; the *species* appear further apart than a comparative analysis of their polypeptide chains might otherwise suggest.[6]

Complexity

Complexity and simplicity, like Yin and Yang, are metaphysical duals; except for a vagrant connection to intuition, it hardly makes a difference what is called which. Mathematicians and philosophers are interested in complexity for their own ends; so are theoretical biologists, who in their better moments are quite capable of evincing a sense of Heraclitian awe when confronted with the intricacies of the protozoan swim bladder. Simple counting principles often seem as if they might provide a general scheme for the measurement of complexity. Suppose that X is a nonempty set of objects and that A, B, C,... are constructed from the elements of X by certain specified operations – concatenation, for example. Can we not then say that the complexity $C(x)$ of any object is a measure of the number of its distinct elements and the separate and specifiable relations between them? $C(x)$ would be a monotonically increasing function of the square of the number of distinct elements in any given construction. Simple, no? And intuitively satisfying?

Apparently not. Label the parts of an ordinary watch in an obvious alphabetic fashion; and the binary relations between its parts as well. The watch when working, let me suppose, has a complexity measured at C; but so, then, does the watch when not working – when not assembled, in fact, binary relations being free for the asking. Examples of this sort, when extended and made precise, suggest ultimately that any complex object belongs to an embarrassingly large equivalence class of objects precisely equal in point of complexity.

Statistical mechanical complexity

A system of identical particles moving within a fixed, bounded, and finite volume of space constitutes a *configuration*; never having seen the blue smoke from a cigar spontaneously collect in but one corner of a warm room, the thoughtful physicist – pipe, slippers, Beagle-eyes, an air of earnest confusion – concludes that not all configurations are equally probable; yet if there are N configurations $\Pr(N_i) = N_i / N$ – this for each i. This incompatibility between what one sees and what one gets is known as *Boltzmann's paradox*, an unhappy name if only because no real paradox is forthcoming; but an unhappiness nonetheless. Distinct configurations, Boltzmann argued, may be grouped into *states*; what the altogether more elegant Gibbs called *ensembles*. Within thermodynamics – statistical mechanics from above – the entropy S of a system appears perpetually in the ascendancy and tends inexorably to a maximum; statistically, Boltzmann reasoned, S is thus proportional to

$$S = k \log W \; ; \tag{9.4}$$

where k is Boltzmann's constant, and W a measure of those configurations compatible with a given state – *complexions* as they are called in old-fashioned texts. Configurations are alike in point of probability: not so complexions; the probability of finding a mechanical system in a given state is proportional to the number of distinct complexions realizing that state. At equilibrium, the complexions are at a maximum; and so, too, the entropy, which functions as a kind of ectoplasmic measure of ***randomness*** or ***disorder***.

Complexity under a classification

Statistical mechanics has a good point to its credit, and implies a second. Certain states of a physical system may be multiply realized; their number, if counted, makes for a measure of sorts. What is measured within statistical mechanics is plainly not complexity; the description of entropy as disorder serves only to explain the whole business to the baffled undergraduate, with the explanation rapidly withdrawn by the time he enters graduate school. Still, I am struck by the extent to which the mathematical definition of entropy is made possible by an enterprising reorganization of the way in which mechanical systems are classified; in assessing complexity, a concept with a brutish family resemblance to disorder, the classification may well come first.

An example? Of course. I shall pass glowing colored slides about shortly. Consider the set of all functions $f: R^n \to R$. Those smooth functions whose critical points are nondegenerate are known as *Morse functions* and are at once open, dense, and locally stable in $C^\infty(R^n, R)$. Any Morse function may be expressed in canonical form: if x is a critical point of f, there exists a number k such that in a neighborhood of x, and after a suitable change in coordinates,

$$f(x) = x_1^2 + \cdots + x_k^2 - x_{k+1}^2 - \cdots - x_n^2 . \tag{9.5}$$

Such is Morse's lemma. Their mathematical docility suggests that the Morse functions are simple, if anything is; but the Morse functions are simple *because* they are Morse functions, and not Morse functions because they are simple; simplicity is a derivative quality, like color, contingent upon a classification, and unremarked otherwise.

The concept of a degenerate singularity makes for a simple classification on the space of smooth functions $C^\infty(R^n, R)$; but a set of objects may be simple under a classification even if the classification is itself unpleasantly complex. Writing some years ago, Smale asked whether there exists a least Baire set U in the space of all dynamical systems Dyn(M) on a compact manifold M, whose elements might be qualitatively described "by discrete numerical and algebraic invariants".[7] The question as posed admitted of a simple answer: no. What is needed, Smale later concluded, is a sequence of nested subsets U_i[Dyn(M)], where k is relatively small, U_i open, and U_k dense. As i increases, more of Dyn(M) is swallowed; as i decreases, stability and regularity properties come to the fore. It is for U_1 that Axiom A is satisfied, nonwandering sets are finite, and the transversality condition is met. U_1 thus consists of "the simplest, best-behaved, nontrivial class of dynamical systems"; but nothing in Smale's organization of Dyn(M) is simple at all.

A set is *absolutely simple* under a classification if it is at once open, dense, and locally stable; under this definition simplicity does not come in degrees. Often, suitable sets turn out to be merely of the first Baire category, the best one can do; sets that are dense need not be stable, and vice versa. First category sets and sets of measure zero coincide in the case of countable sets; but not beyond.

From the point of view of statistical mechanics, simplicity and complexity are concepts that involve configurations; complexity under a classification is a matter of routine: what is complex is singular, unusual. These notions may be brought into alignment – but only for a certain class of objects. An object A is ***dissective*** only when it may be decomposed to a finite stock of parts in a finite number of steps. The mammalian eye is a dissective structure; so is the whole of a mouse, a moose, or a mole; but curves and concepts, the real numbers, the coast of Britain, sea-green sea-waves, and, perhaps, the entire bizarre universe of elementary particles, are indissective. A dissective object is thus composed of its parts taken together under a certain distinctive relationship. Say that A is composed of $a_1, a_2, ..., a_n$ under R. By a relational alternative to R I mean a single permutation of the parts of A. If A, for example, contains but two parts, a and b, say, under the relationship $R(a, b)$, $R(b, a)$ is a relational alternative to R – the only one in fact. Given R, I denote by R^* the full set of all relational alternatives to R. If A is dissective it is R^* that forms its complexion class: the set of all sets of its parts under all and only their relational alternatives.

An elementary partition of a complexion class splits the class as a whole into equivalence classes; relative to a partition, complexity and simplicity are attributes of equivalence classes, and are judged simply by size. To the extent that $[E_i]$ is larger than $[E_j]$, it is simpler as well; and vice versa. Almost all structures in theoretical biology may be dissected to a finite, although very large, base; in this sense, biological complexity and simplicity have pliant finite measures.

The mammalian eye, for example, is a dissective structure. Its parts (on one level of dissection, at least) are proteins, which are arranged in various delicate and precise ways. I am ignoring, now, any dynamic considerations and thinking instead of the mammalian eye as a static object. The complexion class to the mammalian eye consists of all and only those rearrangements of proteins that comprise relational alternatives to the mammalian eye itself.

What makes an eye distinctively an eye, rather than some assembly of jelly-like proteins, is obviously the fact that it is capable of sight. This invocation of function sounds an unavoidably Aristotelian note; but without some concept such as function or purpose, theoretical biology loses much of its point. Let me partition the relational alternatives to the mammalian eye into equivalence classes on the simple basis of function. In the full complexion class, those structures that are capable of sight fall to one side; and those that are blind and stare sightlessly, fall to the other. Complexity and simplicity appear as matters of relative size: the larger the equivalence class, the simpler the structures. Given the delicacy of the mammalian eye, most of its relational alternatives will be incapable of sight; like the Morse functions, these complexions are simple structures; but again, simple because they are sightless, and not sightless because they are simple.

Complexity in strings

Of the 2^n binary sequences of length n, some, such as

$$0, 0, 0, 0, 0, ... \tag{9.6}$$

seem simpler than others,

$$0, 0, 1, 0, 1, \ldots \tag{9.7}$$

for example; yet the most natural probability distribution over the space of n-place binary strings assigns to both the same probability: 2^{-n}. It goes against the grain, mine, at any rate, to reckon (9.6) as likely as (9.7), especially when n is large; but nothing in the sequences themselves indicates obviously the point of distinction.

The goal of science, René Thom has suggested, is to reduce the arbitrariness of description; substitute data for description, and the apothegm gains my assent. A law of nature is data made compact: $F = ma$, said once and for all, the whole of an observed or observable world compressed into just four symbols. A series of observations compactly described is rational; if rational, not random. This curious but compelling chain of deductions prompted Kolmogorov to argue that randomness in binary sequences or strings might be measured by the degree to which such strings admit of a simpler description.[8] In following this line, Kolmogorov took the first step toward severing information theory from its unwholesome connection to the theory of probability. If S is a binary string its length is measured in bits: an n-place binary string is n bits long. By a *simpler description* of S, Kolmogorov meant a string D shorter than S such that D describes S by acting as the input to a fixed computer that generates S. Strings that cannot be compactly described are *complex, random*, or *information-rich*; strings that can, are not; of these adjectives, only the second preserves even a vagrant connection between the concept that it connotes and what is being measured. This rather inelegant idea makes plain the felt difference between a string of n 0s, and a mixed string. Sequence (9.6), for example, may be expressed by a program, speaking loosely, whose length is $\log_2 n + C$. If $n = 32$, $\log_2 n = 5$: the relevant instruction is simply to write or compute 0 2^5 times. C measures what little is needed to carry out the instructions; $32 - 5 = 27$, the compactness of the program. The shortest program that computes a mixed sequence such as sequence (9.7), by way of contrast, may well be close to 32 bits in length: to compute the sequence, the computer must first store it precisely.

The details? They have been changing since Kolmogorov first spoke, oracle-like, on the subject in a note published in 1967; like a snake engulfing an egg, the theory of recursive functions is engaged in swallowing algorithmic information theory, a development that I deplore, but accept as inevitable. Consider the set of all n-place binary strings A^* over a binary alphabet A and let TM be a fixed computer – a Turing machine, say; g is a general input–output function on TM mapping strings onto strings. The *complexity* of a string S of length n is the length of the shortest binary string D that generates S under TM by means of g. Whatever the complexity of S, D will plainly be maximally complex, and, hence, entirely random. Otherwise, it would not be the shortest description of S. All finite length strings quite obviously have a finite measure of complexity; and only finitely many distinct strings of the same length have the same finite measure of complexity. Quite surprisingly, the decision problem for complexity is recursively unsolvable; this result follows almost directly from the unsolvability of the halting problem for Turing machines. Like truth, randomness is a property that remains ineluctably resistant to recursive specification.

If all else fails, a binary sequence of length n may be generated by a binary sequence of length n: there are 2^n such algorithms, and $2^{n-1} - 2$ algorithms shorter than this. On any reasonable interpretation of complexity, algorithms within a fixed integer k of n itself must be reckoned random or complex or nearly so. Thus $2^{n-k-l} - 2/2^n$ algorithms have a complexity less than $n - k$; and are hence nonrandom or simple. If $k = 10$, this ratio is roughly 1 in 1000; of 1000 binary sequences of length n, only one can be compressed into a program more than ten bits shorter than itself. Hence:

Theorem 9.1 The set of random sequences of length n in the space A^* of all binary sequences of length n is generic in A^*.

These random sequences are simple under a classification because they are typical, but complex in a stronger and more absolute sense because they are random or information-rich. In this context, genericity is a ***finite*** measure of size. The number of purely random strings grows exponentially with n, of course. If most binary sequences are random, the appearance of sequence (9.6) prompts a natural stochastic surprise: sequences such as (9.7) are what one expects. The definition of Kolmogorov complexity may be directly extended to recursively enumerable sets; sets of strings especially, and hence languages.

Language-like Systems

When it comes to language, there is syntax and semantics. Phonetics is the province of the specialist; pragmatics remains a pale albino dwarf. To semantics belongs the concept of meaning; to syntax, the concept of a well-formed formula or a grammatical sentence. The reference to logic is happy if only because it highlights the fact that language-like systems go beyond the natural languages. Any language no doubt exists primarily to convey meaning; but meaning in mathematics is a matter of a model – an extrinsic object.

The construction of strings within a language-like system involves concatenating or associating simpler strings; any finite string may be dissected to a finite set of least elements. Going up, concatenation; going down, finite dissection; retrograde motion of this sort suggests that language-like systems on this level be represented algebraically as semigroups. Let A be any nonempty set of objects – words, for example, or letters, or numbers. A has the structure of a semigroup if there exists a mapping $A \times A \to A$ such that for all a, b, and c in A

$$(a \circ b) \circ c = a \circ (b \circ c) \ .$$

In English words go over to sentences from left to right; in Hebrew, from right to left; but in any case, one step at a time. Let A be a finite set of words now, with words understood implicitly as the least elements of a natural language; and let A^* be the set of all finite sequences $(a_1, \ldots, a_n)$ whose elements $\{a_1, \ldots, a_n\}$ lie in A. To endow A^* with the structure of a semigroup, it suffices to define an associative mapping $A^* \times A^* \to A^*$: easy enough. If

$$S_1 = (a_1, \ldots, a_m)$$

and

$$S_2 = (b_1, \ldots, b_n) \ ,$$

then

$$S_1 \circ S_2 = (c_1, \ldots, c_{m+n}) \ ,$$

where

$$c_i = a_i; \ c_{m+j} = b_j \ ,$$

$$i = 1, 2, \ldots, m; \ j = 1, 2, \ldots, n \ .$$

A^* is at once a ***free-semigroup*** over a finite alphabet and a ***universal language***: no sequences are left out.

Almost all language-like systems are large in the sense that they have many distinct strings. Meditating on the matter in the late 1950s, and regularly thereafter, Noam Chomsky argued that every natural language is infinite by virtue of its recursive mechanisms – conjunction and alternation, for example – and, simultaneously, that such mechanisms are recursive by virtue of the fact that every natural language is infinite. Both halves to this argument, taken together, describe a closed circle in space. Whatever the truth, language-like systems, if they are infinite, are countably infinite and no bigger.[9]

Going further toward a definition of a language-like system involves the badlands beyond triviality. Linguistics, the French linguist Maurice Gross once provocatively remarked, admits of but a single class of crucial experiments. Native speakers of a given language are able to determine whether a given sentence is grammatical. Experiments of this sort exist because no language-like system encompasses the whole of a set of strings drawn on a finite alphabet – a curious and interesting *fact*, which the sheer concept of communication might otherwise not suggest. The distinction between grammatical and ungrammatical strings induces a primitive classification on a language-like system; and reflects an even stronger principle of fastidiousness: the vast majority of language-like strings are not grammatical at all and represent syntactic gibberish. The fastidiousness of language-like systems is yet again a fact: it would be easy, if unrewarding, to design an artificial language in which most strings were grammatical. From the point of view of grammar, the strings of a natural language are complex under the classification of strings into grammatical and ungrammatical sets. With the strings arrayed in front of the mind's bleak and rheumy eye, in ascending order, by length, with sets of strings stacked like an inverted pyramid, the grammatical strings in a language-like system appear as nothing more than a thin smudge; they are thus ***complex*** under this classification because they are singular, unusual. The origins of this bit of natural history are to be discovered, no doubt, in the algorithmic properties of the human brain: in order to store a natural language, the brain must first represent it – in the form of recursive rules, for example. This suggests that language-like systems are low in point of Kolmogorov complexity; and from this point of view, ***simple***.

A natural language, I have already observed, realizes two metric spaces (cf. p 240); but the informal example that I gave involved the concept of meaning, and not grammar. No matter: the point carries over to the case at hand – and comprises the third of three queer natural facts that nothing in the concepts of grammar or communication obviously implies. Thus, let T be a typographic metric space of strings under the natural metric; the same set of strings comprises a second metric space under the degenerate distance function $\mathbf{d}^*$: if s and s' are both grammatical, $\mathbf{d}^*(s,s') = 0$; if not, $\mathbf{d}^*(s,s') = \infty$. These are the ***natural*** and (degenerate) ***grammatical*** metric spaces of a language-like system. In a language-like system, natural and grammatical metric spaces are plainly ***not*** in phase.

Two models of generation

Linguistics is a rebarbative, hair shirt of a subject; and grammar a vexing property. Linguists, for reasons of their own, are often interested in the weakest of generative devices that specify all and only the sentences of a natural language.

Representation by grammar

A ***phrase structure grammar*** is a quadruple $G = (A, T, S, P)$, where A is some finite alphabet of symbols; T, a distinguished subset of A – the set of so-called terminal symbols; S, a distinguished initial symbol; and P, a finite set of production rules of the form $u \rightarrow v$; u is a nonempty set of nonterminal symbols, and v some specified string of characters. The set of all strings of terminal symbols constitutes a ***phrase structure language*** – a proper subset of the set of all strings A^* defined over A.

By a *context-free* production rule, I mean one in which u may occur in any context – in effect, a rule in which u figures in isolation. Correspondingly, there are context-free grammars.

Example 9.1 Let $A = (a, b), T = (a, b)$, and P be the two rules $S \rightarrow ab$; $S \rightarrow aSb$. This grammar generates all and only the strings of the form $a^n b^n$.

Representation by systems of equations

Consider the context-free grammar G whose production rules are $S \rightarrow aSa$, and $S \rightarrow c$, where $T = (a, c)$, and S is an initial symbol. Let the variable f_i range over terminal symbols. The action of the production rules may be mimicked by an equation:

$$S = f_1 + f_2 + \cdots + f_n \ ,$$

where addition is construed as set theoretic union. For G,

$$S = aSa + c \ .$$

Replacing S by $S^0 = c$,

$$S^{(1)} = aca + c \ .$$

This process repeated ultimately yields a system of equations

$$S^{(1)} = aca + c$$

$$S^{(2)} = a(aca + c)a + c = a^2ca^2 + aca + a$$

$$\vdots$$

$$S^{(n)} = a^n ca^n + \cdots + aca + c = \sum_{i=0}^{i=n} a^i ca^i .$$

At the limit, the solution $s^{(\infty)} = \sum_{i=0}^{i=\infty} a^i ca^i$ is given by a formal power series in noncommutative variables.[10]

A language-like system has *formal support* when each and every string in the system may be described by a single algorithm; only for context-free languages may grammars and systems of equations be balanced against each other. Elsewhere, the situation is darker. There is a sense, however, in which these two representations exhaust the possibilities for the description of structured and infinitary objects; and correspond, in the Metaphysical Large, to the alternatives confronting an imaginary Deity in creating the observable world.

Weak Theories

The vitalist believes that life cannot be explained in terms of physics or chemistry. In the nineteenth century, in Germany and France, at least, his was the dominant voice before Darwin; and natural philosophers, such as Cuivier or von Baer, or Geoffrey St. Hilaire, dismissed mechanism with a kind of troubled confidence that suggests, in retrospect, a combination of assurance and wistfulness. Orthodoxies have subsequently reversed themselves with no real gain in credibility. David Hull, in surveying this issue, concludes that neither mechanism nor vitalism is plausible, given the uninspiring precision with which each position is usually cast.[11] *D'Accord*. To the extent that the refutation of vitalism involves the reduction of biological to physical reasoning, the effort involved appears to me misguided, and reflects a discreditable, almost oriental, desire for the Unity of Opposites. On the standard view of reduction, the sciences collapse downward until they hit physics: *Rez-de-Chausee*; but our intellectual experience *is* divided: mathematics, physics, biology, the social sciences. Each science extends sideways for some time and then simply stops. The ardent empiricist, surveying the contemporary scene, might well incline to scientific polytheism, with mathematics under the influence of an austere Artin-like figure, and biology directed by a God much like Wotan: furious, bluff, subtle, devious, and illiterate.

Still, the philosopher of science is bound to wonder why so many philosophers have remained partial to the reductionist vision, and hence to mechanistic thought in biology. David Armstrong, J.J.C. Smart, Michael Ruse, and even the usually cagey W.v.O. Quine, call on elegance to explain their attachment. Were the sciences irreducibly striated, one set of laws would cover physics, another biology, and still a third, economics and urban affairs, with the whole business resembling nothing so much as a parfait in several lurid and violently clashing colors. This is an aesthetic argument, and none the worse for that, but surely none the better either. If elegance is inadequate as a motive, intellectual anxiety, realized unconsciously, is not.

Vitalism commences from the conviction that nothing in our experience is much like the life that ripples and bubbles so abundantly over an entire planet, and nowhere else, apparently. Now mathematical physics is not only the preeminent discipline of our time – it is where the laws are. Evolutionary theories in biology are ***weak*** in the sense that they are not directly sustained by the authority of physics; and, worse, weaker still in being ***counterphysical***. Thermodynamic arguments count against the very existence of the structures that they are meant to explain. Fact heavy, law poor, such theories remain surprisingly resistant to confirmation. Were biology an aspect merely of physics, the sceptic would get short shrift: there, the answer to whether what works, works, is simply that it does.

Science is unavoidably general. To say that copper conducts electricity is weakly to imply the counterfactual conditional that were anything much like copper it would conduct electricity as well. It has often appeared to philosophers of science that specifying what it means for something – an x, say – to be much like copper inevitably comes to claiming that, among other things, x conducts electricity. Still, the similarity in structure between two domains of discourse – computer programs and natural languages, for example – may be obvious on grounds other than the fact that they share the same laws.

When I speak of a theory, I follow the logician's lead: a theory consists of a consistent set of sentences in a given language; the set-theoretic or algebraic structures in which a theory is satisfied comprise its models. Two models that share the same structure are isomorphic and hence elementarily equivalent in the sense that they satisfy the same sets of sentences. What I am after is a weaker notion entirely – partial similarity in structure. I know of no way, unfortunately, to define this concept so that the definition applies equably to biology, and, say, geology; I suspect, in fact, that partial similarity in structure will require a definition with indefinitely many separate clauses. Whatever the details, similarity in structure is bound to be a matter of degree, so that it makes sense of sorts to say of two models that they are at a certain distance, one from the other. In this way a family $\{M_i\}$, $i = 1, 2, \ldots$ of (possibly) first-order models may be given an appropriate and empirical metric structure.

Suppose that T is a theory holding in M; and let M^* be a model at some fixed distance from M. By the ***symmetric difference*** T/T^* of T I mean the number of formulas T^* of T that fail to hold in M^* when T is interpreted in M^*.

A theory T is general, I shall say, if for any $\varepsilon > 0$, there exists a $\delta > 0$ (a function of ε, of course) such that

$$\mathrm{d}(M, M^*) < \delta \rightarrow T/T^* < \varepsilon \tag{9.8}$$

Generality in my sense is a kind of stability; and as Dr Johnson remarks, the soul must ultimately repose in the stability of the truth.

To see an analogy between the operations of life, on the one hand, and the operations of language, on the other, is to raise the question whether the laws of biology have a natural and legitimate interpretation in linguistic terms. I am myself indifferent to the fate of the Darwinian theory, and perfectly prepared to believe, along with Wickramasinghe and the luckless Hoyle, that life originated in outer space, or that the Universe-as-a-Whole is alive and breathing stertorously;

but if Darwinian theories work in life, they should work elsewhere – in language-like systems, I should think. Should they fail there, this may be taken as evidence for the inadequacy of Darwinian theories, or as evidence for the inadequacy of the analogy that prompted the comparison in the first place.

I stress this point if only because it has so often been misunderstood.

Life as a language-like system

It was von Neumann who gave to the idea that life is like language a part of its curious current cachet. The last years of his life he devoted to a vast and clumsy orchestration of cellular automata, showing in a partial fashion that when properly programmed they could, like abstract elephants, reproduce themselves. Some years before, McCulloch and Pitts had constructed a series of neural nets in order to simulate simple reflex action; Kleene demonstrated that their nets had the power of finite automata and were capable of realizing the class of regular events; von Neumann's automata had the full power of Turing machines. Michael Arbib, E.F. Codd, G.T. Herman, A. Lindenmayer, and many others, have carried this work forward, with results that asymptotically approach utter irrelevance.[12] Yet the analogy between living systems and living languages has not lost any of its brassy charm. There is information, of course, which is apparently what the genes store; replication, coding; messages abound in the bacterial cell, with *E. Coli*, in particular, busy as a telephone switchboard. So striking has the appropriation of terminology become, that some biologists now see the processes of life, in all their grandeur, as the effort of a badly protected and vulnerable bit of genetic material to keep talking for all eternity.

Unlike an argument, an analogy stands or falls in point of plausibility; good arguments in favor of bad analogies are infinitely less persuasive than bad arguments in favor of good analogies. Certainly the proteins, to stick with one class of chemicals, may be decomposed to a finite base – the 20 amino acids. The precise, delicate, dance-like steps that are involved in their formation suggest, moreover, that they satisfy some operation as abstract as concatenation. On the other hand, the number of possible proteins, although large, is finite; but one of the joys of analogical reasoning is the vagueness with which the line between success or failure may be drawn.

The grammatical strings of a language-like system are low in Kolmogorov complexity, and so are not random. Such is the fastidiousness of a language-like system. What of the proteins? If they are random, it makes little sense to think of them as biological words or sentences. Jacques Monod, whose metaphysical attitude toward biology suggested nothing so much as a kind of chirpy bleakness, drew attention to the random character of the proteins in ***Le hasard et la necessite***; his argument has been gravely accepted by many molecular biologists.[13] In fact, the evidence leading to his conclusion is fragmentary; the standards of randomness to which he appealed, imprecise. Thus it struck Monod that knowing, say, 249 amino acid residues in a chain 250 residues in length, one could yet not predict the last member of the chain; much the same is true for English sentences, of course; it is, in any case, simply untrue that protein strands exhibit such wanton degrees of freedom. Within protein chemistry, there are many instances of what appear to be strong internal regularities: palindromic patterns, for example.

Nonetheless, I am in sympathy with Monod to this extent: it is unlikely that the analogy between life and language will be profitably pursued on the atomistic level of the nucleic acids or the proteins themselves.

PART THREE

Arguments Good and Bad

The theory of evolution is haunted by an image and an observation: the first, that of the hapless chimpanzee, typewriter-bound, endeavoring, quite by chance, to strike off the first twenty lines of Hamlet's soliloquy; the second, the comment of an anonymous Jansenist logician, who remarked, quite sensibly, "that it would be sheer folly to bet even ten coppers against 10000 gold pieces that a child arranging at random a printer's supply of letters would compose the first twenty lines of Virgil's *Aneid*". Image and observation do not quite cohere into a single argument: it is clear in neither case *how* the imagined stochastic experiment is to stop. Still, the notion of randomness yet lies at the center of evolutionary thought, and there it sits, toad-like and croaking. On the simplest and most intuitive conception of probability, what can occur is weighted against the background of what might occur: five diamonds: all other combinations of the cards. In poker, there are 2598960 five-card hands, but only 5148 flushes. It is their ratio that one might expect to observe as cards are actually dealt; but in the longest of long runs, the passage to the limit gives content to the intuitive idea that a number of successive trials will converge to a particular real number: 0.002, for example, if flushes are being counted.

One of the curiosities of the very notion of probability is the inescapability of the improbable. The laws of thermodynamics, to take a notorious example, are anisotropic: they go in one direction; downhill, as it happens, a circumstance with what appears to be overwhelming personal support. Statistical mechanics provides a brilliant and persuasive explanation for thermodynamic laws; yet Poincaré demonstrated, in an absurdly easy proof, that any statistical mechanical configuration, of whatever degree of implausibility – k molecules of gas, for example, occupying $1/V$ of the total volume V of a finite and bounded container – is bound to recur, in all its vividness, poignant symmetry, and complexity, given enough time. Physicists often explain the discrepancy between thermodynamics and statistical mechanics by arguing that the time involved is very long. No doubt.

The evolution of life on this planet is, as Darwin realized, not a hurried affair. Early on, Darwinian biologists got rid of the theological limits set to the age of the Earth by Bishop Ussher and others in the seventeenth century; the scale within which Darwinian evolution might have worked is bounded by perhaps five billion years. Nineteenth century biologists assumed that whatever else one might say about Darwinian biology, it would not fail for lack of time; this thesis twentieth century biologists have carried over intact.

Five billion years is apt to seem long if one is counting the minutes; but it is not long enough to sample on a point by point basis a space whose cardinality is roughly 10^{15} – touching base with a new point at every second, say; and yet there are 20^{250} possible proteins – a number larger by far than the expected life of the universe measured in seconds. In a space of this size, the odds against discovering a specific protein – fishing it from an urn, say – are prohibitive: 1 in 20^{250}.

I spoke hastily just now of a *specific* protein: if *any* protein will do, the odds improve: in a uniform probability space $\{a_i\}$, $\Pr(a_1 \vee a_2 \cdots \vee a_i) = 1$. The distinguished British biologist Peter Medawar has seized upon this point, and commenced happily to trot, but in what I think is the wrong direction.[14] "Biologists," he writes, "in certain moods are apt to say that organisms are madly improbable objects or that evolution is a device for generating high degrees of improbability. I am uneasy about this entire line of thought," he continues, "for the following reason:

> Everyone will concede that in the games of whist or bridge any one particular hand is just as unlikely to turn up as any other. If I pick up and inspect a particular hand and then declare myself utterly amazed that such a hand should have been dealt to me, considering the fantastic odds against it, I should be told by those who have steeped themselves in mathematical reasoning that its probability cannot be measured retrospectively, but only against a prior expectation ... For much the same reason, it seems to me profitless to speak of natural selection's 'generating improbability' ... it is silly to be thunderstruck by the evolution of organ A if we should have been just as thunderstruck by a turn of events that had led to the evolution of B or C instead."

Medawar is roughly right about probability: the fallacy to which he refers is the *error of retrospective specification*; and consists precisely in reading back into an original sample space information revealed only on the realization of a particular event. In poker, a deal distributes n hands of equal probability: 1 in 2598960, as it happens. This sample space is retrospectively specified if one hand in particular is contrasted with the full set of 2598959 hands that remain, and probabilities assigned to the partition so created; what appears initially as one among equiprobable events becomes under retrospective specification an improbable event in a sample space of only two points. It is embarrassing for an author to point such things out. Still, Medawar is wrong in the general conclusions that he draws from this paragraph. Card sharps and statisticians are little interested in the set of all five-card sequences. In poker, sequences are *initially* partitioned into equivalence classes of uneven size: a royal straight flush, of which there are four, a straight flush, four of a kind, a full house, a straight, three of a kind, two pairs, and, then, finally, whatever is left – the vast majority. There are four ways to achieve a royal straight flush; many more ways in which to realize a full house. Since they are specified in advance, partitions in poker carry no taint of retrospection; and plainly, in poker there is only a rough correlation between the internal character of sequences within a partition and their payoffs: what is important here, as elsewhere, is the classification, which is very largely arbitrary.

Medawar's argument, on its face, thus involves rather an uninspiring mistake, but it is not yet a mistake in evolutionary thought. The human eye, a chastened Medawar might argue, turning his back on his own analogy between life and the cards, represents one arrangement of its constituents: any other might have done as well. In admiring the structure that results, we suffer from misplaced awe, like a toad contemplating a dog. Does this argument carry conviction eye-wise? Is it reasonable to suppose that any other arrangement of the eye's constituents would result in an eye? In anything at all? The question sounds an unavoidably Aristotelian note: an eye is an organ with a specific function – sight, most obviously; an eye-like configuration does not count as an eye unless it can see. To frame the discussion thus is to answer the question immediately, at least on the level of intuition; but what I have said must not be confused with an *argument* in refutation.

Viable proteins

Linguistics is possible if only because human beings have strong and reliable intuitions about natural languages. The polypeptides are alien strings, accessible only through an arduous act of the biochemical imagination. Grammar effects a segregation of strings in a language-like system; beyond grammar, aloof, untouchable, there is *meaning*; the two concepts do not coincide. Some grammatical strings, in a natural language, at least, are grammatical and meaningless; others, meaningful but ungrammatical; but meaning and grammar belong together, yoked pairs in the same corner of some dimly understood conceptual space. An algebraic system of strings in which no distinctions of meaning and grammar are recognized is profligate; and pointless because of its profligacy.

In a preanalytic sense, the concept of meaning indicates a kind of coherence; and has a usefulness of application in domains other than language. A life well-spent is meaningful: its parts and patterns are ordered; full with life, biological creatures are filled with meaning, a kind of blunt, irrefrangible purpose; in death, this meaning disappears, and what is left, the corpse and its grim constituents, appears all at once to lose the integrity of the creature itself, and becomes, instead, a thing among other things, an object merely. To the vitalist, living creatures instantiate some unique property that remains stubbornly unseen elsewhere – in the domain of objects studied by mathematical physics, for example; in death, this property vanishes, like a fluid evaporating. In mechanistic thought, the passage from life to death is rather like a phase transition, a singularity of sorts in the trajectory of the organism, a disabling and permanent catastrophe, that reflects, as it must, only a change in the constituents of the organism, a variation in its underlying pattern. The concept of a complexion, which figures in statistical mechanics, provides a useful measure of meaning. The complexion set to a biological organism represents those relational alternatives of its biological parts that correspond to living systems. The unalterable fact that living systems die and hence do not persist indicates that some of their complexions fail to preserve life and hence meaning; in fact, the number of meaningless complexions must be significant: most of the arbitrary rearrangements of a complex organ – a mammal, say – result in nothing more than a botch – a circumstance with which every surgeon is familiar. The Central Dogma of molecular biology establishes a relationship between strings of nucleotides and strings of proteins; to the extent that the whole of a biological organism may be resolved into its protein-like parts, the Central Dogma establishes a larger, more indirect, relationship between molecular biological order and order in the larger sense of life. This relationship has an inverse: if only certain forms of life have meaning, this, too, is reflected, as it must be, in the universe of molecular biological strings – on the level of string *ensembles*, for example. If certain protein ensembles are meaningful, and not others, this suggests, but does not imply, that the same distinction is palpable on the level of the individual proteins themselves. The term *viable* I mean as a biological coordinate to the Siamese concepts of meaning and grammar; a protein is viable only when it achieves a certain minimum level of biological organization and usefulness. What level? What kind of organization? Usefulness in what respect and to what degree? Who knows?

Full loads, fair loads, fair samples

In a natural language, sentences decompose to words; words to letters. Grammatical constraints hold weakly at the level of English words. The set of all word-like combinations of English letters of fixed length n, I shall say, make up a ***full load***; the set of all grammatical words, a ***fair load***. Within molecular biology, a full load corresponds to all possible proteins of normal length: a set whose cardinality is 20^{250}. To the fair loads in English correspond the viable proteins in molecular biology. How large is the biological fair load? Again, who knows? Whatever its ultimate size, those proteins that have already been synthesized in the course of biological history are viable if anything is: nothing succeeds like success. This set is a ***fair sample*** of a fair load. Its size Murray Eden calculates at 20^{52}. The task that he sets himself is the infinitely delicate one of drawing inferences about the fair load from its fair sample.[15]

Between the fair sample of a fair load, and the fair load itself, is the difference between what is and what might be; between the fair load and the full load, the difference between biology and mathematics. In English, the difference between the fair load and the full load is as absolute as death. Any two words of English thus resemble each other more than they are likely to resemble a word generated at random from the letters of the English alphabet. In the case of the polypeptides, Murray Eden writes:

> Two hypotheses suggest themselves. Either functionally useful proteins are very common to this space, so that almost any polypeptide one is likely to find has a useful function to perform, or else the topology appropriate to this protein space is an important feature of the exploration: that is, there exists certain strong regularities for finding paths through this space.

In asking whether the viable proteins are common in the space of all polypeptides, Eden is asking, in effect, whether the fair sample is marked by discernable statistical regularities. "We cannot now discard the first hypothesis," he adds, "but there is certain evidence which seems to be against it: if all polypeptide chains were useful proteins, we would expect that existing proteins would exhibit very different distributions of amino acids." Statistical tests appear to show that pairs of proteins are drawn from a common stock. His example involves the alpha and beta human hemoglobin chain. One form of hemoglobin has 146 amino acid residues, the other 140. The two chains may be set down, side by side, and matched, residue by residue. They agree at 61 points; there are 76 points at which they differ, and 9 points at which no match is possible because the chains are not of the same length. It is plausible that one chain was derived from the other, or that both were derived from a common ancestor. What is curious about these pairs of proteins, however, is the fact that even though the chains do not agree completely in the order of their amino acids, they do agree in their *distribution*; reason enough, Eden argues, to suppose that the proteins themselves are drawn from a statistically significant fair sample.

The criticism of this historically important argument, I leave as an exercise.

Delicate inferences

In ***What is Life*?**, Schroedinger argued that living systems must have recourse to what he dubbed an "aperiodic crystal" in order to store information. Crystals are repetitive, regular, and information poor; the order of a living system is specific, irregular, information rich. There is a certain splendid effulgence to the vocabulary of theoretical biology that it would be uncharitable not to cherish. H.P. Yockey identifies order with Kolmogorov complexity; and so does R.M. Thompson, a mathematician who in writing on theoretical biology alternates between information theory and a pious endeavor to communicate to the reader his appreciation for the many faces of Krishna.[16] On the other hand, G.J. Chaitin and R.M. Bennett identify biological order with algorithmic simplicity. A division of intuition on so fundamental a point may suggest a degree of conceptual confusion approaching the schizophrenic.

If biological words are characterized by a high degree of Kolmogorov complexity, could time and chance have combined to discover a structure comparable, say, to cytochrome c or any of the modern hemoglobin chains? This is the question raised by the redoubtable H.P. Yockey: the problem as posed has but two parameters.[17] In the beginning the primeval soup, which I always imagine as rather a viscous, Borscht-like fluid, contained perhaps 10^{44} amino acid molecules. There is, inevitably, an element of fantasy to all quantitative calculations of this sort. At each second, over the course of 1×10^9 years, an indefatigable stochastic Deity arranges and then rearranges the 10^{44} amino acid residues in sequences whose length $N = 101$. There are

$$20^{101} \text{ or } 2.535 \times 10^{131} \tag{9.9}$$

such sequences. The odds against discovering any one in particular thus stand at 1 in 2.535×10^{131}. Not all residues, however, are equally probable. Save for a very large set of strings of small probability, the number of sequences of length N is

$$a^{NH} , \tag{9.10}$$

where

$$H = -\sum_{j=1}^{n} p_j \log_a p_j . \tag{9.11}$$

Here p_j measures the probability of the jth residue, and $a = 2$, so that H is measured in bits.

In the end – the details are not important to my argument – Yockey concludes that

$$H = 4.153 \text{ bits/residue} ; \tag{9.12}$$

the number of 101 place sequences is

$$2^{4.153 \times 101} = 1.8067 \times 10^{126} . \tag{9.13}$$

"Information theory," he remarks, "shows that, in this case, the actual number of sequences is smaller than the total possible number by a factor of 10^5". Now there are, in all, 3.8×10^{61} families of cytochrome c sequences; in order to obtain any one of them by chance, Yockey argues, it would be necessary to repeat an elementary stochastic experiment 3.15×10^{58} times on 10^8 separate planets "in order to have a reasonable expectation of selecting at least once a member of the ensemble of 3.8×10^{61} cytochrome c sequences in only ten of them".

From nothing, nothing, the Darwinian doubters have always claimed; and I have been there with the best of them; but *this* argument, couched as it is largely within the algorithmic theory of complexity stands on what seems to me dubious ground:

(1) A binary string is random to the extent that its shortest program is roughly of the same length as the string itself; this definition trades only in counting bits. Now, the impulse to assert that contemporary proteins are random owes much, I think, to the rather primitive idea that life, if complex, requires complex constituents or atoms; I have suggested something similar in arguing that the proteins inherit a grammatical distinction from the structures that they constitute. Kolmogorov complexity, however, is ill-defined on any level of biological organization past the molecular; but even if a mammal or a mollusk could be represented as a binary string, nothing suggests that those strings would be high in Kolmogorov complexity. Quite the contrary. Life in the large, on the level of the organism itself, is organized with what appears to be brisk algorithmic efficiency. Living creatures are simple in the sense of Kolmogorov complexity; but complex under the classification of their complexions. In this sense, they behave much as a language-like system. This observation is compatible with the thesis that protein strings are, nonetheless, high in Kolmogorov complexity; but it is compatible, too, with the contrary thesis that protein strings reflect the complexity of life by means of their *organization* and not their complexity. Nothing in the concept of Kolmogorov complexity measures the algorithmic organization of a string or set of strings; two equally complex strings may well differ in their *time complexity* to the extent that only one is polynomially bounded.

(2) The difference between the space of available proteins, and the small subset actually chosen by evolution, makes for a trite contrast; yet what lends to cytochrome c its position of statistical distinction? "Because of the very fundamental function of the cytochromes," Yockey writes, " ... the histones and other proteins, which are believed to be of very ancient and even precellular origin, one cannot relax the *specificity* requirement derived from cytochrome c" [emphasis added]. In generous conversation, Yockey has amplified this point by suggesting that the specific protein chains necessary for life correspond to the set of words in a language – fair and not full loads; a curious remark inasmuch as words in a natural language are low, and not high, in Kolmogorov complexity. Still, I am sympathetic to the drift of this line; but the difficulty goes beyond the problems of an imperfect analogy. Certain classes of proteins, Yockey argues, are necessary for life. Such are the information-rich, complex strands; other strands are specific in the limited sense that they are statistically unlikely: "only a tiny fraction of the (available) sequences will carry specificity." It follows by **Theorem 9.1** (p 245) that specificity and complexity are not the same thing: the *set* of complex strands (of a given length) is in the majority; their emergence is probabilistically favorable, indeed, unavoidable. Cytochrome c, considered simply as a complex protein, is no more likely to appear than any other complex protein; but no less likely either. Having discovered cytochrome c, quite by chance,

Life might have made do with any other protein of comparable complexity. If by specificity, Yockey means statistical unlikelihood in a uniform sample space – the space of all complex proteins, for example – his surprise at the emergence of cytochrome c is attributable to retrospective specification; if not, what then is specificity, the mysterious middle term to his argument? If the specific proteins have some independent description, Yockey does not provide it; and their size, apart from suggesting that it is low, he does not calculate.

PART FOUR

Der Prozess

The evidence in favor of the thesis that proteins are random sequences of amino acids is exiguous; and random words may well be grouped into nonrandom sequences. This suggests that the close study of the statistical properties of certain proteins may involve a kind of dense conceptual myopia, something that reflects a passionate absorption in minutiae. The *process* by which evolution in strings takes place, on the other hand, is macroscopic and global, an energetic probabilistic swarming over sample spaces that are never specified by means of mechanisms that are never clarified.

Biological paths

Life loiters over two metric spaces. The first is alphabetic; the second, zoological. Evolution comprises a drama in the large, at the zoological level; but the Central Dogma requires that any change in the large be mirrored by an alphabetic change, and so the process is doubled as it is divided. To talk blithely of evolution in strings is to assume the completion of the two first steps in biological evolution: the emergence of life-like systems from inorganic matter; and the adventitious creation of the modern biological system of replication and genetic information. An explanation of these steps I cede to the forces of the Night: my more limited concern is with evolution as a process that takes place once the genetic machinery is throbbing moistly. In evolution at the molecular level, one amino acid is dropped from a protein string, another is inserted: *make way!, move over!, get out!, get lost!*, to cast the operations in easily understood terms; even if the process is more complicated, it may mathematically be resolved into discrete and finite steps. Whatever the details, proteins change over time; and the changes leading to their creation may be regarded as a *path* $P = p_1, p_2, \ldots, p_n$ or protein *sequence*. Suppose that A comprises the full stock of 20 amino acids; $A/$, the set of all *words* of amino acids precisely 250 points in length; and A^*, the set of all finite sequences drawn over $A/$. I assume – an *assumption* note! – that A^* has the structure of a language-like system under the binary and associative operation of protein *concatenation*, where concatenation has precisely its usual linguistic meaning.

Stochastic processes

Let S be a system and X the set of its states or configurations. State transitions are represented by a transformation $T: X \to X$, an artifice expressing the action of the system's laws of evolution. If $T_{s+t} = T_s T_t$, $[T_t \in R]$ is a *flow*, or *group action* of R on X.

On the Darwinian theory, evolution is at its secret heart stochastic; it is natural, therefore, to specialize the concept of a process to the case in which X is a measure space, T a measure preserving transformation. This is the domain chiefly of ergodic theory. Its underlying, indeed, fundamental, object is a probability space (X, B, u), where X is a set of states, B a σ-algebra of measurable subsets of X, and u a countably additive nonnegative set function on B. $u(X)$ is, of course, 1. Let T be an invertible injection from X onto X; if $u(T^{-1}E) = u(E)$ for all E in B, T is a *measure-preserving transformation*; the system (X, B, u, T), a *basic probability space*.[18]

By the *orbit* of a measure-preserving transformation T, I mean the extended history of a single point x under T from the infinite past to the infinite future: a trajectory from void to void. Artificially truncated at x, the system is in an initial state or condition. A real valued function $f: X \to R$, whose values correspond to $f(x)$, $f(Tx)$, $f(T^2x)$, ... acts to measure a system along its orbit; the class of such measurements is defined only to the extent that f is itself measurable:

$$\bar{f}(x) = \lim_{N \to \infty} 1/N \sum_{k=0}^{N-1} f(T^k x) \tag{9.14}$$

is thus the *time mean* of the system;

$$\int_X f \, du \tag{9.15}$$

its *space mean*: systems in which the two coincide for every measurable function are *ergodic*.

Example 9.2 Let A be an alphabet of n symbols $a_1, a_2, \ldots, a_n$, with probabilities $p_1, p_2, \ldots, p_n$, such that $p_i > 0$, and $\sum p_i = 1$. The product space n^z consists of the set of all two-sided sequences in n; the various probabilities assigned to each sequence induce a measure u on n^z. The *shift transformation* $(fx)_n = x_{n+1}$ is measure preserving; the system that results is a finite-valued stationary stochastic process with identically distributed terms.

Example 9.3 Let $M = (a_{ij})$ be an $n \times n$ stochastic matrix. Let $p = (p_1, \ldots, p_n)$ be a row probability vector fixed by M:

$$pM = p \ .$$

Keep the product space and shift transformation from **Example 9.2**; U_M may be extended to a countably additive measure on the algebra generated by cylinder sets; by the Caratheodory–Hopf theorem, U_M thus forms a measure on the Borel Fields of n^z.

Example 9.2 models, say, a doubly infinite series of coin flips, each with probability of one-half; **Example 9.3**, a regular Markov chain, where p measures the a

priori probability of each symbol, M, the transition probabilities from one symbol to another.

Entropy

Consider a *source* consisting of a finite alphabet A and an associated string of symbols, ... $x_0 x_1 x_2$ ... , where each x_i is an element of A. Symbols appear in sequence with a fixed probability p_i; if the probabilities are independent, the *average entropy per symbol* is

$$H = -\sum_{i=1}^{n} p_i \log_2 p_i \quad . \tag{9.16}$$

H is at its maximum if each $p_i = 1/n$. In general, the probability that a particular symbol appears in sequence may depend on symbols that have gone before. This is true if the source is a finite-state Markov device. Let A^I be the ensemble of all doubly infinite sequences drawn on A; the cross section on A^I of sequences that coincide at a finite number of points $a_i = x_{t_i}$, where t_i represents any set of integers, is a *cylinder set*. Now if A contains k letters, the number of n-term sequences over A is k^n; and each sequence is a cylinder in the larger space A^I. It is the cylinder that has a fixed probability Pr(C): the set of all n-term sequences represents a finite probability space, k^n points in size. The ***average amount of information per symbol*** sent out by a source of this sort is

$$H_k = -1/k \sum_{C \in C_k} \Pr(C) \log_2 \Pr(C) \ ; \tag{9.17}$$

the entropy of the source itself

$$H = \lim_{k \to \infty} -1/k \sum_{C \in C_k} \Pr(C) \log_2 \Pr(C) \ . \tag{9.18}$$

The concept of a source may be specialized to the case of a measure-preserving system under ergodic constraints.[19]

The Shannon–Macmillan theorem

A source puts out sequences; at any given time, there will be only finitely many – A^n in fact, if A is a finite alphabet, and n is the length of each sequence. Finite length sequences are cylinders in the infinite probability space determined by the source; they inherit a probability structure. If n is sufficiently large, there exists an arbitrarily small ξ and $\delta > 0$ such that the n-term sequences may be separated into two groups. For the first

$$|\log_2 \Pr(C)/n + H| < \xi \ ;$$

for the second,

$$\sum \Pr(C_i) < \delta \ .$$

This is Shannon's theorem, a result in mathematics that appears to add an author in regular periods. In any case, sequences of the first group are characterized by the fact that $(1/n)\log_2\Pr(C)$ is arbitrarily close to $-H$. The probability of any such sequence C_i is thus 2^{-nH}; the number of such sequences is 2^{nH}, and comprises a very small share of the total number $a^n = 2^{n \log_2 a}$ of available sequences: a happy result. In coding a channel of communication, attention need be directed only to a tiny sample of the output.

The Stochastic Structure of an Evolutionary Source

In considering evolution as a stochastic process, the object of study becomes biological paths; and not biological words – stray proteins, say, or bits of nucleic acids. The full set of paths in evolutionary space comprises an infinitely large set of strings, if only because evolution appears unbounded as a natural language. The sheer stress on the notion of randomness in popular accounts of evolutionary thought suggests at first that something like a pure Bernoulli process may underlie the whole business, an extended coin flip by means of a coin with 20^{250} separate faces. This is obviously absurd. Evolution is a process by which an ensemble of strings changes over time. Each string is composed of points – amino acids, in fact; the probability that any particular point will change is arbitrary but low; there is little likelihood that ***all*** points in a string will change simultaneously. Transition probabilities in a neighborhood N of a set of proteins E are thus ***concentrated*** in that neighborbood.

If an ensemble of proteins occupies a certain finite set of states A_i, its evolution comprises a finite state ***Markov process*** – a stochastic source satisfying the hypothesis of the Shannon–Macmillan theorem.

Trapping problems

The entropy of a source is a measure of its stochastic character: H at its maximum represents a high degree of uncertainty; all messages are equally probable. The hypothesis of the neo-Darwinian theory is that evolutionary sources are largely random. What this means is, in fact, not entirely clear; but it surely implies that H is relatively large. Let H_{Max} thus be the ***imagined entropy*** of an evolutionary source. "If the process of manufacturing messages", Chomsky and Miller remark, "were completely random, the product would bear little resemblance to actual utterances in a natural language".[20] Going backward, if the utterances of a natural language are regular, their source is not random. To the extent that a fair sample of evolutionary paths is regular, the fair load is regular as well. A source ***specifically designed*** to generate the fair load of protein paths has thus an entropy H

$$H < H_{\text{Max}} \, . \tag{9.19}$$

In itself, this is neither controversial nor surprising; if the degree of protein regularity is small, the difference between H and H_{Max} is negligible; if large, an evolutionary source ***over-generates***. The real issue is a matter of degree, a question of finesse. Linguistics, of course, suggests that if H is very much lower than H_{Max}, over-generation becomes inordinate; a stochastic source cannot, in general, converge on any ***natural*** language whose complexity is beyond the recursive capacity of finite-state automata; but while life may be a language-like system, it is not necessarily like a natural language, Chinese, say, or even Esperanto.

Hemoglobin chains

Statistical entropy is a measure of uncertainty; and a measure, too, of the number of alternative messages – my use of the word is metaphoric – that a stochastic source may generate. H at a relative maximum indicates that a source may send out multiple messages, a kind of energetic babble; at a relative minimum, H is ***constrained*** – by the rules of grammar, for example, or the laws of logic. In the case of life, Murray Eden observes, path lengths between proteins are most obviously limited by time, and evolution must be achieved within bounds set by the number of generations in the history of an organism. Meandering paths between proteins are temporally inaccessible. The alpha and beta hemoglobins, Eden argues, were derived by a process of evolution, one from the other; a path between the two sequences must thus exist. Eden calculates that this path at its shortest requires something like 120 separate steps, where each step involves a specific point mutation.[21] The population size of hemoglobin proteins – the fair sample – is, he estimates, 10^6; the rate of mutation 10^{-6}. Each step in this path corresponds to a positive gain in fitness: movement upward along a local gradient of relative perfection. In a conservative sense, it would take roughly 2700000 generations to convert a population of 10^6 alpha hemoglobin chains to a population of beta hemoglobin chains. So far, so good.

If certain paths, of whatever length, are inaccessible to life, a stochastic source is *occluded*. This is certainly what the hypothesis that life is a language-like system implies; it is implied, too, by the fact that contemporary hemoglobin chains exhibit relatively little ***variance***: certain possible paths are deficiently viable, ungrammatical in a sense. Nature, in passing from one chain to another, has evidently rather a small target in mind.

2700000 generations for the evolution of a protein is short: twice that number is long. Let k be a point midway between these numbers. If HG comprises an initial set of 10^6 hemoglobin chains, P_{HG} is the full load of protein paths $[P_{ij}]$, whose initial terms P_{1j} all lie within HG. The number of ***targeted*** protein paths in P_{HG} is small: so much to do, so much to see. The number of targeted protein paths that reach their target within k generations is vastly smaller: so much to do, so little time. If the entropy of a stochastic source is great, the untargeted meandering paths are apt to be favored; by contraposition, this implies that the source entropy for an evolutionary system is rather low; constrained, in fact, by the choice of time and targets; but what expresses these constraints, the Darwinian theory does not say.

Weizenbaum Theory [22]

It is a peculiarity of molecular biological strings that, like the elements of a natural language, they realize two spaces. These are spaces with distinct and ***different*** metrics: there is no reason to suppose that they are in phase. Evolution as a process works most directly on biological organisms, which must perish or persevere in the face of circumstance. To the extent that evolution is a process by which organisms converge over time to some local (or global) optimal, the processes of convergence that are sketched broadly in life must have some substantial echo at the molecular biological level, where words and strings hold sway. The relationship between metric spaces that this pattern exemplifies is quite general – the province, in fact, of Weizenbaum theory. Thus let M and N be two metric spaces, each with its own natural metric; points in M are labeled $t_1, t_2, \ldots, t_n$; points in N, $e_1, e_2, \ldots, e_n$; $f: M \to N$ is a mapping between points in M and points in N – a bijection, to make matters trivially simple. M and N are arbitrary, and admit of obvious specification:

(1) M is a typographic metric space; N, the space of biological organisms (see p 240).
(2) M is a typographic metric space under the natural metric on words; N, the same space under distance defined in terms of meaning or grammar (see pp 245–248).
(3) M is a typographic metric space; N, a space of algorithms.

Thus f might map linear sequences of DNA or proteins, or sets of such sequences, onto organisms, or sets of organisms; equally, f might map a linear string of letters onto a sentence, with a fixed meaning in a natural language; or onto an algorithm in a given computer language such as Algol; then, too, f might map fixed strings in an assembly language onto a computer program. In each of these cases, f does not preserve metrics; M and N are not necessarily in phase.

In addition to the natural metric on M, there exists an ***induced metric*** $d_{N(i)}$ on M defined by the following relationship:

$$d_{N(i)}(f^{-1}e, f^{-1}e') = d_N(e, e') . \tag{9.20}$$

The Weizenbaum experiment

To specify a Weizenbaum experiment, it is necessary to provide M with a ***probability transition system*** Pr determining for each point t in M the probability that t will change to t'; and an ***initial probability distribution*** Pr_0. A ***distinguished element*** $e^* \in N$ is fixed from the first. Within the context of molecular biology, transition probabilities are focused on relatively nearby strings – this because point mutations result in string-like changes of a short typographic distance. In a biological Weizenbaum experiment, this fact is respected to the extent that the typographic metric space and the probability transition system are mutually in accord: probabilities follow typographic neighborhoods. Elsewhere, probabilities and distances are adjusted accordingly.

A point t_0 is selected in accordance with the initial probability distribution Pr_0 over M. The distance $d_{N(0)}$ from $f(t_0)$ to e^* is measured; the system engaged for $i = 1, 2, 3, \ldots$; as t_{i-1} moves to t_i, the distance $d_{N(i)}$ between $f(t_i)$ and e^* is recorded. The outcome of the Weizenbaum experiment is the sequence

$$d_{N(0)}, d_{N(1)}, \ldots, d_{N(n)} .$$

The Weizenbaum experiment is *successful* if:

Condition W For $d_{N(0)}$ at an ***average distance*** from e^* the sequence $\{d_{N(i)}\}$ converges to a neighborhood of 0.

Condition W, when met, implies that $\{d_{N(i)}\}$ is both stable and oriented. The graph of a sequence of points constitutes a trajectory; the set of trajectories in N that are at once stable and oriented is of measure zero. A successful Weizenbaum experiment thus establishes that $\Pr(M)$ ***cannot*** be arbitrary with respect to its ***induced*** metric structure. In particular, points that are far in the ***induced*** metric have small transition probabilities: those probabilities that count must be concentrated on nearby objects – nearby in the sense of the induced metric. On the other hand, transition probabilities over molecular biological strings are, on the neo-Darwinian theory, focused on neighborhoods that are nearby in a natural metric.

It is perhaps for this reason that, with the exception of life itself, no one has ever seen a successful Weizenbaum experiment.

Eigenvalues of natural selection

In Darwinian thought, the effects of randomness are played off against what biologists call the *constructive* effects of natural selection, a mechanism that philosophers have long regarded with sullen suspicion. Wishing to know why a species that represents nothing more than a persistent snore throughout the long night of evolution should suddenly (or slowly) develop a novel characteristic, the philosopher will learn from the definition of natural selection only that those characteristics that are relatively fit are relatively fit in virtue of the fact that they have survived, and that those characteristics that have survived have survived in virtue of the fact that they are relatively fit. This is not an intellectual exercise calculated to inspire confidence.

Natural selection is a force-like concept; and, as such, acts locally if it acts at all. Mathematicians often assume that evolution proceeds over a multidimensional fitness surface, something that resembles a series of hills and valleys; a great deal that is theoretically unacceptable is often hidden in a description of its topology. But I am anticipating my own argument. In speaking of locality, I mean to evoke the physicists's unhappiness at action at a distance. Strings that are far apart should be weak in mutual influence; this is a spatial constraint. Then again, no string should be influenced by a string that does not yet exist. This is a temporal constraint, a rule against ***deferred success***. The historical development of a complex organ such as the mammalian ear involved obviously a very long sequence of precise historical changes. Comparative anatomy suggests that the reptilian jaw actually migrated earward in the course of evolution. It is very difficult to understand why each of a series of partial changes in the anatomy of the reptilian jaw should have resulted in a net increase in fitness ***before*** the advent of the mammalian ear. Certain genes within the bacterial cell, to take another example, "are organized into larger units under the control of an operator, with the genes linearly arranged in the order in which the enzymes to which they give rise are utilized in a particular metabolic pathway".[23] The genetic steps required to organize an operon cluster do not "confer any selective advantage to the phenotype so that individual steps are independent".[23] The rule against deferred success functions as a prophylactic against the emergence of teleological or Aristotelian thought in theoretical biology.[23]

I have pictured evolution on the molecular level as a process involving paths; natural selection acts to induce a statistical drift on some paths, and not others; those paths involving a positive gain in fitness are favored. At any particular time, at any particular place, one has an ensemble E of protein strings, embedded, so to speak, in an underlying probabilistic structure, a measure-preserving system, to keep to the concepts already introduced. To this structure, natural selection is grafted, and acts, presumably in virtue of a property that may be represented by the action of a real-valued, measure-theoretic function: thus $f(x)$, $f(Tx)$, $f(T^2x)$, ... are successive local calculations of fitness under the action of the system's transformation, the ***eigenvalues*** to the system. Suppose now we consider a finite-state system consisting of an alphabet of 26 letters; and the set of sequences k places in length. There are, of course, 26^k such sequences. Each letter $a_i \in A$ occurs with a fixed and independent probability p_i. The shift transformation moves a given string one place to the left. In effect, this system is simply the finite-state stationary process with identically distributed terms mentioned in the example already discussed; and may be represented as a linear array of k squares. An initial probability distribution fixes the configuration of the system for the first (integral) moment; at each subsequent step, every square changes: the odds in favor of any particular letter appearing are 1/26. If doubly infinite in extent, this system models the play of k 26-sided dice continued from the indefinite past to the indefinite future.

What are the chances, one might ask (with a marked lack of breathlessness in my own case), that a system of this sort – a pure Bernoulli process – could converge on a particular sentence of English? Following Mannfred Eigen, let us suppose that the sentence in question is TAKE ADVANTAGE OF MISTAKES, so that k is 23; this is the ***target sentence*** – S.

Even here, poised between irrelevance and imprecision, delicate and important biological questions arise.[24] Thus, while it makes sense of sorts to say that for every string, there exists a target – there would be many target sentences – it makes far less sense to say, as Eigen does, that there exists a target for every string – just one, in fact. Fixed in advance, a target so singular would seem suspiciously like a goal and hence ***streng verboten*** in evolutionary thought. How might such a target be represented and by what means might its influence be transmitted to strings? These are not trivial questions.

In any event, nothing in Eigen's own example quite indicates why a stochastic system with a target sentence, however defined, should stop when it has reached its goal. This, however, is a trivial defect, easily made good by the construction of an ***evaluation measure***. Suppose, for the sake of simplicity, that fitness involves only a mapping from strings to 0 and 1: at S, $f(S) = 1$, elsewhere, f is 0. An evaluation measure serves to size up strings in point of fitness as they appear: at S, where $f(S) = 1$, it orders the system simply to stop; at all other strings, the command is to mush on.

Stochastic device, target sentence, fitness function, and evaluation measure, taken as a quartet, comprise an ***Eigen system***. The enterprising Professor William R. Bennett Jr has calculated that an Eigen system would require a virtually infinite amount of time to reach even a simple target sentence – a number roughly a trillion times greater than the life of the universe In the same spirit, Murray

Eden has figured that life would require something like 10^{13} blubbery tons of *E. Coli* "if one expected to find a single ordered gene pair in 5 billion years". The trouble is not simply one of finding the right letters: it is also the problem of not losing them once they are found.

What more, then, is needed? The opportunity, Stephen Jay Gould remarks, for the system to capitalize on its ***partial successes***. Curiously enough, this is Eigen's answer as well, a bizarre example of independent origin and convergent confusion. As Eigen works though his example, his system is designed to retain those random changes that fit the target sentence. Looking at the record of Eigen's own simulation, we see that quite by chance the letter A appears in the first generation in the right place on the sequence. It stays intact, A'ish so to speak, for the rest of the simulation. When an E pops up, it, too, gets glued to the system.

The result is an ***advanced Eigen system***, and an improvement over the hopelessly slow Eigen system already described. Under the advanced Eigen system, fitness is no longer an all or nothing affair; f thus takes values, let us say, between 0 and 1. Scanning every new string, the evaluation measure selects those strings s_i such that $f(s_i) > f(s_{i-1})$. These the system retains until it finds a string superior in point of fitness. The result is a sequence of strings the ***ascends in fitness***. At S, as before, the system stops.

An advanced Eigen system may well reach a target system in rather a short time: unfortunately, in theoretical biology, as elsewhere, the question is not whether but ***how***. To the extent that fitness is purely a local property, it is difficult to understand why ***every*** ascending sequence should necessarily converge to a neighborhood of 1, and hence indirectly toward S. A string that only partially conforms to S is locally no fitter than a string that remains resolutely unlike S. On the other hand, if each of the ascending sequences converge to S it is very hard to see that fitness is a local property, and hard thus to understand what it is that an evaluation measure manages to measure. What is unacceptable is the obvious and tantalizing idea that an evaluation measure judges fitness by calculating the ***distance*** between random strings and a target sentence: distance is not a local property; an evaluation measure so constructed would plainly be responding to signals sent from the Beyond, a clear case of action at a distance. The problem of discovering a target sentence remains unchanged, hopeless. In fact, this is precisely what the advanced Eigen system actually measures, since an arbitrary sentence in which A appears in the second position is judged fit only because it is closer to the target sentence than it might otherwise be. When the matter is carefully explained, theoretical biologists understand at once that the very concept of a target sentence constitutes a beery and uninvited guest in evolutionary thought. I have taken the argument a step further by insisting that evaluation measures themselves be purely local.

Need I insist that the situation is made no better if instead of a specific target sentence I talk of systems set for success when they reach any sentence whatsover? I suppose, since it may at first appear easier to design a system that by randomly changing letters, in what Eigen hopefully calls the evolution game, approximates an arbitrary English sentence instead of just one. The illusion of ease is ill-gotten, of course: a target sentence is a minor stand-in for a major concept. If no particular target sentence is fixed in advance, then any sentence of English, once reached, makes for success. Simply to stop, the system must have an abstract characterization of all the English sentences. Of these, there are infinitely many. A system bouncing briskly from one set of random permutations to another, no less than the linguist or logician, thus requires nothing less than a grammar of the English language if it is not to keep babbling forever.

I have described grammars in terms of the notion of formal support; these concepts receive no definition in Darwinian theory.

Notes

[1] See my review of Michael Ruse's *The Philosophy of Biology*, in *Philosophy of Science* 41, Number 4, December 1974, for a discussion of Smart's position, and related issues.

[2] See, for example, Thomas Kuhn (1970) *The Structure of Scientific Revolutions* (Chicago: University of Chicago Press), a book which has prompted a vast secondary literature and much merited soul-searching among analytic philosophers of science.

[3] "Those who are oppressed by their own reputations," Dr Johnson remarks, "will perhaps not be comforted by hearing that their cares are unnecessary." Francis Crick (1966) *Of Molecules and Men* (Seattle: University of Washington Press).

[4] I discuss reductionism from the perspective of atomistic theories in Berlinski (forthcoming) *The Rise of Differential Topology* (Boston: Birkhaeuser Boston). See also Kenneth Schaffner (1967) Approaches to reduction. *Philosophy of Science* 34 (1): 137–47.

[5] Michael Ruse has argued for his thoroughly incoherent position in Ruse (1973) *The Philosohy of Biology* (London: Hutchinson). The concept of evolution was, of course, in the European air for at least a century before Darwin wrote. European biologists are yet unreconciled to Darwin. In this regard, see Pierre Grasse (1977) *Evolution of Living Organisms* (New York: Academic Press). The facts of molecular biology, it is worth stressing, are not in dispute: it is their interpretation that remains clouded. The central role of DNA, in particular, has troubled many thoughtful observers. "To attribute such powers to a single substance", Grasse remarks "however complicated its molecular structure, is in my view aberrant."

[6] M.C. King and A.C. Wilson (1975) Evolution at two levels in humans and chimpanzees. *Science* 88 (4184).

[7] S. Smale (1980) *The Mathematics of Time* (New York: Springer).

[8] A. Kolmogorov (1967) Logical basis for information theory and probability theory. *IEEE Transactions on Information Theory* IT - 14 (5). I have patterned my discussion on: G.J. Chaitin (1974) Information-theoretic computational complexity. *IEEE Transactions on Information Theory*, IT - 20 (1). The interested reader should consult Chaitin's other papers, and relevant papers by Solovay. Chaitin's bibliography may be consulted for details.

[9] See, for example, Noam Chomsky (1972) *Language and Mind* (New York: Harcourt, Brace Javonovich).

[10] The idea of representing context-free languages by means of a system of equations in noncommutative variables is due to M.P. Schutzenberger. See M. Gross (1972) *Mathematical Models in Linguistics* (New Jersey: Prentice-Hall) for details. I am inclined to think that the Deity, in creating the observable world, hesitated between programming or painting the whole business. As a programmer, he would have chosen a set of recursive rules; as a painter, a system of simultaneous equations.

[11] David Hull (1974) *Philosophy of Biological Sciences* (New Jersey: Prentice-Hall), although staid, contains a competent discussion of many of these issues.

[12] See, for example, L. Lofgren (1975) On the formalizability of learning and evolution, in Suppes, Henkin, Joja, and Mosil (Eds) *Logic, Methodology and Philosophy of Science* (Amsterdam: North-Holland).
[13] J. Monod (1971) *Chance and Necessity* (New York: Alfred Knopf).
[14] See Peter Medawar (1977) *The Life Sciences* (London: Wildwood House).
[15] Murray Eden (1967) Inadequacies of neo-Darwinian evolution as a scientific theory, in P. Moorhead and M. Kaplan (Eds) *Mathematical Challenges to Neo-Darwinism* (Philadelphia: The Wistar Institute Press).
[16] R.M. Thompson (1981) *Mechanistic and Non-Mechanistic Science* (Lynbrook, New York: Bala Books).
[17] H.P. Yockey (1977) A calculation of the probability of spontaneous biogenesis by information theory. *Journal of Theoretical Biology* 67.
[18] See K. Petersen (1983) *Ergodic Theory* (Cambridge: Cambridge University Press) for details.
[19] My discussion follows that of A.I. Khinchine (1957) *Mathematical Foundations of Information Theory* (New York: Dover Publications).
[20] N. Chomsky and G. Miller (1963) Finitary models of language use, in Luce, Bush, and Galanter (Eds) *Handbook of Mathematical Psychology* (New York: John Wiley & Sons).
[21] Eden, *op. cit.*
[22] The idea of the Weizenbaum experiment is due to M.P. Schutzenberger.
[23] Eden, *op. cit.*
[24] See, for example, Eigen (1971) Self-organization of matter and the evolution of biological macromolecules. *Die Naturwissenschaften* 10. Together with Ruth Winker, Eigen has recently (1981) published a popular account of his thought under the title *The Laws of the Game* (New York: Harper & Row).

References

Grantham, R. (1974) *Science*, **185**, 62.
Schroedinger, E. (1945) *What is Life*? (New York: Macmillan).
Smart, J.J.C. (1963) *Philosophy and Scientific Realism* (London: Routledge and Kegan Paul).
Watson, J.D. (1965) *Molecular Biology of the Gene* (New York: Benjamin) p 67.

CHAPTER 10

Universal Principles of Measurement and Language Functions in Evolving Systems

H.H. Pattee

The ability to construct measuring devices and to predict the results of measurements using models expressed in formal mathematical language is now generally accepted as the minimum requirement for any form of scientific theory. The modern cultural development of these skills is usually credited to the Newtonian epoch, although traces go back at least 2000 years to the Milesian philosophers. In any case, from the enormously broader evolutionary perspective, covering well over three billion years, the inventions of measurement and language are commonly regarded as only the most recent and elaborate form of intelligent activity of the most recent and elaborate species.

In this discussion I argue that such a narrow interpretation of measurement and language does not do justice to their primitive epistemological character, and that only by viewing them in an evolutionary context can we appreciate how primitive and universal are the functional principles from which our highly specialized forms of measurement and formal languages arose. I present the view that the generalized functions of language and measurement form a semantically closed loop which is a necessary condition for evolution, and I point out the irreducible complementarity of construction and function for both measuring devices and linguistic strings. Finally, I discuss why current theories of measurement, perception, and language understanding do not satisfy the semantic closure requirement for evolution, and I suggest approaches to designing adaptive systems which may exhibit more evolutionary and learning potential than do existing artificial intelligence models.

My approach is to generalize measurement and linguistic functions by examining both the most highly evolved cognitive systems and the simplest living systems that are known to have the potential to evolve, and abstracting their essential and common measurement and linguistic properties. I want to emphasize that when I speak of molecular language strings and molecular measuring devices I am not constructing a metaphor. Quite the contrary, I mean to show that our most highly evolved languages and measuring devices are only very specialized and largely arbitrary realizations of much simpler and more universal functional principles by which we should define languages and measurements.

Generalized Measurement

The classical scientific concept of measurement requires a distinct physical measuring device that selectively interacts with the system being measured, resulting in output that has a symbolic interpretation, usually numbers. Most scientists regard the output of a number as an essential requirement and, indeed, numbers are required if the language of science is restricted to mathematics. If the laws are expressed by equations of motion, then the initial conditions must be numbers if we are to use the equations to predict other numbers. However, without questioning the enormous advantages of numbers and formal mathematical representations of laws, it is obvious that measurements are possible without numerical outputs (e.g., Nagel, 1932). For example, timing, navigating, surveying, weighing, and even counting were once accomplished by iconic, mimetic, or analog representations. Today the trend is away from the outputs of traditional laboratory measuring devices with visible numerical scales and toward transducers that feed computers and robots directly. In all cases the type of output from a measurement is chosen according to the particular functional requirements of the system as a whole.

The essential point is that while the selection of *input patterns*, the choice of *output actions*, and the *relation* of input to output in any measuring device is largely arbitrary, the only fundamental requirements for useful measurements are the *precision and reproducibility, or local invariances, of the input–output relation*, and the *functional value* of the entire operation to the system doing the measuring. The requirement of reproducibility means that the measuring device must be *isolatable* from the system being measured, and *resettable*, so that the measurement process can be repeated an arbitrary number of times to give the same output for the same input pattern. However, such an abstract description of measurement is incomplete, since it omits the crucial requirement of *system function*, or the *value* of the measurement.

From the abstract definition of measurement alone we would conclude that any relatively fixed or constrained set of particles in a physical system qualifies as a measuring device if we interpret pattern as simply the initial conditions of the free particles and action as the alteration of their free trajectories after collision with the constrained set. Thus, we might say that a rock in a stream maps the input flow pattern to the output action of turbulence, or say that in crystal growth the constraint of a dislocation on a crystal surface maps the patterns of molecular collisions to the specific action of binding more of its own constituents. However, we do not normally call these cases measurement processes. By contrast, the pattern recognition required for specific substrate binding and catalytic action of cellular enzymes I would call a measurement, even by the most rigorous definitions that apply to highly specialized, artificial devices. How do I justify this? Clearly, the enzyme's action is more complicated than crystal growth, but I do not see the level of complexity of the measuring device as the only criterion; for example, calipers are a simple, artificial constraint, that we may use to measure size. The only distinction I find convincing is that of ***system function*** or, more specifically, that of pattern–action mapping that supports the persistence or survival of the system, and subsidiarily of the measuring constraints that make up the system. In other words, there must be ***functional closure***. It is necessary that the enzyme serves a function in the cell for its pattern recognition and catalytic action to be called a measurement. This is still too broad a definition, since it gives no clues as to the characteristics of function, other than survival, that are required of measurement. We must specify some further conditions on this mapping from patterns to actions that are necessary for efficient or effective measurements. Are there also conditions on the way that successful *systems* of measuring devices interact? Let us consider what is common to some extreme examples of successful measuring devices.

Measurement as a Classification

The most important, and yet the most deceptive, aspect of our highly evolved artificial measurements is the feeling we have as intelligent observers that we know what attribute we are measuring independently of the measuring constraints. This is a half-truth. We usually have an abstract concept of what attributes we wish to measure and design the constraints of the measuring device so that its output action expresses these attributes and minimizes all others. Since the output action is designed to be very simple, we often tacitly assume that the corresponding input patterns are very simple. For example, we think of temperature as a simple property of a gas, but our thinking does not change the complex molecular collisions of the gas. This is actually a useful deception in building classical models, although it leads to erroneous results in quantum mechanics. In fact, the measuring device necessarily interacts physically with all of the system's innumerable degrees of freedom, and it is precisely because of the innumerable internal constraints of that particular device that only a few degrees of freedom are available for the output actions.

It is primarily this property of mapping *complex input* patterns to *simple output* actions that distinguishes useful measurement functions from merely complex physical constraints. Without this complex-to-simple or many-to-one mapping process we would not be able to identify equivalence classes of events and, consequently, we would not be able to construct simple models of the world. I would go further and claim that *this classification property of measurement is an epistemological necessity*. Without classification, knowledge of events would not be distinguished from the events themselves, since they would be isomorphic images of each other. This also implies symmetry in time, and measurement must be an irreversible process.

From a broad biological perspective, the entire nervous system has evolved for the principle function of quickly and reliably mapping the ineffably complex configurations and motions of the environment to a very few vital actions; that is, run, fight, eat, sleep, mate, play, etc. Although these actions can be decomposed into complex subroutines, the decision is still which of only a small number of actions to employ. The entire organism can therefore qualify as an extreme case of a generalized measuring device. Let me return now to the other extreme of evolution and consider measuring devices at the molecular level.

At the cellular level we have the example of the single enzyme molecule. The action of an enzyme, like the action of an artificial measuring device, may be described very simply. Generally, it is the catalysis of one particular covalent bond and, consequently, we might think of the corresponding input pattern simply as one particular substrate molecule. But this would miss the essential property of an effective measuring device, which is to reduce the complexity of input interactions by means of its internal constraints. When we speak of an enzyme as highly specific it is another way of saying that it recognizes or distinguishes very complex input patterns.

This ability to recognize complex input patterns and, as a consequence, execute a simple action requires physical constraints of a special type. Since the many-to-one mapping is arbitrary, the constraints must arbitrarily couple the *configurations* available for fitting the input pattern to the *motions* of the device that produces the output actions. In physics these are called nonholonomic or nonintegrable constraints. A holonomic constraint is a restriction on the configurations of a set of particles, such as occurs in forming a crystal from a solution of molecules. This freezing-out of configurational degrees of freedom necessarily freezes-out the corresponding motions of the crystallized molecules, so that we see the constrained system as a rigid solid. A nonholonomic constraint may be defined as a restriction on the motions of the particles *without* a corresponding restriction in the particle configurations. In other words, a formal expression of a nonholonomic constraint appears as a peculiar equation of motion for selected velocity components, where certain configurational variables serve as initial conditions. However, we cannot generally eliminate any configurational variables of the system by using these relations because of the nonintegrability of the equations of constraint. This results in a flexible or allosteric configuration. What we call machines are made up of holonomic, rigid parts that are coupled by nonholonomic, moving linkages. In such machines more configurations of the parts are allowed kinematically than are allowed in the dynamic motions of the parts (e.g., Pattee, 1972b). In proteins it is these nonholonomic constraints that couple the complex configurations or patterns of the substrate to the allosteric motions causing catalytic actions.

The complexity of patterns that can be usefully distinguished clearly depends, in part, on the complexity of the internal constraints of the measuring device that fits the pattern. What is not so clear, but equally important for recognition, is that the output action must be simple and repeatable. In fact, we can imagine a complex fit that requires complex constraints without any corresponding simple action, as in a pile of gravel. We also speak of complex actions resulting from complex constraints, as in the weather. But it is only when complex interactions result in simple, repeatable actions that we speak of recognizing patterns. Enzymes require hundreds of amino acid residues to fold into a structure which we say fits the substrate; but any solid has molecules that physically fit each other just as well, yet we do not generally picture solids as pattern recognizers. It is only the simple catalytic action that establishes the fit interaction as a pattern candidate; but I would again argue that the only objectifiable existence of patterns is ultimately established by some form of system closure. That is, *the distinguishing property of measurement constraints is that their pattern–action mapping supports the system that is necessary to synthesize these constraints.* Since this is such a fundamental condition, let me discuss it in more detail. We shall see that for evolution to be possible, functional closure must be more complex than just autocatalytic cycles.

Function Requires Construction

Returning now to the human level we can say that the primary function of measurement is to map the ineffably complex interactions of the physical world into attributes which are necessary for our survival in this world. To realize this function, it is obviously necessary for us to pay attention to these attributes. This justifies the epistemological illusion of thinking about the world in terms of these measured attributes; that is, in terms of the simple *outputs* of the measuring devices rather than the complex inputs. In the everyday use of observations and measurements there is no survival value in analyzing the inner details of measuring devices. In other words *performance* of measurements does not benefit from *analysis* of the constraints of the measuring device.

In fact, if one analyzes the measurement constraints using a microphysical description, the measurement *function* unavoidably disappears into a measurement-free physical system with more degrees of freedom. On the other hand, *it is from this more detailed physical system that the complex measurement constraints must have been synthesized in the first place.* This means that we must have *control* over physical details of constructing measurement devices even though we do not want or need knowledge of these details while we actually perform measurements. The measurement activity therefore requires both *functional primitives*, in the sense that any analysis of the constraints of the measurement device necessarily obliterates the essential classification action, and *constructional primitives*, in the sense that knowledge of the function of the device can result in no necessary rules for synthesizing the device's constraint.

This apparently improbable interrelation between genes and enzymes is the simplest case of what I call *semantic closure* (Pattee, 1982). By general semantic closure I mean the relation between two primitive constraints, the generalized measurement-type constraints that map complex patterns to simple actions and the generalized linguistic-type constraints that control the sequential construction of the measurement constraints. The relation is semantically closed by the necessity for the linguistic instructions to be read by a set of measuring devices to produce the specified actions or meaning. The semantic closure principle is supportable from several levels:

(1) As an empirically based generalization from the facts of molecular biology.
(2) As a theoretical requirement based on the logic of heritable systems (e.g. von Neumann, 1966; Polanyi, 1968).
(3) As an epistemological condition necessary for the distinction between matter and symbol (Pattee, 1982).

It may also be stated, as a complementarity principle, that the properties of measurement and language cannot be adequately defined individually, but form an irreducible, complementary pair of concepts.

Generalized Language

There is common agreement on many of the universals of language ***structure*** (e.g., Hockett, 1966). Natural and formal languages are discrete, one-dimensional (1-D) strings of elements from a small alphabet. The strings are further constrained by lexical and syntactic rules which may be very simple or very complex. These rules may be precise and explicit, as in formal languages, or ambiguous and difficult to formulate as in natural languages. Language strings are constructed and read sequentially, although all natural languages also have the essential metalinguistic ability to reference themselves out of sequence; that is, to construct strings that refer to other strings in the language. From what is known of the structure of the gene it appears to qualify fully as a natural language system (e.g., Pattee, 1972a).

When it comes to language ***function*** it is more difficult to find simple generalizations, let alone common agreement. Language unquestionably has many functions; for example, memory, instruction, communication, modeling, thought, problem-solving, prediction, planning, etc. What I am proposing is not inconsistent with any of these functions. However, my criteria for functions in both measurement and language are based on the most ***primitive*** conditions for evolvable systems. These include:

(1) The ability to construct and coordinate measuring devices and other functional structures under the control of a heritable description (i.e., genetic control).
(2) The ability to modify function by changing the description (i.e., mutability).
(3) A heritable process for evaluating the description–construction system as a whole (i.e., natural selection).

The impressive techniques of molecular biology have shown us in some detail how present cells accomplish these processes, so in a phenomenological sense they are no longer considered problems by biologists. However, there remain the essential mysteries of how such cellular systems came to exist and how multicellular systems develop. That is, how does such a coordinated set of linguistic instructions and measuring constraints evolve from a nonliving physical world and how are such intricate multicellular morphologies constructed and maintained by these linguistic and measurement devices? I comment on approaches to these problems in the last section but, since I have no solutions, for the present I assume the existence of cells and simply generalize from the structure of linguistic constraints and how they function in the cell system.

We considered the basic function of measurement devices as mapping complex patterns to simple actions. In cells, these actions are typically the catalysis of a single bond; in effect, the smallest change that can be made in the constraints of a molecular system. However, this small change is only made if a complicated set of other constraints is satisfied, namely the recognition of the substrate molecule. In a linguistic device the functions are quite different. The function of the linear sequence is to control the sequence of actions necessary to construct the measuring device, but it does this through the simplest possible type of constraint, the chain of single bonds. At the other extreme of language constraints, we find that one principle function of our spoken and written languages is to give instructions;

and it is an impressive fact, often taken for granted, that by forming discrete strings from about 30 types of simple marks we can effect the construction of almost any conceivable pattern, whether it is in the brain, in the actions of the body, or in the construction of artifacts. How these transformations take place from the simple, 1-D string of constraints to the physical structures and actions represented by these strings is almost a total mystery for natural language, even though we find we are able to know the meaning from the strings. But by contrast, at the molecular level we know in great detail how the genetic strings are transformed into the structures and actions represented by the strings, but we have no way of deriving the meaning of any string; that is, of how to tell from the genetic message alone what the function is of the protein it describes.

A generalized language might therefore be characterized as a simple chain of constraints that controls the construction of complex patterns. If we try to formalize this further, as we did with measurement, we might be tempted to say that linguistic devices map a domain of 1-D constraints to a range of n-D patterns. However, this would be a misleading abstraction. In the case of the measuring device, it is the actual constraints of the device itself that recognize the input pattern and are physically responsible for the output action. Therefore, by saying that the device *maps* the input pattern to output action we mean that it is responsible for dynamically *executing* the mapping. On the other hand, a string of constraints in a language is dynamically inactive. Language strings are pure configurations; that is, they have no significant motions or velocity components. Thus, symbol strings are rate independent in the sense that their meaning, or what they control, does not depend on how fast they are read.

Semantic Closure

We explained earlier how the *action* of measurement constraints is functionally primitive, since analysis of the details of the constraints interferes with the measurement function. In a similar way the *meaning* of a linguistic string is functionally primitive, since analysis of the mechanisms of production of the string interferes with the meaning. In practice, we look at the results of a measurement and do not confuse ourselves with the constructional details of the measuring device. Similarly, when we generate linguistic strings we focus on the meaning, not the mechanics of production.

This complementary primitiveness of measurement and linguistic meaning is not only an observable fact of biology but also, I believe, an epistemological necessity. As we said earlier, it is essential that we be able to directly picture the world from what we perceive or from the outputs of measuring devices without having to also know the physical details of the perceiving or measuring constraints as parts of the nonmeasuring interactions of the physical world. This requirement of semantic primitiveness of perception and measurement accounts for what I call the epistemic illusion of the reality of the world, which is not involved with the complex and largely arbitrary constraints that execute perception and measurement. The alternative possibility, that we must analyze these measurement constraints, only leads to irrelevant details at best or an infinite regress at worst. In a complementary sense it is essential that we be able to directly grasp the meaning of linguistic strings without becoming involved with the complex and largely arbitrary details of the constraints that generate and interpret strings. Just as in the case of measurement, this requirement of semantic primitiveness of language accounts for the epistemic illusion that strings have an intrinsic meaning independent of the dynamical constraints that generate them or that they ultimately control. Only through semantic closure do these two primitives complement each other and form an autonomous, evolvable system. The semantic closure principle allows us to treat the ***action*** of a measuring device as primitive because the details of its construction are accounted for by a linguistic string, while the ***meaning*** of the linguistic string can be treated as primitive because the details of interpretation are accounted for by a set of measuring devices. For me, it is this fundamental relation between the ***relative primitives*** of measurement and language constraints that distinguishes evolvable or epistemic systems from normal physical systems. In the final sections I elaborate on why this closure principle offers more promise for models of evolvable systems than other approaches. Before doing this, let me summarize the properties of generalized measurement and language.

Properties of generalized measurement

(1) Measuring devices are localized, isolatable, resettable structures with repeatable actions.
(2) Measuring devices have no intrinsic output actions, but may be triggered to simple actions by specific input patterns (nonholonomic constraints).
(3) Measurement constraints obey all physical laws, but are not derivable from laws (generated by system function).
(4) Measuring devices are constructed sequentially under the control of linguistic constraints, but a complete, finite set of measuring devices is necessary to read linguistic strings (semantic closure).
(5) Measuring devices execute a many-to-one mapping from complex input patterns to simple output actions (classification).
(6) Measuring devices do not occur in isolation, but form functional, coherent sets within a system.
(7) The value and quality of any measurement is a system property determined by the survival of the system in which it functions.
(8) Beyond these properties, the domain of input patterns, the range of output actions, the choice of mapping, and many other aspects of measuring devices are largely arbitrary.

Properties of generalized language

(1) Language structures are discrete, 1-D strings made up of a small number of types of elements.
(2) Language strings have no intrinsic actions, but may trigger action in measuring devices (nonholonomic constraints).

(3) Linguistic strings obey all physical laws, but are not derivable from laws (generated by system function).
(4) A complete, but finite, set of measuring devices is necessary to read and interpret linguistic strings.
(5) Linguistic instructions are necessary to control the synthesis of this interpreting set (semantic closure), as well as the synthesis of other functional components of a system.
(6) Language strings are transcribed sequentially, independently of rate, but they may reference themselves out of sequence (metalanguage).
(7) The value and meaning of any linguistic string is a system property determined by the survival of the system that it controls.
(8) Beyond these properties, the physical structure, the choice of alphabet, the units of meaning, and many other aspects of language strings are largely arbitrary.

Models of Evolution

How can these primitive closure requirements for measurement and language be incorporated into a model of an evolving or learning system? How would such a model differ from previous models? Let us begin with the second question. Many more or less literal simulations of genetically controlled, self-reproducing systems have been studied, beginning with von Neumann's self-reproducing automaton in which he first explicitly recognized the need for a genetic description as well as a universal constructor that must read and execute this description if evolution is to produce increasingly complex systems. However, von Neumann (1966) was more interested in the logical or linguistic aspects of the model than in the physical aspects of pattern recognition and measurement. He was well aware of this neglect of the physical aspects of the problem ("...one has thrown half the problem out the window and it may be the more important half."), but at that time (ca. 1948) the Turing concept of computation was well-developed, while molecular biology was still a great mystery.

Many later simulations of evolution have been attempted for the purpose of improving the adaptation or optimization process in formal or artificial systems (e.g., Bremermann, 1962; Fogel *et al.*, 1966; Klopf and Gose, 1969; Holland, 1975; Barto, 1984). Only a few models of evolution have been constructed to help conceptualize and test the postulates of neo-Darwinian theory (e.g., Moorehead and Kaplan, 1967; Conrad and Pattee, 1970). In all but one of these models, the process of natural selection is accomplished by fitness criteria which are explicit and preestablished by the programmer. In the Conrad and Pattee model no explicit fitness criteria were introduced. Instead, a set of general conditions or rules of interaction between organism and environment were defined, such as conservation of metabolic resources. However, the nature of the environment with respect to the organisms was preestablished; that is, no genetically modifiable measurement constraints were introduced in this model. Thus, in the existing models of evolution the environment has been represented as a fixed, objective framework that produces the selection pressures on the populations of organisms. Our present complementary view of language and measurement requires the epistemic condition that the organism can only respond directly to the simple output of measurements of the environment. As we have seen, these simple outputs are a consequence of complex constraints resulting from genetically controlled syntheses. However, there is no explicit relation of the gene string to the input–output mapping of the measuring device. Gene strings that construct measuring devices cannot be thought of as programs that manipulate data structures in a computer. In the latter case, every program instruction must be completely explicit. Explicit actions require that all types of inputs, outputs, and hardware operations be preestablished. By contrast, in the organism it is the genetic instructions that construct the hardware that determines all the inputs, outputs, and actions. Genetic consequences are therefore entirely implicit. One cannot assign an element of the gene to an element of action, yet this is the central requirement of a program or effective procedure in computation. Furthermore, simply to say that the architecture of present computers is totally unlike the architecture of organisms is a misleading understatement, since even the concept of architecture plays an entirely different role in organisms to that in computers. For these reasons any form of computational metaphor for organisms must be treated with skepticism.

Up until quite recently the predominant view of genetic control has been very much like the view of computation as an explicit program control of data strings in memory. The alternative view that morphogenesis depends both on autonomous dynamics (archetypes) and internal constraints (chreods) for which genes provide only local switching forces is well known (Waddington, 1968), but for many years lacked empirical evidence and a conceptually clear, formal model. Currently, such topological and dynamical models of morphogenesis are more popular largely because of the application of elegant mathematical formalizations of the singularities, bifurcations, degeneracies, and instabilities of dynamical systems. These mathematical and physical theories of continuous systems arose from completely distinct concepts and methodologies to those of the computational models of morphogenesis, yet they have also led to models for the growth of many types of biological patterns as well as impressive claims for more general powers (e.g., Thom, 1975; Prigogine, 1980; Eigen and Schuster, 1979; Haken, 1981). However, in spite of these significant contributions to mathematical and physical theory, biologists usually perceive the excitement over these formal models as coming more from the physicists and mathematicians who are impressed with the complex patterns that can be generated from such simple equations and boundary conditions. The problem is that molecular genetics is itself so well-established at the foundations of biology that dynamical models are not likely to be useful until they can incorporate these linguistic constraints into their models of evolution and development. While some of these dynamical models have helped clarify measurement constraints (e.g. Prigogine, 1980), none of them has directly contributed to the genotype–phenotype closure relation that is necessary for evolution.

It is also instructive to review current theories of cognitive activities at the other end of the evolutionary scale where the subjects of interest are perception, action, learning, language, knowledge, and other forms of intelligent activity. It is significant that here also we find two opposing schools, one based on explicit linguistic strings and the other on implicit measurement dynamics. The first school arose from logic and computation theory, and is now dominated by the paradigm of the computer as the universal symbol system that can model cognitive tasks such as pattern recognition, classification, learning, and understanding natural language. It is the claim of the computationalists or information processors that these tasks can be understood as purely linguistic or string-processing activities without reference to measurement or any physical dynamics, except as preestablished input and output transducers for the strings. These computational modelers appear to have a principled commitment to the epistemic illusion characteristic of linguistic constraints that strings contain implicate meaningful information, and that by processing these strings with a sufficiently clever rewriting of the rules, this meaning can be explicated (e.g., Newell, 1980; Pylyshyn, 1980). In a somewhat less principled way, the information processors are committed to the complementary epistemic illusion of measurement that only the simple output action need be entered into their models, and that the origin of the complex dynamical constraints that generate these simple outputs need not be considered as a part of their cognitive process.

The opposing school, which arose from the ecological physics approach of J.J. Gibson (1979) takes the other extreme of basing their models on a principled avoidance of linguistic constraints, which they argue are neither essential for mapping perception to action nor for the construction of measurement constraints. Ecological physics models are based on extensions of the dynamical singularity theories of physics (e.g., Turvey and Carello, 1981; Turvey and Kugler, 1984), and understandably emphasize perception–action models rather than genetic control or language understanding.

Both the information processing and the ecological physics schools of cognitive modeling appear to have committed themselves to their exclusive methodological principles without serious consideration of the empirical facts of development and evolution. In effect, the information processors are committed to the principle that discrete strings possess intrinsic meanings independent of the physical dynamics that generate the strings, while the ecological physicists are committed to the principle that physical dynamics possess intrinsic meanings independent of the genetic strings that have constructed the dynamical constraints. One simple, but very frustrating, fact of evolution is that natural selection does not follow the physical or logical principles of most other scientific models, but operates only through opportunistic and even haphazard experiments. Survival depends on balancing many highly interrelated, qualitative system properties such as speed, reliability, efficacy, recovery from error, efficiency, and adaptabiity. Thus, although it may be technically efficient for us to recognize shape by computation on a string of data obtained by an arbitrary scanning of the shape, the enzyme is much quicker using direct 3-D template recognition with no computation whatsoever; and although it is technically possible to cast a machine from a 3-D template with no string processing, the enzyme is constructed more reliably by sequentially processing a gene string. At the cognitive level why should this opportunistic strategy be different? We can recognize the number of rocks in a pile directly if there are less than 6 or 7, but we must count them sequentially if there are more. In a fraction of a second we directly recognize our complex friends in a crowd, but may have to follow long strings of inductions to identify a simple mineral in a rock. The brain, like the cell, has clearly evolved the power both to directly perceive patterns (measure) and to process strings (compute).

To me, the effort to model the brain as exclusively one or the other type of constraint may be useful engineering – in principle it can be done – but that is not our problem. Our problem with the nervous system is to understand the functional interrelation of direct perceptions and language necessary for efficacious action and learning, just as the problem with the cell is to understand the functional interrelations of gene strings and cellular dynamical constraints necessary for development and evolution. These interrelations are certainly very complex and largely unknown, but what is perhaps the most fundamental evolutionary fact we already know, and that is the meaninglessness of strings or dynamics taken in isolation. From the evolutionary perspective it is only the semantic closure of genotypic language strings and phenotypic measurement dynamics that defines any biological organism in the first place. Whether any physical strings or dynamical constraints can be said to form a language or a measuring device, or whether either has function or meaning can only be decided in terms of its origin and function in the life of the organism.

Conditions for Artificial Evolving Systems

I now come to the question of how this semantic closure property of measurement and language can be incorporated into an artificial system. Although language and measurement are complementary primitives they do not relate symmetrically. We pointed out that measurement constraints are dynamically active without linguistic inputs, even though they may have been constructed under linguistic constraints. Measurement devices physically execute the mapping from input patterns to output actions. This means that a system of measuring constraints, once constructed, can perform complex dynamical tasks without further linguistic control. In other words, the specific actions of measurement systems do not require a program to run them. By contrast, a linguistic constraint has no intrinsic dynamics, it is rate independent, and it therefore can execute no rule or action by itself. Every action of a linguistic system must therefore have an external rule or program step to execute it. In effect this is how computation is defined. Only a string that is mapped into another string ***by means of*** an effective procedure can qualify as formal computation; but a measurement by itself is not an effective procedure since it has no explicit input. Conrad and Hastings (1985) have proposed naming such direct transformation a new computational primitive, but since there is no explicit input, they must use a nonstandard definition of computation. Gibbsonians often refer to measurement constraints as "smart machines" that accomplish their function without computation, to contrast them with string processing that requires smart programming if any useful output is to result.

I am proposing that any model of an evolutionary process must clearly represent and functionally distinguish language and measurement constraints (i.e., the genotype and phenotype) and must preserve the properties and relations of each. This includes the construction of the measuring devices under the constraints of the linguistic strings and the reading of these strings by measuring devices. It must also include the ability of the strings to gradually or suddenly modify the inputs, outputs, and mappings of the measuring devices and must allow the representation of measuring devices to function under an autonomous dynamics once they have been constructed. This latter condition is difficult to fulfill in an artificial model since the function of a measuring device depends on its interactions with an environment. If we try to simulate the natural environment, the model becomes very complex and yet is incomplete. On the other hand, if we invent too simple an artificial environment, the measurement mapping becomes trivial. The engineering approach is to have the model adapt to the real natural environment, but this requires the construction of real measuring devices under genetic control. This may be practical, but one could question its status as an explanatory model, or even a model at all, since it would appear to be a real evolving system. One more pedagogic-type model might utilize an artificial environment that could be gradually modified in the hope of inducing new measurements by the organism. What are the simplest conditions under which we can expect such emergent behavior?

It appears obvious that the simulation of language constraints on a computer is simpler than the simulation of measurements. However, there is an enormous difference between natural languages and artificial programming languages, which is easily recognized, but not understood. Typically, computer languages do not tolerate mutations or recombinations, whereas genes and natural languages depend on such changes for evolution and creative expression. One difference which may be significant is the lack of complementary measurement constraints in current computer architectures. Since linguistic constraints have no intrinsic dynamics, the computer does nothing unless given a program step. Furthermore, this step must be explicit; that is, the mapping from the domain of program steps to the range of output actions must be unconditionally defined in advance. This total dependency on linguistic inputs results in a total intolerance to the absence of inputs or to inputs with syntactical error. It also follows from the requirement of explicitness in the program steps that errors are also explicit; that is, changes in input–output mappings cannot be gradual. Natural systems, on the other hand, operate with measurement constraints under autonomous dynamics that do not require linguistic inputs for their function. Furthermore, this function depends only implicitly on the linguistic strings that controlled their construction; that is, the mapping from strings to measurement function cannot be specified as a sequence of unconditionally defined steps as in a program. Each linguistic step contributes to the final function only in conjunction with the contributions of other steps so that no single linguistic input step can be assigned an unconditional consequence in the output action. This input–output relation can be observed most directly in the folding transformation that converts the linguistic string constraints of the polypeptide's primary structure into the 3-D globular structure of a functioning enzyme. The significant result of this transformation is that a mutation or recombination of the linguistic string may result in all degrees of functional change, from virtually no change, to gradual or continuous change, to discontinuous change, to a new function. This same variability in meaning occurs in natural language where a single change in a letter or word may result in no change of meaning, a shift of meaning, or an entirely new meaning.

The nature of this relation between description and function or between language and meaning is certainly the most crucial and yet the most puzzling aspect of any epistemic or evolutionary system. It is a problem as old as philosophy and even now it is not clear that a complete explanatory model is possible. My only conclusion from this discussion is that unless an artificial system contains representations of the constraints of both generalized language and generalized measurement, as well as the complementary relations between them that I have described as semantic closure, the model is not likely to evolve similarly to living systems or to contribute significantly to the theory of evolution.

References

Barto, A. (1984) *Simulation Experiments with Goal-seeking Adaptive Elements. Final Report, June 1980–August 1983* (Avionics Lab., Air Force Wright Aeronautical Lab, Wright-Patterson Air Force Base, Ohio 45433).

Bremermann, H.J. (1962) Optimization through evolution and recombination, in M.C. Yovits, G.T. Jacobi, and G.D. Goldstein (Eds) *Self-Organizing Systems* (Washington, DC: Spartan Books).

Conrad, M. and Hastings, H.M. (1985) Scale change and the emergence of information processing primitives. *J. Theoret. Biol.* (in press).

Conrad, M. and Pattee, H.H. (1970) Evolution experiments with an artificial ecosystem. *J. Theoret. Biol.* 28: 393–409.

Eigen, M. and Schuster, P. (1979) *The Hypercycle: A Principle of Natural Self-Organization* (Heidelberg, Berlin, New York: Springer).

Fogel, L.J., Owens, A.J., and Walsh, M.J. (1966) *Artificial Intelligence Through Simulated Evolution* (New York: Wiley).

Gibson, J.J. (1979) *The Ecological Approach to Visual Perception* (Boston: Houghton-Mifflin).

Haken, H. (1981) Synergetics: is self-organization governed by universal principles?, in E. Jantsch (Ed) *The Evolutionary Vision*, AAAS Selected Symposium 61 (Boulder, CO: Westview Press).

Hockett, E. (1966) The problem of universals in language, in J.H. Greenberg (Ed) *Universals of Language* (Cambridge: MIT Press) pp 1–29.

Holland, J. (1975) *Adaptation in Natural and Artificial Systems* (Ann Arbor, MI: Michigan University Press).

Klopf, A.H. and Gose, E. (1969) An evolutionary pattern recognition network, *IEEE Trans. of Systems Science and Cybernetics* 5: 247–50.

Moorehead, P.S. and Kaplan, M.M. (Eds) (1967) *Mathematical Challenges to the Neo-Darwinian Interpretation of Evolution* (Philadelphia: The Wistar Institute Press).

Nagel, E. (1932) Measurement. Reprinted in Danto, A. and Morganbesser, S. (1960) *Philosophy of Science* (New York: World Publishing Co.).

von Neumann, J. (1966) *Theory of Self-Reproducing Automata*. Edited and completed by A.W. Burks (Urbana, Ill; University of Illinois Press).

Newell, A. (1980) Physical symbol systems, *Cognitive Science* 4: 135–83.

Pattee, H. (1972a) The nature of hierarchical controls in living matter, in R. Rosen (Ed) *Foundations of Mathematical Biology*, Vol. 1 (New York: Academic Press) pp 1–22.

Pattee, H. (1972b) Physical problems of decision-making constraints. *Int. J. Neuroscience* 3: 99–106.

Pattee, H. (1982) Cell psychology: an evolutionary approach to the symbol–matter problem. *Cognition and Brain Theory* 5(4): 325–41.

Polanyi, M. (1968) Life's irreducible structure. *Science* 160: 1308–12.

Prigogine, J. (1980) *From Being to Becoming: Time and Complexity in the Physical Sciences* (San Francisco: W.H. Freeman & Co).

Pylyshyn, Z. (1980) Computation and cognition: issues in the foundations of cognitive science. *The Behavioral and Brain Sciences* 3: 111–69.

Thom, R. (1975) *Structural Stability and Morphogenesis* (Reading, MS: W.A. Benjamin).

Turvey, M. and Carello, C. (1981) Cognition: the view from ecological realism. *Cognition* 10: 313–21.

Turvey, M. and Kugler, P. (1984) An ecological approach to perception and action, in H.T.A. Whiting (Ed) *Human Motor Actions: Burnstein Reassessed* (Amsterdam: North Holland Publishing Co).

Waddington, C.H. (1968) The basic ideas of biology, in C.H. Waddington (Ed) *Towards a Theoretical Biology I. Prolegomena* (Edinburgh: Edinburgh University Press) pp 1–32.

Springer